The Physics Around You

Dale D. Long
Department of Physics
Virginia Polytechnic Institute and State University

Wadsworth Publishing Company
Belmont, California

A Division of Wadsworth, Inc.

Photo Credits

Part-opening photos: I, II, III: Historical Pictures Service, Inc., Chicago; IV: Culver Pictures, Inc.; V: United Press International Photo.

Figure 1.6, Edwin S. Roseberry; 1.9, Robert Brent Corp., Healdsburg, CA; 1.11, Wide World Photos; 1.12, Photo by Wm. A. Bernhill/New York Public Library Picture Collection; 1.13, Wide World Photos; 2.2, Wide World Photos; 2.14, Dr. Harold Edgerton, M.I.T.; 2.16, Pfaff & Kendall, Newark, NJ; 2.17, Wide World Photos; 4.15(a), ©Raimondo Borea, (b), Willa Percival Photo/Photo Researchers, Inc.; 4.17, Mano Photo; 4.19, Matt Berg; 4.20, Bob Gomel/DPI; 4.28, Sikorsky Aircraft; 4.30, Busch Gardens, Williamsburg, VA; 6.9, Wide World Photos; 6.10, Port Authority of New York and New Jersey; 8.12(a), Leon M. Stith/Black Star, (b), Stock, Boston; 8.15, American Cancer Society; 10.3, JOEL, Ltd., Tokyo, courtesy of Prof. H. Hashimoto; 10.9, Union Carbide; 11.18, Wide World Photos; 12.36, Sekai Bunka Photo; 13.12, General Electric Company; 14.10, Research-Cottrell, Inc.; 14.12, Daniel R. Rubin/DPI; 15.10, Wide World Photos; 15.22, Donald E. Hall, 15.24, C. G. Conn, Ltd., Oakbrook, IL; 15.25, Dr. Marc S. Lapayowker, Temple University; 15.27, Virginia Department of Conservation and Economic Development; 15.29, Education Development Center; 15.31, ©1976 Memorex Corp.; 16.13, Schaeffer & Seawell/Black Star; 16.16, George Resch/Fundamental Photographs; 16.27, Polaroid Sunglasses Inc.; 16.28, Polaroid Sunglasses, Inc.; 17.1, Darwin Van Campen/DPI; 17.5, Loren Long; 17.7, Rosemary Eakins, Research Reports; 17.9, Loren Long; 17.11(b), George Resch, Fundamental Photographs; 17.18, ©1979 Ian Lucas/Photo Researchers; 17.36, ©Joel Gordon 1979; 18.6, Don Leonard; 18.12(a), ©1976 Joel Gordon, (b), (c), St. Vincent's Hospital, New York; 18.22(a), Wide World Photos, (b), University of Chicago.

To Lou, Don, and Doug

Physics Editor: Marshall Aronson

Production: Greg Hubit Bookworks

Design: Joe di Chiarro

Cover photo: Gregg Mancuso

Illustration: Ayxa Art; Darwen and Vally Hennings

Chapter-opening art: Carlos Perez

Picture research: Research Reports, New York

Printed in the United States of America

1 2 3 4 5 6 7 8 9 10—84 83 82 81 80

Library of Congress Cataloging in Publication Data

Long, Dale D
　　The physics around you.

　　Includes bibliographical references and index.
　　1. Physics.　I. Title.
QC23.L85　　　530　　　79-22727
ISBN 0-534-00770-8

Contents

Part V   **The Atom**

Preface

The purpose of this book is to acquaint nontechnically oriented readers with the general principles of physics by focusing on the physics around them. It uses everyday experiences and devices to illustrate the principles and to assist the readers in understanding them. The book can serve as a text for either a purely conceptual physics course, or one with minimal mathematics, of duration from one quarter to one year.

Energy and energy transformations are the unifying concepts of the book, and form the bases throughout for explaining many physical processes. This approach permits an immediate discussion of topics relevant to the readers and allows flexibility in choosing or ordering topics for study.

Simple algebra is the most advanced mathematics used. You may treat the several numerical examples in each chapter at any of three levels: (1) for a conceptual course, read them to gain further insight into the meaning of the principles under discussion, but with no expectation of doing similar calculations; (2) for a more quantitative course, read them as illustrations of how to do similar problems; or (3) completely ignore them without loss of continuity in the text. Numerical problems at the ends of the chapters are included for use in quantitative courses, but are not essential for conceptual study.

The style is deliberately informal and conversational. Questions throughout the chapters are intended to promote dialogue between the text and reader. Many of the qualitative questions both within and at the ends of the chapters are intentionally left open to various interpretations in order to foster thought and discussion concerning the real-life factors that influence the answers. Sections marked with the ** symbol describe applications only, and contain no new physics.

Finally, let's consider the important question: How can you use this book in your course or for your reading needs? The book is designed for you to choose the topics you want to study. To a large extent the parts, and to some extent the individual chapters, are independent of each other. Chapter 1 is the only complete chapter required for all readers. I also strongly recommend Sections 2.5, 2.11, and 4.3.

The Instructor's Manual contains a Chart of Section Prerequisites that shows which sections are prerequisite to any given section in the book. By using this chart, you may choose for your course or reading those topics most appropriate to your needs. You may cover them in any order consistent with the prerequisites shown. The first chapter in each part contains sections that are needed for later chapters in that particular part. Other chapters are relatively independent of each other, but it is important to refer to the chart to determine the specific interdependence. When the text makes references to earlier material not specified as prerequisite, they are not fundamental to the particular discussion and can be overlooked when the earlier section was omitted.

Acknowledgments Several people provided the initial support and encouragement to undertake this book. Ed Jones, as advisor to elementary education majors, audited my course from which the text grew, read and commented on the sample chapters, and provided the interest and enthusiasm I needed to get started. Vic Teplitz, my department head at that time, provided extensive support and encouragement. My wife Lou, both initially and throughout the project, has provided the faith that this book would become a reality, during my many periods of doubt.

The development from random scribblings to workable book came about with the help of many. Autumn Stanley has been of immense help in developing an informal and readable writing style. Michael Snell was of continual assistance throughout the writing stage. The physicists who reviewed the book—Russell Cloverdale, Andrew Kowalik, Ronald Brown, Stanley Yarosewick, Joseph Aschner, Jean Moore, H. H. Forster, Roger Creel, William Paske, and Jack Wilson—provided innumerable constructive comments and suggestions that I have tried to incorporate as completely as possible. My many colleagues in the V.P.I. & S.U. Physics Department whom I have cornered in the halls have helped to clarify my thinking on many subtleties of physical interactions. Sonje Hinich's thoughtful comments have added clarity to several of the chapters. My grammar consultant, Frieda Bostian, was always available to simplify the complexities of the language. My octogenarian neighbor, Willamay Dean, kept my nose continuously "to the grindstone." Beth Burch and her staff have been most helpful in typing the manuscript, always in time to make up some of my tardiness. Larry Bolling has provided several of the photographs. Janet Manning has assisted in copying and preparing the manuscript for submission. I thank my many students who have used the manuscript as a class textbook and provided useful feedback. My special appreciation goes to my sons Don and Doug for their suggestions of illustrations, for providing the kind of curiosity that has inspired my desire to try to explain physics to nonscientists, and for tolerating my writing on this book when they desperately needed me as a pitcher for their softball games.

I thank Greg Hubit for his able guidance during the process that turned the manuscript into a published book. The staff at Wadsworth Publishing Company has provided the support I have needed throughout the effort. I particularly appreciate the work of Marshall Aronson, who inherited in midstream the shepherding of this project.

Dale D. Long
Blacksburg, Virginia

1

Energy

Energy makes our lives go. It moves us along roads, through the air, over the water. It scrambles eggs in the morning, warms houses in winter, cools stores in summer. It brings us ice in July and strawberries in January. Energy gives us music to lift our spirits and light to illuminate this page at night. Most important, it keeps our bodies working.

Yet, before 1973 most Americans thought little, if at all, about energy. In our minds, the gasoline pump would never run dry, a truck would deliver fuel oil as often as we liked, and the power company was happy to supply us with all the electricity we could use. But with the 1973 oil crunch, people the world over began to recognize the importance of energy and to wonder about what would happen if it ran out.

Bombarded as we are by warnings that our energy *is* running out, it is no wonder that we ask: Just what is energy? Why is it so vital to our very existence?

Energy is one of those things that are best defined by what they do. So first let's consider some of what energy *does* and how it does them. Then we can say what it *is*.

1.1   What Does Energy Do?

We've already mentioned some specific jobs that energy does for us. Now, let's take an overall look at the way it works for us, and try to develop some general categories of energy use.

First, there are **mechanical** jobs. These involve setting something in motion, or keeping it going against either the resistance of the air or the resistance of the surface over which it moves. Some examples would be a car moving along the road, a train pulling a load of coal up a mountain, and an elevator lifting people. Many mechanical jobs involve rotation: the turning of an egg beater as it whips the eggs, the spinning of a circular saw blade as it cuts a board, the rotating of a fan blade as it stirs the air, the whirling of a lawn mower blade as it cuts the grass. Such mechanical jobs all require machines. But sometimes *you* are the machine using energy to move something—as, for instance, when you throw a bowling ball down the lane, or lift a stack of books onto a shelf, or pull a bobsled up a hill. All these jobs represent mechanical uses of energy.

A second category of energy use is the production of **heat.** Familiar examples from this group include the heating of houses, water, and food. We typically do these jobs by burning some fuel or using electricity to produce the heat.

A third category is that of **cooling:** refrigerating, freezing, air conditioning. We most often do these with electricity.

The final categories we'll mention include the production of **sound** and **light.** Both sound and light are usually produced electrically, and we see the results in radios, televisions, light bulbs, and other special devices.

These categories cover most of the everyday ways that we use energy. Obviously this energy—whatever it is—is very handy stuff. Think about how you use fuel or electricity each day, and try to fit those uses into the general categories we have just established.

Our examples indicate that energy for everyday use always seems to come in the form of either fuel or electricity. We use fuel to run engines or to provide heat. Even electricity (as we'll see in detail in later chapters) is usually generated by burning some kind of fuel. The generating plant may burn a fossil fuel—coal, oil, or gas—or use a nuclear fuel—uranium or plutonium—in nuclear reactions to produce heat. In either case, the plant uses heat to produce steam that mechanically turns an electrical generator. Of course, some power plants use mechanical sources to turn the generators. Hydroelectric plants, for example, use water falling over a dam; windmills use moving winds.

Figure 1.1 summarizes the sources and the everyday uses of energy we've discussed. It also shows the usual intermediate stages between source and use. Notice that electricity is often the intermediate stage, and that mechanical processes can be at either the starting or the receiving end, or both. The burning of fuel produces heat that is used either directly or to do some other job. Though some energy sources or uses do not fit into our diagram, the main ones are included.

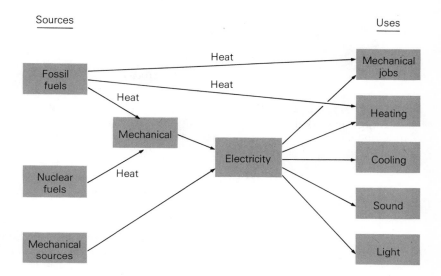

...

Figure 1.1 A schematic diagram showing the sources, intermediate stages, and final everyday uses of energy.

1.2   **Energy: Its Definition and Its Forms**

Now that we've taken an overall look at energy and what it can do, we can get serious about *defining* it. We're still going to use what energy does or what it can do, focusing first on mechanical processes. We'll use specific examples, but keep in mind that these examples illustrate general ideas. Before we can get to our definition, however, we'll need to describe two other terms: force and work.

Work and Mechanical Forms of Energy

Suppose you're planning to hang up your grandmother's picture with a thumbtack. Your objective is to *push* on the tack, and have it *move* into the wall. That push, whether or not the tack moves, is an example of what a physicist would call a **force.** When you push or pull on an object, whether or not it moves, we say you exert a force on it.

Now suppose that you not only exert a force on the thumbtack, but that the force causes it to *move* into the wall, as in Fig. 1.2. The force from your thumb then does *work* on the tack. **Work** is done when a *force moves* something through a *distance*. When you pull on a desk drawer, you exert a force on it; if that force moves the drawer, we say that the force does work on the drawer. If a locomotive pulls on a rusty old freight train, it is exerting a force on it; if that force moves the train, it does work on it. When your car stalls,

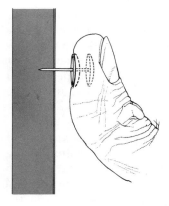

Figure 1.2 Your thumb does work on the tack only if the force moves it into the wall.

Figure 1.3 The force on the car does no work if the car doesn't move.

and you get out to push it, the force on the car does work only if you move the car (Fig. 1.3).

Kinetic Energy   Now let's return to grandmother's picture, which has fallen down because the tack wasn't strong enough to hold it up. You decide to put a nail, rather than a thumbtack, into the wall. Pushing with a thumb can't handle this project; the job calls for a hammer. Why not just hold the hammer against the nail and push? Having a handle to hold, you can exert much more force than by pushing directly with your thumb. But it still isn't enough; you have to *swing* the hammer. The *motion* of the hammer gives it the ability to exert a force on the nail and move it through a distance. The hammer's *motion* gives it the *ability to do work*.

This *ability to do work is energy.* The moving hammer has energy *by virtue of its motion.* We call this form of energy **kinetic energy.**

Here are a few other examples of kinetic energy at work. A moving bowling ball does the work of pushing the pins aside—all ten of them at once if it hits just right. A 260-pound linebacker moving at full speed does work on a quarterback as he tackles him. The moving air of the wind does work on the blades of a windmill used to generate electricity. In a hydroelectric power plant, water falling over the dam does work in turning a turbine (a special kind of waterwheel) that turns the generator. In all these examples, the *motion* of the object gives it energy—the ability to do work.

Gravitational Potential Energy   We'll illustrate our next form of energy in Fig. 1.4 with an old-fashioned waterwheel. Water flowing onto the wheel fills the upturned cups. The weight of the water in these cups causes the wheel to rotate, thereby operating the mill attached to the wheel. In other words, the force exerted on the wheel because of the weight of the water does work on the wheel. The water's position in the gravity of the earth enables it to do work. We call this form of energy, by virtue of position in the earth's gravity, **gravitational potential energy,** or sometimes just *potential energy* for short.

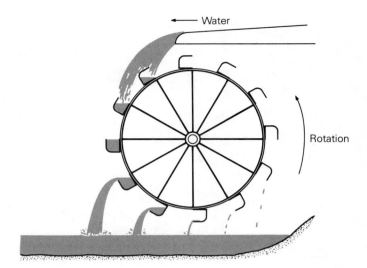

Figure 1.4 The gravitational potential energy of the water does the work of turning a waterwheel.

We mentioned earlier that the kinetic energy of the water gives it the ability to turn a modern waterwheel—a water turbine. It is the potential energy of the water at the top of the dam that does work on the water, speeding it up as it falls, and giving it the kinetic energy it needs to turn the turbine (Fig. 1.5).

The farther an object moves downward, the more work its potential energy can do. In other words, the amount of potential energy an object has increases with its height. But its height has to be specified relative to some particular, arbitrarily chosen, level.

If you have ever visited Thomas Jefferson's home, *Monticello,* you have probably been fascinated by his many ingenious inventions. Among them is his "seven-day" clock mounted in the entrance hallway (Fig. 1.6).

Jefferson wanted a clock that would run for a week. He suspended two sets of heavy weights from cables, ran the cables over pulleys in a corner of the room, and wound the other end of the cables around wheels inside the clock. As the weights gradually descended, they gave up their potential energy in doing the work of running the clock. But Jefferson ran into a snag when the weights hit the floor and the clock stopped before the end of the seven days.

Rather than increase the weight and redesign the clock, Jefferson cut holes in the floor to let the weights pass into the room below. He then had available the potential energy of the weights descending through two stories of height rather than just one. That got his clock through the seven days without anyone having to rewind the weights to the ceiling. In other words, by cutting holes

Figure 1.5 In a hydroelectric power plant, gravitational potential energy of the water does work that gives the water kinetic energy. The kinetic energy does the work of turning the turbine.

in the floor he was choosing a different reference level for measuring the potential energy of his weights.

Elastic Potential Energy   Screen doors on older houses sometimes have a metal spring to pull them shut after someone goes through (Fig. 1.7). The stretched spring, by virtue of being stretched, has potential energy that can do work in closing the door. We call this *elastic potential energy* because the spring is elastic: It returns to its original shape once the stretching force is removed. We store energy in the springs of many everyday devices: squeezing the grip of a heavy-duty staple gun stores potential energy in a spring that then fires the staple into wood or other tough material; winding a watch compresses a coil spring, thus storing elastic potential energy used later to turn the hands of the watch. Some

Figure 1.6 Thomas Jefferson's "seven-day" clock over the door at his home, *Monticello.* The weights are the steel balls you see in each corner. The dark strips on the wall behind the right-hand cable have the days of the week written on them; Friday and Saturday are in the basement. (Photograph courtesy of Thomas Jefferson Memorial Foundation.)

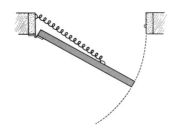

Figure 1.7 Elastic potential energy provides the energy to close this screen door.

devices store elastic potential energy in compressed air. If you cock a BB gun, for example, you compress air that stores energy used later in firing the BB.

Rotational Kinetic Energy Look at the rotating wheel in Fig. 1.8 and notice that kinetic energy is associated with the motion of each part of the wheel. Each chunk of matter (for example, the one marked "x") moves in a circle with some speed, and therefore has kinetic energy. Rather than considering the separate kinetic energy of each of these particles, we can lump the energy all together and think of it as kinetic energy of rotation of the entire body. This kinetic energy can do work through the forces exerted in slowing or stopping the rotation. When we need to distinguish this *rotational kinetic energy* from that due to motion in a straight line, we call the latter *translational kinetic energy.*

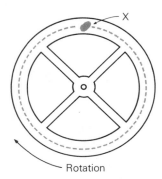

— X

→ Rotation

Figure 1.8 As the wheel rotates, the chunk of matter labeled "x" moves in a circle shown by the dashed line.

Rotational kinetic energy has many practical uses. For example, energy is often *stored* in flywheels—massive wheels that, once started rotating, need a lot of work to be stopped. The New York City Transit Authority has experimented with a system in which the work done in stopping a subway train goes toward increasing the rotational kinetic energy of a set of flywheels. This energy is then used to get the train back up to high speeds.* Flywheels can also help to produce a smooth motion in the face of uneven inputs of force. A potter's wheel, for example, contains a flywheel that is turned with the feet (Fig. 1.9). Because very little of the rotational

Figure 1.9 A potter's wheel has a flywheel to store rotational kinetic energy and maintain smooth rotation.

kinetic energy stored in this wheel is used up as the potter works on a bowl, the wheel maintains a smooth rotational speed. Automobile engines have flywheels that prevent rapid fluctuations in engine speed.

---

* Alden P. Armagnac, "Flywheel Brakes," *Popular Science,* Feb. 1974, p. 70; April 1976, p. 81.

Energy—The Result of Doing Work

The forms of energy we've discussed so far might be called *mechanical,* since using them involves mechanical activity (that is, the doing of work as bodies exert forces *while* they move). Think again about the examples we've used for mechanical energy, and ask yourself how the various devices got their particular kind of energy. You'll notice that in every case, energy was obtained as the *result* of something doing work, of something exerting a force through a distance.

The hammer got its kinetic energy as the force of your hand did work on it. Mr. Jefferson gave the weights of his clock potential energy by lifting them to the ceiling. The spring of the staple gun got its potential energy as you squeezed the handle and compressed the spring. Even the water at the top of the dam got there by the long tortuous route of evaporating, condensing in the clouds, falling as rain, and then flowing to the dam. In every case, the device or object got its energy by work being done on it. We might then equally well define energy as the *result of doing work.*

Question	What does the work that gives

    1. a thrown baseball its kinetic energy?
    2. a rock from a slingshot its kinetic energy?
    3. a set mousetrap its potential energy?
    4. a roller coaster its kinetic energy after passing down a steep grade?
    5. a book on the top shelf of the library its potential energy?

Other Forms of Energy

Figure 1.1 shows that mechanical forms of energy are only a small part of the overall energy picture. We'll now look briefly at the other forms, which we'll discuss in more detail in other parts of the book.

The term *heat* appears in several places on our energy diagram. **Heat** is a form of energy because it can do work. For example, the heat produced as gasoline ignites inside the cylinders of an automobile engine causes the air in the cylinders to expand, pushing the pistons down. This downward motion of the pistons gets converted to rotation of the wheels much as the downward push on a bicycle pedal causes the bicycle's rear wheel to turn. On a larger scale, the heat produced from burning fuel in a power plant generates steam that turns the generator. This is the "mechanical" step between fuels and electricity shown in Fig. 1.1.

Heat can also be the *result* of doing work. Rub your hands together quickly. You feel heat. When one object slides over another, there is a resistance to the sliding that we call *friction*. The work done against the friction force opposing the motion always produces heat.

Question   Your car brakes work on the basis of friction between the brake shoes and the brake drums that rotate with the wheels. Why will the brakes of your car overheat if you keep them on all the way down a high mountain?

Electrical energy obviously plays a major role in our lives. The basis of electrical energy resides in the fact that, associated with the parts of the atoms composing all matter, is a property called electrical charge. The force between these charges holds the individual atoms together, and also holds the atoms that make up matter to each other. When these charges move relative to each other, there is a potential energy change, similar to the potential energy change when a weight moves relative to the earth. During a chemical reaction, such as the burning of a fuel, these charges rearrange themselves to give a different (lower, in the case of burning) potential energy. This change in potential energy, along with any associated changes in kinetic energy of the atom's constituent particles, we call **chemical energy.** Thus fuels—coal, oil, natural gas, wood—store chemical energy that is released as heat as they burn.

There is also electrical energy associated with the *motion* of electrical charges. We call this motion of charge an electrical current. This form of electrical energy is in some ways analogous to the kinetic energy of material objects—that energy associated with their motion. It's this form of electrical energy that operates most of the electrical devices we routinely use.

Electrical processes and their accompanying energy changes are important not only in our useful everyday gadgets. Every part of the human body is controlled by electrical signals and uses electricity for its functioning.

Sound and **light** are both types of waves. Any kind of a wave transmits energy, as we'll see in Part IV. For now, notice that both sound and light have the ability to do work. Sound, for example, can do work on your eardrum as it pushes it in and out, initiating the process by which your ear sends a message to your brain, a message that you interpret as sound. Light can do work wiggling the atomic particles in the receptors in the back of your eyes, enabling them to send messages to your brain about the light received. Thus, sound and light are forms of energy.

Finally, there is **nuclear energy,** a topic currently much in the news. A particular type of force—called the nuclear force—binds the nuclei of atoms together. This force can do work on nuclear particles as it rearranges them within, or ejects them from, a nucleus. Thus, the rearrangement of nuclear particles to form different nuclei can result in the conversion of nuclear energy into other useful forms.

1.3   **Units of Work and Energy**

If you were having an interview for a job, and the interviewer said to you, "We intend to pay you *every* month," you would probably feel comforted. But unless you heard exactly *how much* they intended to pay every month, you might not leave with a very warm feeling.

Or suppose you were visiting the Washington Monument and were considering walking to the top rather than riding the elevator. If you ask a woman just returning from the walk how long it takes, and she says, "twenty," you still won't know very much. She might mean 20 minutes, 20 hours, or 20 days, depending upon her physical condition.

These illustrations point out two facts about understanding quantities of something. When you have to answer the question "How much?" you need both a number and some particular units: dollars, minutes, or whatever is appropriate. Information about energy is no exception. We've now gone about as far as we can in talking about energy, unless we can talk about quantities of energy. To do that, we need to have some units for it. Since energy is the ability to do work, and also results from doing work, we can express both energy and work in the *same* units.

The basic unit of work or energy for scientific purposes is the joule (abbreviated J, named after James P. Joule, a nineteenth-century English physicist noted for helping establish the relationship between heat and other forms of energy. Rather than define the joule now, let me give you a feel for it by saying that it is just about the amount of work you would do if you lifted a baseball from the floor to your dining room table.

Among the many other units used for work and energy in specialized situations are the foot-pound (used with the British system of units); the kilowatt-hour (often used for electrical energy); the British thermal unit, Btu, and the calorie (used mostly for heat); and the Calorie (with capital C, used by nutritionists, who are now switching to kilocalorie since 1 Calorie equals 1000 calories). You can find the conversion factors between these units in

Appendix II. We will define and use these units where appropriate later in the book. For now, let's look at one example of what a specific quantity of energy can do in its various forms.

Let's see how much work can be done with a piece of boysenberry pie. A typical small piece of pie might contain 200 kilocalories, or about 800 000 joules (J) of energy. This is enough to lift an 850-kilogram Volkswagen Rabbit as high as a football field is long! Does that give you an idea why you have to exercise so hard to work off that extra piece of pie? This same amount of energy would heat enough water for you to wash your face twice on a cold morning, and is about 200 times the kinetic energy Pete Rose loses when he slides into second base. It would light a 100-watt bulb for 2¼ hours. For comparison, 17 grams (about 1¼ tablespoons) of gasoline burned completely will release the same 200 kilocalories or 800 000 joules of energy. These unequal-sounding comparisons should begin to appear more reasonable as you progress through this book.

1.4 Conservation of Energy

Today, we are often told to "conserve energy": to turn off the lights when we're not in the room, to set thermostats low in winter and high in summer, to drive a light car at moderate speeds, to use energy-efficient appliances. Admirable as these practices are, they have nothing to do with what physicists mean by conservation of energy. From the point of view of the physicist, *energy is conserved*, and whether or not Aunt Minnie forgets to turn off her basement light has no influence on the matter one way or the other.

We're referring now to what we call the *principle of conservation of energy*. The principle says that *the total amount of energy in the universe, or in any isolated system, remains constant.* By an isolated system, we mean one that no energy enters or leaves. To put it another way, *energy can be neither created nor destroyed; it can only be transformed from one form to another.* Physicists accept this principle as an "article of faith," for no one has ever reported an experiment that violates it. In every experiment that at first seemed to show a violation, something has always turned out to be wrong with either the experiment itself or the interpretation of its results. Someday we may find circumstances under which energy is not conserved. But right now the principle seems as solid as anything we know about our physical world.

If energy is conserved anyway, what does poor Aunt Minnie conserve by shivering through the winter with her thermostat set

**PHYSICIST PROPOSES
STRANGE NEW PARTICLE**

TÜBINGEN, GERMANY, (December 7, 1930)—A group of physicists gathered here for a conference heard today a proposal for the existence of a strange, new particle, as yet unobserved. Wolfgang Pauli, a noted Austrian physicist, made the proposal in a letter from Zürich, Switzerland, to the conferees.

Pauli suggested the particle because a certain kind of radioactivity, called beta decay, seems to violate the principle of conservation of energy and certain other accepted conservation principles. When an atom beta-decays, it emits a particle called an electron, and in the process becomes an entirely different kind of atom. The problem is this: When experimenters measure the total energy of the electron and new atom, it adds up to less than the energy of the original atom. Rather than abandon the hallowed principle of energy conservation, Pauli suggested that another particle is emitted along with the electron, and that it carries away the rest of the energy. He admits that the particle would need to have some strange properties: very small mass, probably less than that of the electron; no electrical charge, since it otherwise would have been detected already; and very high speeds. The name "neutrino" has been suggested for this proposed particle.

Most of the physicists at the conference remain highly skeptical of Pauli's proposal. Although they do not like the idea that energy might not be conserved, they find Pauli's suggestion outlandish and highly questionable.

Pauli himself was unable to attend the conference because of social commitments. He said in his letter, "Unfortunately, I cannot personally appear in Tübingen, since I am indispensable here on account of a ball taking place in Zürich on the night from 6 to 7 of December."

on 20°C rather than a cozy 25° (where she kept it back in the "good ole days")? She conserves energy in a *usable* form. What the energy conservation principle talks about, on the other hand, is *total* energy. When Aunt Minnie saves fuel by not burning it, she is saving energy in a form that can be used for practical purposes. Once she burns it, its chemical energy is converted to heat that eventually dissipates in the environment. This energy is no longer available for practical use. However, *the total amount of energy in existence has not changed by the burning of fuel.*

Total energy is conserved. So what? For one thing, it gives us a handy way of analyzing what happens in a system. Let's use an analogy. Suppose you have a friend who suspects his roommate of having sticky fingers—especially when they pass near his billfold at night. Your friend might use a conservation principle to check out his suspicions. Suppose in a given month he has $510 to spend, and he keeps an accurate tally of where it goes ($194.50 for rent, $121.76 for food, $39.18 for utilities, $26.40 for books, $41.28 for miscellaneous itemized expenses) but has only 88¢ left. Using the

SCIENTISTS CONFIRM EXISTENCE OF NEUTRINO

AIKEN, SOUTH CAROLINA. (July 20, 1956)—American physicists reported today in the journal *Science* that they have confirmed the existence of the neutrino, the particle proposed by Wolfgang Pauli nearly 26 years ago. The scientists, Clyde Cowan and Frederick Reines and their co-workers from the University of California's Los Alamos Scientific Laboratory in New Mexico, detected a small fraction of the billions and billions of neutrinos given off during the beta-decay processes inside a nuclear reactor. They carried out the experiment at the U.S. Atomic Energy Commission's Savannah River Laboratory in Aiken, South Carolina.

Pauli, who won the Nobel Prize in 1945 for his discovery of what is now called the Pauli exclusion principle, proposed in 1930 that the neutrino must exist. Otherwise, beta-decay processes would violate the principle of conservation of energy, as well as other conservation laws of physics. Even though this elusive particle interacts so weakly with matter that it had always escaped detection, physicists have accepted its existence for some time. Nevertheless, directly observing it has reaffirmed their faith in the energy conservation principle.

principle of conservation of dollars—the total number of available dollars stayed constant—he can conclude that $86 disappeared under mysterious circumstances.

A conservation principle is one of the most powerful tools we have for analyzing physical processes. As we go through this book, we will find that several physical quantities are conserved in isolated systems. We then can use the appropriate conservation principle to analyze what's happening in a particular system. We'll now use the energy conservation principle.

1.5   Energy Transformations

Since energy is neither created nor destroyed, but only transformed, the study of energy is obviously a study of its transformations: changes from one form to another. Looking back at Fig. 1.1, you can see that what happens between a source of energy and its use is merely a transformation, or series of transformations, from one energy form to another. In some cases, this is a direct, one-step process. In others, there are various intermediate stages. We'll now look at a few general types of energy transformations and illustrate them with specific everyday examples.

Transformations between Potential and Kinetic Energy

Many everyday processes involve transformations between purely mechanical forms of energy—potential and kinetic energy—while other forms of energy remain nearly unchanged. In such a system it must be true that

$$\text{potential energy} + \text{kinetic energy} = \text{constant}.$$

An apple hanging on a tree has potential energy. If it falls from the tree, it loses potential energy as it falls. But it gains kinetic energy. Until it hits the ground (at which time it is no longer isolated) the kinetic energy gained equals the potential energy lost.

A clock pendulum swinging back and forth is continually undergoing changes between potential and kinetic energy. At the highest point in its swing, where it reverses direction, it has no kinetic energy. As it swings downward, its height above the earth decreases, lowering its potential energy. The loss in potential energy is matched by the gain in kinetic. As it swings up the other side, it gives up its kinetic energy to potential energy. The same thing happens as your little brother swings in his swing. Energy is continually changed back and forth between potential and kinetic energy.

A Yo-Yo provides an interesting example of energy transformations. After the string is wound completely and the Yo-Yo released to fall, the string unwinds, and as it does, some of its potential energy changes to translational kinetic energy, but most of it changes to rotational kinetic energy. When the Yo-Yo gets to the end of the string, it possesses rotational kinetic energy that it has no way of losing. Since it must continue to rotate, it winds back up the string, converting this rotational kinetic energy to potential energy again.

One final example: The water that goes over the dam in a hydroelectric plant has its potential energy converted to kinetic energy for use by the turbine.

In every case, the energy is transformed as work is done. The potential energy of the apple just as it is released from the tree is able to do work on the apple in moving it. This work done on the apple gives it kinetic energy. If you caught the apple before it hit the ground, and threw it back toward the tree, the kinetic energy it had when you released it would give the apple the ability to do the work of increasing its potential energy. One form of energy *does* the work; the work done *increases* the other form of energy. *But the total energy remains constant as long as the system is isolated,* which is *not* the case *during* the catching and throwing processes.

Question Assume you throw the apple as just described. What gives the apple its kinetic energy? For each of the energy interchanges in this action, describe the work done by one form of energy, and how this work increases the other form.

Frictional Effects

You know from experience that if you're swinging and you stop "pumping," and no one pushes you, you gradually slow down. In any one swing, you don't go as high as you did before. Similarly, when you're "yo-yoing" and you hold your hand still, the Yo-Yo never comes all the way back up to your hand after you release it. The subway train, whose flywheels store energy while stopping, never quite has enough energy in the flywheels to get the train back to its original speed. In every case, there is a loss of *mechanical energy.*

Since *total* energy is conserved, the mechanical energy lost must go to some other form of energy. It goes to *heat* as the parts of the system rub against each other and against their surroundings, thereby doing work against the resulting frictional force. In swinging, you push the air aside to get through, and the chains rub against the hooks holding them. The Yo-Yo meets air resistance and the sides of the groove rub against the string. All the moving parts of the train experience frictional resistance. The work done against friction decreases the total mechanical energy of the system unless some energy is added from an outside source. We can then write an energy balance equation of the form:

(initial kinetic energy) + (initial potential energy)

= (final kinetic energy) + (final potential energy)

+ (work done against friction).

Question   Why do you have to wind a pendulum clock? Note: The unwinding spring gives the pendulum a little kick during each swing. Compare this kick with the push somebody has to give you to keep you swinging at the same pace.

Question   When you put the brakes on in your car, where does the kinetic energy of the car go? If you put on the brakes gradually, where is this heat produced? If you slam on the brakes and lock the wheels so they can't turn, where is the heat produced?

Figure 1.10 The appearance of a smooth metal surface as seen under strong magnification.

Before leaving friction, let's mention lubrication. The main reason for friction between solids is that any surface, no matter how smooth, is jagged when looked at with enough magnification (Fig. 1.10). When two such surfaces are in contact, the jags interlock to provide resistance to relative motion. If the parts rub together continually, friction will wear them out. Oil added around the parts fills the holes and forms a thin layer of liquid between the pieces of solid. Then each solid part rubs only on liquid, thereby greatly reducing friction.

Other Energy Transformations

To say very much about the other kinds of energy transformations illustrated in Fig. 1.1, we need to have an understanding of the special areas whose energy we are describing: light, electricity, and so on. That's what the various parts of this book are all about. For now, we'll just mention a few everyday examples of nonmechanical energy transformations.

When you turn on a light switch, you let electrical charge flow through the filament of the bulb. As the filament resists the flow of this charge, it gets heated by an effect similar to friction. In that way, electrical energy is converted to heat energy. Some of this heat energy is converted to light when the filament gets hot enough to glow.

An automobile engine is a device for using the chemical energy stored in gasoline to do work in moving the car along the road. In other words, the engine converts this chemical energy to: (1) kinetic energy of the car and its contents; (2) potential energy of the car and its contents if it moves uphill; and (3) heat generated as the car meets frictional resistance. The engine cannot convert all the stored chemical energy to work. It is first converted to heat in the engine. Most of this heat (typically about 90%) is either carried by the water and antifreeze to the radiator, where it is given up to the air flowing past the radiator fins, or is expelled through the exhaust pipe. If the energy lost through the radiator and exhaust by a typical "full-sized" car at highway speeds could be converted to electrical energy, it could operate 600 100-watt light bulbs! The remaining small fraction of the energy does the work of moving the car.

Where does the energy come from that provides the gravitational potential energy your body gains in walking up a flight of stairs? It comes from the chemical energy stored in food you have eaten. Your body is able to use this chemical energy to do work in walking. When walking upstairs, you have to do more work than on a level because you have to increase your body's potential energy. Walking or riding a bicycle uphill is harder than walking or riding on level ground because additional work must then be done to increase the potential energy.

Try to "think energy" for a day. As you go about your daily routine, look for ways that you cause energy to be transformed from one form to another. What is the form before and after the transformation? Is there an intermediate form?

1.6   Efficiency

Most of the time when we deliberately transform energy from one form to another, we don't get all of it converted into the form we want. Due to the friction in the pipes carrying water from the top of a power plant dam to the turbine, not all of the potential energy lost by the water gets converted to kinetic energy on the way to the turbine. Due to the friction between the chain and sprocket wheels, and between the other moving parts on a bicycle, all the work you do on the pedals isn't converted into the work of moving the bike. In the case of a light bulb, some of the electrical energy used is given off as heat, not light. When Mr. and Mrs. Warmbody burn natural gas in their furnace, some of the heat produced gets lost from the ducts carrying hot air to their rooms; they do not benefit from all of the energy in the fuel. We use the term *efficiency* to describe how much we get out of a system relative to what we put into it.

We define efficiency as follows:

$$\text{efficiency} = \frac{\text{output energy (or work)}}{\text{input energy (or work)}} \times 100.$$

The output energy is the amount of useful energy we get out of a system in the form we want. The input is the amount of energy we put in, in some other form. If we are talking about a mechanical machine—such as a bicycle—on which we do work on one part, and another part of it does work in return, we use work instead of energy in the definition. In some devices—such as automobiles— we put chemical energy in and get mechanical work out; the efficiency is then the ratio of work out to energy in. In all cases, we multiply the ratio of energies by 100 in order to express the efficiency as a percentage. From the energy conservation principle, efficiency must always be less than (or equal to) 100%.

The difference between energy (or work) input and the energy (or work) output is usually considered "lost" energy. It may be lost from doing useful work, but it must appear *somewhere* in *some* form. In a mechanical system where the loss is due to friction, the lost energy goes to heat. We'll discuss a few examples of the use of efficiency.

Suppose you ride your bike up a certain hill and, in doing so, the potential energy of your body and the bike increases by 200 joules. If you must do 115 joules of work to move against the wind resistance, and you consider this to be part of the useful output work, the total output work is 315 joules. If your feet have to do 350 joules of work on the pedals, the mechanical efficiency of the bike is

$$\text{efficiency} = \frac{\text{work out}}{\text{work in}} \times 100 = \frac{315}{350} \times 100,$$

or 90%.

The average home heating system that burns oil to heat water which then circulates through radiators or baseboards to heat the house is about 50% efficient. This means that half the chemical energy supplied in oil to the heating system does a useful job, and the rest is wasted. A system properly designed and functioning well is capable of about 70% efficiency; some heat must be carried up the flue if air is to enter to sustain combustion.

A large electrical generator can produce over 1 billion joules of electrical energy *per second*. For it to do this, the plant needs to burn fuel at a rate that supplies 2.7 billion joules per second of chemical energy. In other words, the overall efficiency is around 37%.*

Question If you have a choice of heating your room with a 50% efficient oil burner or with a 100% efficient electric heater (that is, one that converts all the electrical energy supplied to it to heat) which way would you use more fuel? Assume your electrical power company is 37% efficient and operates on oil.

In some situations the "lost" energy isn't lost from usefulness. An ordinary 100-watt light bulb is only 2.5% efficient at producing light. That is, only 2.5% of the electrical energy supplied to it gets converted to light. The rest goes to heat, most of which is radiated into the room. If it happens to be winter, the load on your heating system is reduced by the amount of heat supplied by the bulb.

---

*The laws of thermodynamics tell us that it is not *even in principle* possible to convert all the heat to work for turning the generator. The 37% is, therefore, not as bad as it seems.

1.7   **Physics and Physical Laws**

You would probably guess from the title of this book that it is supposed to be a *physics* book. Yet all we've talked about is energy. We have said nothing about physics or what it is. I want to assure you that this *is* a physics, not an energy, book. It discusses all the traditional topics usually covered by a "general physics" book.

I suggest that you compare the Table of Contents with the energy diagram of Fig. 1.1. You will see a strong similarity between the blocks in the figure and the various parts of this book. You won't find the cooling block explicitly covered as a major section because we classify it with heat. The blocks concerning fuels are special considerations within the study of atomic and nuclear interactions. The prominent role played by electricity in this diagram is closely related to the dominant place of electricity in the center of this book.

Physics deals with a description of the interactions of matter in the categories indicated by the headings of the parts of this book and, directly or indirectly, by our energy diagram. It also deals with the energy interchanges within and among these various categories. I have left this discussion of what physics is about until late in this chapter, in hopes that by now you see that physics *is where you are,* that it deals with the things going on around you and within you, as well as the things physicists have tucked away in their secluded laboratories. In this book we're going to focus on the physics *around you,* while at the same time giving you a peep at the physics "done" by physicists in their laboratories.

The Meaning of Physical Laws

A physicist describes the way parts of matter interact with each other in terms of what we call *laws.* These are not laws of the type passed by Congress, but are principles or rules telling the way physical quantities behave or interact with each other under certain circumstances. Physical laws cannot be invented. As statements describing the behavior of the physical world, they have to be *discovered.*

So far we've talked about one law: the law of conservation of energy. (We called it a "principle" earlier, but the two words mean basically the same thing and can usually be interchanged.) We will encounter other laws as we go through the book.

Physical laws say nothing about *why* parts of our universe interact as they do. Rather, they describe the nature of these interactions. For example, in the next chapter we will discuss the law of gravitation discovered by Sir Isaac Newton. This law tells us that a force of attraction exists between any two pieces of matter, and it tells us how that force depends on both the amount of matter present and its relative separation. But it gives no explanation of *why* the objects attract each other.

When we study magnetism, we will find that when a particle with an electrical charge moves through a magnetic field, a force acts on this particle. The force acts in a direction that is perpendicular to both the magnetic field and the direction of motion of the particle. Most people, when they first learn this, find the direction of the force to be somewhat illogical, and they often ask "Why?" The appropriate physical law describes what happens, but does not say why it happens. All we can say about "why" is that that's the way nature is.

You may find this state of affairs a bit disappointing if you were looking forward to having a lot of your whys answered by reading this book. I suggest that you withhold that disappointment. I think you will find it rather fascinating that you can understand many of the complex processes that go on around you in terms of a few physical laws. Most of these laws should seem reasonable in terms of your own intuition and experiences.

The job of the physicist is to explain the processes of nature in terms of as few physical laws as possible. These laws should be as simple as possible and still describe the interactions to which they apply. Often laws must be modified as new facts are learned, or expanded to include situations not considered when the laws were discovered.

Usually you can describe the behavior of a given event or process from the viewpoint of more than one law. In this book, I have chosen the energy conservation law as the unifying theme throughout, since it applies to all areas. We will try to understand the various physical processes in terms of energy and energy transformations. That's the reason for discussing energy first.

Physicists usually write physical laws in mathematical language. That way the law is more precise and leaves no room for misunderstanding. Let me illustrate with a simple example. Suppose a fourth grader tells you that she rode with her father for 2 hours while he drove continuously at a speed of 80 kilometers per hour, and she wants your help in figuring out how far she rode. You could tell her that the faster and longer she rides, the farther she will go. This information would not be very helpful. But if you tell her to multiply her speed by the time, you are giving her a quantitative mathematical relationship from which she can get a precise answer. In expressing the laws of physics mathematically, we gain the same advantages as we do when the distance in the fourth grader's problem is expressed mathematically.

You've probably heard that physics "has all that math," and you may feel a little unsure about what lies ahead in that respect. In this book you will not find any math more complicated than simple algebra. We will find that when we express physical laws and properties quantitatively, it not only gives us a more concrete understanding, but lets us figure out some useful information. (Remember, you want to know how much you're to be paid in your job each month, not just that you will be paid.) This means that we'll often be using numbers as well as units that give meaning to the numbers. The next section deals with the way we handle numbers and units in this book.

Working with Numbers and Units

When we specify the value of some quantity, we will usually round it off to two significant digits: for example, 7800 rather than 7764; or 0.041 rather than 0.04129. Also, when we do a calculation using two-digit numbers, we will usually round off the answer to two digits. When one answer will be used in another calculation, we will keep an extra digit in the first answer to avoid confusion in the later one. If it seems to you that this rounding off is throwing away all those good numbers your calculator gives you, I suggest you read Part A of Appendix I. It describes the use of significant digits and how they give you information about how accurate a number is.

For some topics, we will need to use very large or very small numbers. It will save you and me both a lot of trouble if we use the "powers-of-ten" notation for such numbers. That means expressing $2\,900\,000$ as 2.9×10^6 and $0.000\,071$ as 7.1×10^{-5}. If you're rusty on, or unfamiliar with, this notation when we start using it, spend a few minutes with Part B of Appendix I and brush up on it.

In Section 1.3 we illustrated that a number for a physical quantity means nothing without specifying the units used. We will stick with metric units, with an occasional comparison in the old "British" system to aid your intuition when the British unit may be more familiar. Our basic units will be a particular subgroup of metric units known as the SI system, where SI stands for *Système International*. This system has been internationally adopted for scientific purposes. In it, we express distance in meters (m), time in seconds (s), and energy in joules (J), where the accepted abbreviations are in parentheses. As we define new quantities, we will give their appropriate SI units.

Appendix II gives you more information on metric units, as well as conversion factors between various units. It also includes a definition of the metric prefixes that express various multiples of 10 of the basic units. These prefixes are very handy, and we'll use them where appropriate. It's much easier, for example, to say that a uranium nucleus has a radius of 7.4 fm than to say its radius is 7.4 $\times 10^{-15}$ m.

1.8   Power

The main terms that we've talked about in this chapter—force, work, energy—we also use in ordinary conversation unrelated to physics. In ordinary use, these terms imply something similar to the more restricted and precise meanings we have used. In like manner, another popular term, *power* (whose meaning relates to the ability to act), seems similar to our definition of energy. However, in physics, "power" has a precise definition. The concept of power relates to energy, but has its own distinct meaning.

Power *means the rate at which work is being done* or *the rate at which some form of energy is being used.* In other words, power is the amount of work done, or energy used, in each unit of time.

Suppose you're spending the weekend with a friend who lives on the fourth floor, and you do 2400 joules (J) of work in carrying your suitcase up the three flights of stairs. You might do this work at a leisurely pace, taking 60 seconds (s) from bottom to top. But if you're in a hurry, you might make it in 20 s. The amount of work done on the suitcase is the same either way, but the rate of doing work, the power, is different.

In the 60-s case

$$\text{power} = \frac{\text{work}}{\text{time}} = \frac{2400 \text{ J}}{60 \text{ s}}$$

or 40 J/s, where J/s means joules per second. When you do the work in 20 s,

$$\text{power} = \frac{2400 \text{ J}}{20 \text{ s}}$$

or 120 J/s. The power is three times as much when you do the work in one-third the time.

We could express power in joules per second, but in the SI system the unit for power has the special name *watt* (W) after James Watt, the inventor of the steam engine. That is,

1 watt (W) = 1 joule/second (J/s).

When we specify the "wattage" of a light bulb, for example, we specify the electrical power; that is, the rate at which it consumes electrical energy. A 100-W light bulb uses electrical energy at the rate of 100 joules per second. This is the rate at which you would do work if you lifted 120 5-kilogram (11-lb) bags of sugar from the floor onto a 1-m-high table in one minute.

The *horsepower* (hp) is a unit of power often used. When James Watt invented the steam engine, he wanted to be able to tell people how many horses his engines could replace. By suspending a weight in a deep well, and hooking a horse to a mechanism that could lift the weight, he found that a typical horse could do 746 joules of work per second, or 746 watts. Watt defined the horsepower to be that amount of power. The Watt engine of 1778 had 14 hp.

The horsepower of an automobile engine tells you the rate at which that engine is capable of doing work. If a 100-hp engine could accelerate a particular car from rest to a speed of 80 kilometers/hour in 20 s, a 200-hp engine could give the same car that speed in roughly 10 s. I say "roughly" because several factors could influence things somewhat: The capabilities of the two engines might vary differently with speed, the frictional effects might be different, and the more powerful engine would probably make the car weigh more. If all these extraneous influences were negligible, the time would be cut in half.

Let's do one more thing with our definition of power. Since

$$\text{power} = \frac{\text{energy}}{\text{time}},$$

we can rewrite this equation as

$$\text{energy} = \text{power} \times \text{time}.$$

In other words, if your electric heater uses a power of 1000 W, and you use it for 4 hours (h), the electrical energy used is

$$\text{energy} = 1000 \text{ W} \times 4 \text{ h}$$

$$= 4000 \text{ watt-hours}.$$

We see from this example that the watt-hour (Wh) is an *energy* unit that results from multiplying the power in watts by the time in hours. Often we use a larger unit, the kilowatt-hour (kWh), which equals 1000 Wh or 3.6×10^6 J.

When you pay your electrical "power" bill, is your payment based on the power used or the energy used? This question is like asking whether you are paying for how much energy you use or for how fast you use it. Hint: look at the units used on your (or somebody else's) latest electric bill.

Key Concepts

A push or a pull, whether or not it produces motion, is a **force.** **Work** is done when a force moves something through a distance. **Energy** is the ability to do work, or the result of doing work.

The main forms of energy are: **kinetic**—energy by virtue of motion; **potential**—energy by virtue of position; **heat; electrical,** including **chemical; sound; light;** and **nuclear.**

The **principle of conservation of energy** states that the total amount of energy in the universe, or in any isolated system, remains constant; energy can be neither created nor destroyed, only transformed.

Any work done against friction produces heat.

$$\text{Efficiency} = \frac{\text{output energy (work)}}{\text{input energy (work)}} \times 100.$$

Power is the rate at which work is done, or the rate at which some form of energy is used.

The laws of physics describe the interactions of matter and the energy interchanges involved in these interactions. These laws describe what happens, but not why it happens.

Questions

Force and Work

1. We said in this chapter that a push or a pull is a force. To have a force, must you have motion?

2. List as many as possible of the objects you have exerted a force on in the past 24 hours.

3. The weightlifter of Fig. 1.11 is holding the weights stationary over his head. Are his hands doing any work on the bar?

Figure 1.11 The weight lifter is holding the weights still.

Forms of Energy

4. Do you have kinetic energy as you are sitting still in your car while it's traveling at 85 km/h?

Figure 1.12 An old-fashioned spinning wheel.

5. Why do experienced carpenters advise beginners to hold the hammer at the end of the handle, as far as possible from the hammer head?

6. A stretched slingshot is capable of doing work on the pebble in it. What form of energy does the slingshot have?

7. Why is the wheel of an old-fashioned spinning wheel (Fig. 1.12) so large? What form of energy is stored in this wheel as it turns?

8. My son has a toy truck that can *slowly* creep a long distance across the floor after you press down on it and give a slow, steady push. It can even climb over small objects in its path. What kind of energy gets stored in the push that gives is this ability? Hint: The toy truck emits a gradually diminishing whine as it eases across the floor.

Conservation of Energy

9. Does "conserving energy" by turning off unnecessary lights and using less hot water have anything to do with the principle of conservation of energy?

10. To describe conservation practices such as those mentioned in Question 9, would it be better to use the term "fuel conservation" than "energy conservation"?

11. Space capsules are equipped with heat shields to protect them from the tremendous amount of heat developed when they reenter the earth's atmosphere. Where does this heat come from?

12. Why does it "burn" to catch a fast-moving baseball?

13. When a train moving at a speed of 80 km/h stops quickly, what happens to the kinetic energy it had while moving?

14. A car initially moving at a speed of 80 km/h coasts on a perfectly level road until it stops. What happens to the kinetic energy it had?

15. When Joe Morgan slides into second base, what happens to his kinetic energy?

16. An apple hanging on a tree suddenly falls. Just before it hits the ground, what has happened to the potential energy it had while hanging on the tree? What has happened to it after it hits?

Figure 1.13 Modern pole vaulters, such as Dave Roberts, use a flexible pole.

17. Can you explain *why* the total amount of energy in an isolated system remains constant?

Energy Transformations

18. Pole vaulters now use flexible fiberglass poles (Fig. 1.13) rather than the rigid poles once used. Why does a flexible pole help them transform their energy to the kind they want in order to win? What kind of energy is that?

19. A roller coaster has no engine or motor inside, but starts its run after being pulled to a high point on the track. Can it ever get that high again on its own? What gives it the energy that makes it go?

20. Why do you need to wind a grandfather clock? Doesn't the pendulum keep it going?

21. What various forms of energy are present as you shoot an arrow with a bow?

22. Some people still use the method the colonials often used to close the gate in the fence around their yard (Fig. 1.14). The weight that rises when the gate opens, pulls the gate shut later. What actually provides the energy to close the gate?

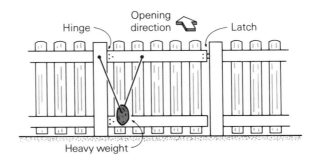

Figure 1.14 A cheap but effective gate closer.

23. What eventually happens to the electrical energy supplied to operate an egg beater? An electric fan? An electric shaver?

24. Describe the energy interchanges involved as you jump on a trampoline.

25. Circus performers sometimes bounce a partner high into the air by jumping onto the high end of a seesaw on which that

Before

After

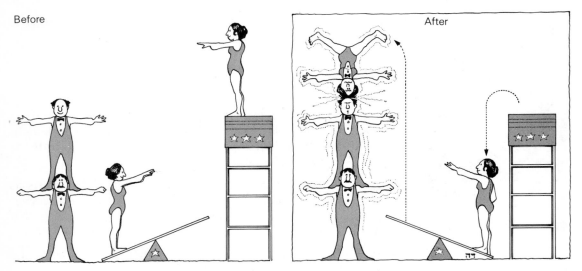

Figure 1.15 Energy transformations at the circus.

person is standing (Fig. 1.15). Explain the energy interchanges. Why doesn't the first person need to be situated as high as the second one goes?

26. What forms of energy are associated with hammering a nail into a board?

27. Why do some A.A.U. track records say "wind-assisted"? From an energy/work viewpoint, why is it harder to run against a wind than to run when there is no wind? Why is it harder to run uphill than on level ground?

28. The heater on most cars works by circulating the water and antifreeze (that normally goes to the radiator) through the heater to give up heat inside the car. Usually, a fan blows air through the heater to speed delivery of this heat to the passengers. Do you get lower gas mileage when you use the heater on such a car?

29. A car equipped with radial tires gets better gas mileage than a car that has bias ply tires. What does this say about the amount of friction that opposes the rolling of a radial tire compared with the rolling of a bias ply tire?

Power and Efficiency

30. A rototiller salesman claims that his chain-drive rototiller with a 3-horsepower engine has 3½ or 4 horsepower at the tiller blades after passing through the chain-drive mechanism. Is this possible? Why or why not? (An example of a chain-drive mechanism is found on bicycles.)

31. If you push a lawn mower across a yard in 10 s, how does the work done compare with pushing it across the same yard in 20 s? How does the power compare in the two cases?

32. Contrary to what you might predict, if you stop up the inlet hose of a working vacuum cleaner, it takes less power to keep the cleaner running. Why? Would the same be true if you stopped up the air-output end? (If you aren't convinced of this power reduction, try to persuade your instructor to connect a vacuum cleaner via an electrical current meter, and check it out.)

Problems

1. The average American eats about 3000 kcal (12 500 000 J) in one day. If all this energy could be converted to mechanical work, how many 25-kg sacks of potatoes could this average person lift from the ground to a 1-m-high truck in a day? It takes about 250 J to put one sack on the truck. Since no one can do this much mechanical work, where does the rest of this energy go? **Ans:** 50 000

2. Show that 1 kWh = 3.6×10^6 J.

3. A commuter train and 300 passengers might have a kinetic energy at full speed of 16 MJ (M = 10^6). If electrical energy to operate the train costs 5¢/kWh, how much does it cost to get this train up to full speed? Assume 100% efficiency in use of the electrical energy. (See Problem 2.) **Ans:** 22¢

4. Suppose a train such as the one of Problem 3 could store in flywheels and reuse ⅓ the energy dissipated in stopping. Estimate how much money New York City might spend equipping a train with flywheels and still save money. Assume, for example, the train lasts five years, and estimate the number of stops it might make per day. **Ans:** At 500 stops/day, $66,900

5. A truckload of eggs has 60 MJ more gravitational potential energy at the top of a 600-m-high mountain than at the bottom. If 1 kcal = 4190 J, how many kcal of heat are generated in the brakes of the truck if it maintains a constant speed going down the mountain by using the brakes only? Neglect wind resistance and other frictional effects. **Ans:** 1.4×10^4 kcal

6. It takes about 120 kcal of energy to heat a gallon of water from 13°C (55°F) to 43°C (110°F). At 20 gallons per bath, how many baths could you take on the amount of heat generated in one trip down the mountain of the truckload of eggs in Problem 5?

   Ans: 6

7. If you lose 1500 J of potential energy in sliding down a sliding board, and you hit the ground with a kinetic energy of 1000 J, how much heat was generated in your posterior and the sliding board?   **Ans:** 500 J

8. A liter of water loses 100 J of potential energy in falling 10 m. If 2000 liters of water pass over a 10-m-high dam and into a water turbine per second, what is the mechanical power output of the turbine? Assume the system to be 60% efficient at using the water's potential energy.   **Ans:** 120 kW

9. Complete combustion of fuel oil releases about 9500 kcal of energy per liter of oil. If the Jones family used 1200 liters of oil to supply 6.5×10^6 kcal of heat to their house last winter, what was the efficiency of their heating system?   **Ans:** 57%

10. If gasoline costs $1.00 per gallon and your car is 15% efficient overall, what amount are you paying per gallon of gasoline that does useful work?   **Ans:** $6.67

11. A particular power plant converts 35% of the chemical energy in the oil it burns to electrical energy delivered to a house. If this energy is used to heat the house with electrical heaters that are 98% efficient in converting electrical energy to heat, what is the overall efficiency of this method of burning oil to produce heat?   **Ans:** 34%

12. A typical large city might produce 1 billion kg of garbage each year. If half of this is burned for producing heat to generate electricity, how much electrical energy in kWh is produced by a 32% efficient power plant if 1 kg of garbage releases 10 million J of energy when burned? If this energy is sold at 5¢ per kWh, what is the gross yearly income from the plant?

    Ans: 440 million kWh; $22 million

13. If running up a flight of stairs increases your gravitational potential energy by 2100 J, and you make it in 3 seconds, what is your horsepower?   **Ans:** 0.94 hp

14. At 5¢ per kWh, how much does it cost to operate your 180-W electric blanket for a month, assuming you use it 8 hours a day, and the heating element is actually operating an average of 50% of the time the blanket is being used?   **Ans:** $1.08

Home Experiments

1. Drive a nail into a hard piece of wood, or pound on a small piece of metal with a hammer or other heavy object. Then feel the nail or piece of metal and notice that it is hot. Interpret this result in terms of energy conservation.

2. Study the action of a Yo-Yo by letting it unwind without moving your hand. Describe the energy transformations.

3. In a large empty parking lot or other area with no traffic, find out how a car's braking distance changes with speed. Start with a low speed, then double and triple it. Put on the brakes at the same spot each time so that it will be easy to compare distances. Assuming the same stopping force each time, the work done in stopping the car is proportional to the stopping distance. Show that the kinetic energy is proportional to the square of the speed.

References for Further Reading

Fowler, J. M., *Energy and the Environment* (McGraw-Hill Book Co., New York, 1975). A readable but comprehensive discussion concerning all aspects of energy in useable form, including sources, relative consumption for various purposes, technology of supplying useable energy, problems, and future projections.

Johnston, W. D., Jr., "The Prospects of Photovoltaic Conversion," *American Scientist,* Nov.-Dec. 1977, p. 729. Discusses the factors involved in converting solar energy directly to electrical energy.

Marion, J. B., *Energy in Perspective* (Academic Press, New York, 1974). Basically the same type of book as Fowler's, but less technical and less detailed.

Post, R. F. and S. F. Post, "Flywheels," *Scientific American,* Dec. 1973, p. 17. Discusses the modern techniques used in one portable method of energy storage.

Scott, D., "Flexible Rafts Harness Wave Power," *Popular Science,* Sept. 1978, p. 94. Describes an interesting approach to generating electricity by an unusual source. Every month for the past several years *Popular Science* has had at least one article concerning what individuals can do to conserve or provide useable energy, and usually at least one article on large-scale methods being used or tried to supply energy.

Starr, C., "Energy and Power," *Scientific American,* Sept. 1971, p. 36. A history of the technological use of energy, and a discussion of present and future alternatives in this area.

Walker, J. *The Flying Circus of Physics with Answers* (John Wiley & Sons, Inc. New York, 1977). Hundreds of fascinating examples of how "Physics and physics problems are in the real, everyday world that we live, work, love, and die in" (quotation from preface of the book).

Walker, J. "Ideas for the Amateur Scientist," *The Physics Teacher,* Nov. 1978, p. 544. Summarizes several ideas for experiments on everyday physics that amateurs can do. Based on articles from "The Amateur Scientist" department of *Scientific American.*

Part I

Mechanics

Every body perseveres in its state of rest, or
of uniform motion in a right line, unless it is
compelled to change that state by forces
impressed thereon.

Isaac Newton

Isaac Newton, born at Woolsthorp, Lincolnshire,
England in 1642, received a BA degree from
Cambridge University in 1665, an MA degree in
1668, and became professor of mathematics
there in 1669. Because of the Plague, he spent
the years 1665 and 1666 at his home in
Woolsthorp, where he laid the foundation for
his later publications in optics, mechanics, and
mathematics. His *Principia,* published in 1687,
contained his renowned laws of motion and the
law of gravitation, which he could verify only
after inventing calculus. His many nonacademic
accomplishments included serving briefly as a
Member of Parliament and as Master of the
Mint in London for the last 27 years of his life.

2

Forces and Their Effects

In the first chapter we pointed out that, when you push or pull on something, you apply a *force* to it. That general statement is not a definition of force. We can define force only by describing *what it does.* In this chapter we will discuss the nature of forces, and the different kinds of things forces can do.

One important thing that forces can do is change the state of motion of an object. As we discuss how forces influence motion, we will need to define a new quantity: *mass.* We will find that the mass of a body is related to, but completely distinct from, its weight. The weight of a body is a gravitational force. The force of gravity instigates the motion of falling.

2.1 Units of Force: the Newton

In the United States we've been accustomed to measuring forces in pounds (lb), whether to express our own weight, the weight of a sack of potatoes, or the thrust of a rocket on its way to the moon. But in the SI system of units, the unit of force is the *newton* (N). One newton is equivalent to 0.225 lb. To get a feel for its size, you can remember that it's about ¼ lb, or about the weight of a McDonald's "Big Mac."

To convert in the other direction, notice that 1 lb weighs 4.45 N. Thus a 10-lb bag of sugar weighs 44.5 N; a 125-lb math major weighs 556 N.

2.2   Contact Forces

Most forces we experience in our everyday routines are what we might call *contact* forces; that is, those forces resulting from the contact of one object with another. When you close a door, your hand exerts a force on the door through contact with it (Fig. 2.1). A bulldozer pushing against a tree applies a force to that tree as its blade makes contact with the trunk. In the famous boxing match where he won the world heavyweight title for the third time, Muhammad Ali's glove made contact with Leon Spinks's face, thereby exerting a force on it (Fig. 2.2). Of the hundreds of ordinary forces that you exert or experience each and every day, only one is *not* a contact force. We'll talk about that exception later in the chapter. Can you pick it out now?

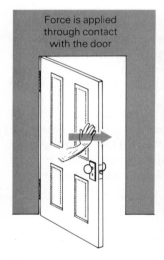

Force is applied through contact with the door

Figure 2.1   Most ordinary forces are contact forces.

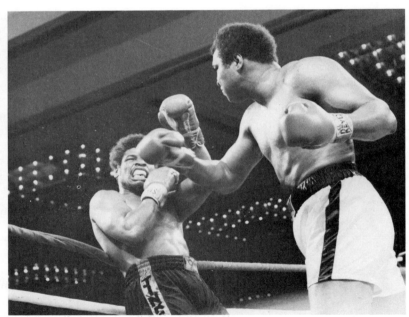

Figure 2.2   Muhammad Ali applies a contact force.

Contact forces are fundamentally electrical in nature. When your hand makes contact with the door, the electrical charges in the atoms of your skin interact with the charges in the atoms on the surface of the door. The result is a force between the atoms of your hand and those of the door. All contact forces come about in this way, including the forces that hold solid bodies together. Adjacent parts of solid bodies are in contact, and are held together by the interaction of the charges of neighboring atoms.

2.3   **The Law of Inertia**

In discussing the concept of "work" in Chapter 1, we talked about forces as well as the motion of the bodies on which those forces act. But we said nothing about *how* forces influence a body's state of motion. We now will consider that important topic—the relationship between force and motion.

A golf ball at rest on a tee does not start to move until it is struck by the club. Nor does this book lying untouched on your desk suddenly begin to move. In general, any object that is not moving does not start to move unless some force acts on it.

On the other hand, a baseball leaving a pitcher's hand continues to move until it is hit by the batter or caught by the catcher. If you apply your brakes on an icy road, your car may continue to move at the same speed. The ice offers very little frictional force to slow your motion. In these latter two examples (and in any other situation) a force is needed to change either the speed or the direction of a moving object.

Sir Isaac Newton's *Mathematical Principles of Natural Philosophy*, published in 1687, included three laws that describe the relationship between force and motion. These are now called **Newton's laws of motion.** The first law covers the situations we have just described, and may be stated as follows: *Every body at rest stays at rest unless acted on by an outside force; every body in motion stays in motion in a straight line with the same speed unless acted on by an outside force.* (Outside force means a force that originates outside the body.)

The first part of this law should make sense to you at once. The second part may not seem quite right. You know, for example, that the pitched baseball will eventually curve down and hit the ground even if it is neither hit nor caught. The car sliding on ice will eventually stop, even if it never hits clear pavement. A moving car taken out of gear on a smooth level highway will slowly come to a complete stop without the brakes being applied. Satellites in orbit around the earth or planets in orbit around the sun move in curved rather than straight paths.

Each of these apparent conflicts with Newton's law involves one or more outside forces. The pitched baseball curves down and eventually hits the ground only because the earth pulls on it with a force—the baseball's weight. Even slick ice offers some frictional resistance that slows the car. The car moving in neutral on the level

highway experiences friction from the air, from the reaction between the tires and the road, and from around the wheel axles as
they rotate.

An object traveling in a circular path can continue to do so only
if there is a force acting on it that continually changes its direction.
For example, if you twirl a rock on a string (Fig. 2.3), the string
exerts a force on the rock that keeps it going in a circle. If the string
breaks, the rock goes off in a straight line in the direction of its
motion at the instant the string broke. Satellites move in curved
paths around the earth because of the gravitational force exerted
on them by the earth.

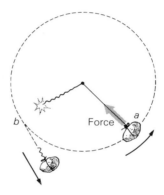

Figure 2.3   At point *a*, the string exerts a force on the rock that keeps it in
circular motion. If the string breaks when the rock is at *b*, the rock moves
off in a straight line because there is no longer a force acting on it.

The property that makes matter tend to continue in a straight
line with both existing speed and direction unchanged is called
inertia. This term means essentially the same here as it does in
everyday use—resistance to change. Because Newton's first law of
motion deals with the inertia of matter, it is sometimes called the
law of inertia. If an object is to *change* its speed or direction, or both,
there must be an outside force acting on it. Our next job, then, is to
describe how such a force changes the motion.

2.4   The Description of Motion

In order to say precisely how forces change the motion of an object,
we need to define the quantities that describe motion: *speed, velocity,*
and *acceleration.*

Speed

Speed means how fast an object is moving. More precisely, speed is the rate at which an object changes position:

$$\text{speed} = \frac{\text{distance traveled}}{\text{time required}} \, ,$$

or

$$\text{speed} = \frac{d}{t} \, ,$$

where the symbols d and t are shorthand for distance and time, respectively.

If you are traveling at a steady speed of 80 kilometers/hour (km/h), you travel a distance of 80 kilometers in one hour. Knowing any two of the three quantities in this equation, you can calculate the third.

Example 2.1 (a) Natalia Maracescu of Rumania currently holds the world women's record of 4 min:23.8 s for the mile run. What was her speed in meters/second (m/s)?

From the definition of speed,

$$\text{speed} = \frac{d}{t}$$

$$= \frac{1609 \text{ m}}{263.8 \text{ s}}$$

$$= 6.099 \text{ m/s}.$$

(b) If Natalia could maintain this speed, how long would it take her to run the 375 km from New York City to Washington, D.C.?

Rearranging our speed equation, we get

$$t = \frac{d}{\text{speed}}$$

$$= \frac{375 \ 000 \text{ m}}{6.099 \text{ m/s}}$$

$$= 61 \ 500 \text{ s} \quad \text{or} \quad 17.08 \text{ h.}$$

Velocity

If you are moving, you have to be moving in some direction. To describe your motion completely, you have to say which way you are moving. The term **velocity** includes both speed and direction. Velocity is the *rate of change of position, with the direction indicated.* If you say you are traveling at 80 km/h, you have given your speed. If you say you are traveling at 80 km/h due north, you have given your velocity. If you change either your speed *or* direction, you change your velocity. Speed, then, is just the "size" of the velocity, without any direction given. We will use the symbol v for velocity.

Acceleration

Velocity can change, of course, and we describe a change in velocity by the term **acceleration.** An acceleration can involve a change in speed, a change in direction, or both. When we talk about motion along a curved path, we will consider accelerations that are changes in direction. For now, we will concentrate on accelerations that are changes in speed only. That is, for the time being, we will consider only motion in a straight line. For this situation only, since the direction is then either unimportant or implicitly understood, we can use the terms speed and velocity interchangeably.

Acceleration is, by definition, *the rate of change of velocity:*

$$\text{acceleration} = \frac{\text{change in velocity}}{\text{time required}},$$

or, in symbols,

$$a = \frac{\text{change in } v}{t}.$$

This equation means that if you start to move your car from a complete stop and reach a speed of 80 km/h in 10 s, your acceleration is given by

$$a = \frac{\text{change in } v}{t}$$

$$= \frac{80 \text{ km/h}}{10 \text{ s}}$$

$$= 8 \text{ (km/h)/s},$$

or 8 kilometers per hour per second. This measurement means that in each second the speed changes by an average of 8 km/h: After 1 s your speed is 8 km/h; after 2 s, it is 16 km/h; after 3 s, it is 24 km/h; and so on.

Example 2.2   (a) A manufacturer claims his car is capable of an acceleration of 10 (km/h)/s. How long would it take this car to reach a speed of 130 km/h, starting from rest?

Since the speed increases by 10 km/h each second, it would be 10 km/h after 1 s, 20 km/h after 2 s, on up to 130 km/h after *13 s*. Alternatively, we could find the answer by using the definition of acceleration,

$$a = \frac{\text{change in velocity}}{t},$$

by rearranging it to get

$$t = \frac{\text{change in } v}{a}$$

$$= \frac{130 \text{ km/h}}{10 \text{ (km/h)/s}}$$

$$= 13 \text{ s}.$$

(b) What speed would this car reach in 4 s?

We can rearrange the equation that defines acceleration to give

$$\text{change in } v = a \times t$$

$$= 10 \text{ (km/h)/s} \times 4 \text{ s}$$

$$= 40 \text{ km/h}.$$

Example 2.3   When a flea jumps, it takes about 0.0008 s to push off. In this time it reaches a speed of about 1 m/s. What is this flea's acceleration? We use the equation

$$a = \frac{\text{change in } v}{t}$$

$$= \frac{1 \text{ m/s}}{0.0008 \text{ s}}$$

$$= 1250 \text{ (m/s)/s}.$$

This acceleration of 1250 meters per second per second is about 30 times that felt by Neil Armstrong during the launch of the *Saturn 5* moon rocket.

2.5   **A Comment about Units**

From our experience with cars, most of us are probably used to measuring speeds in units such as kilometers/hour or miles/hour. For the same reason, it's probably easiest to think of acceleration in units such as kilometers/hour per second. However, speed can be expressed in any units of length per any unit of time—for example, miles/hour, meters/second, centimeters/second, or furlongs/fortnight. Acceleration can be expressed in any units of length per unit of time per unit of time.

Any one physical quantity alone can be written in whatever units are most convenient. But when that quantity is combined with other quantities in an equation, we have to be very careful about the units we use. All the quantities must be expressed in the same **system of units.** For example, to find the distance traveled in 30 minutes at a speed of 80 km/h, you know that you do *not* multiply 80 km/h by 30 min. You must either change the time to hours or the speed to km/min before multiplying.

To avoid errors caused by using mixed units, we will usually express and calculate quantities in what are called *SI units* (The International System of units. See Appendix II, Section B). When we use metric prefixes (see Appendix II, Section A) to specify the decimal point—for example, 3 km rather than 3000 m—we need to convert to the actual SI unit before using the quantity in a calculation. Appendix II, Section D shows how to use the units algebraically to convert from one unit to another and also how to use them as a check on the correctness of your calculations.

In Example 2.3 we found the flea's acceleration to be 1250 (m/s)/s. We can express these units more compactly. Recall from algebra.that

$$(x/y)/y = x/y^2.$$

Units can be treated in the same way. Thus,

$$(m/s)/s = m/s^2.$$

Then the flea's acceleration can be expressed as $a = 1250 \text{ m/s}^2$. You do not have to worry about what a "square second" is. The m/s² is merely a shorthand way of saying meters per second per second. This possibility didn't arise in using the "mixed" units (km/h)/s.

Exercise   In the 1976 World Series some of Don Gullett's pitches were clocked at 150 km/h (94 mi/h). What is this speed in m/s?   **Ans:** 42 m/s

---

Exercise   Assume it takes 0.1 s to throw the ball (that is, the time from backmost position of the ball in the windup to its release). What was the ball's acceleration during the throwing process? Give your answer in (km/h)/s and m/s².

Ans: 1500 (km/h)/s; 420 m/s²

2.6   **Changes in the State of Motion**

Newton's first law describes the behavior of a body that has no outside forces acting on it. According to this law, such a body either

$F = 50$ N

$a = 7$ m/s^2

$F = 100$ N

$a = 14$ m/s^2

Acceleration ∝ force

Figure 2.4 The acceleration, a, of a body is directly proportional to the force, F, applied to it.

stays at rest or continues moving with constant speed in a straight line. That means its acceleration is zero. The law implies that if there *is* an outside force, the body will accelerate. We can now consider how the force and acceleration are related.

Suppose you want to throw a bowling ball; that is, you want to accelerate it from rest to some velocity. You do this by applying a force with your hand. The harder you throw (or, to say it another way, the more force you apply), the higher the speed of the ball when you release it; that is, the greater the acceleration during the throwing process. This is an example of a general law of nature: *The acceleration of an object is directly proportional to the force applied to it.* In symbols,

$$a \propto F,$$

where "∝" means "proportional to." In other words, if a force of 50 N gives the ball an acceleration of 7 m/s^2, a force of 100 N gives it an acceleration of 14 m/s^2. Figure 2.4 illustrates this point.

You probably know from experience that the acceleration of an object depends upon more than just the force applied to it. Suppose you're throwing a softball, and you apply the same 50-N force. The acceleration of the softball will be much higher than that of the bowling ball. The property of the two objects that accounts for this difference is their *mass.*

Mass is a quantitative measure of the inertia of a body. The greater a body's mass, the greater its inertia—that is, the greater its resistance to acceleration. The SI unit of mass is the kilogram (kg).

For a given force, the *acceleration is inversely proportional to the mass* of the object. We write this relationship mathematically in symbols as

$$a \propto \frac{1}{m},$$

$F = 70$ N

$a = 10$ m/s^2

$F = 70$ N

$a = 350$ m/s^2

Acceleration ∝ $\frac{1}{mass}$

Figure 2.5 The acceleration, a, is inversely proportional to the mass for a given force, F.

where m stands for the mass. As Fig. 2.5 illustrates, a force that gives a 7.0-kg bowling ball an acceleration of 10 m/s^2, would give a 0.20-kg softball an acceleration of 7.0/0.2 times as much, or 350 m/s^2.

We can combine the effects of force and mass together, and write

$$\text{acceleration} \propto \frac{\text{force}}{\text{mass}},$$

or

$$a \propto \frac{F}{m}.$$

When one quantity is proportional to another, we can write the first as being equal to a "proportionality constant" times the other.

Therefore,

$$a = \text{(proportionality constant)} \times \frac{F}{m}.$$

But the SI unit of force is defined so that the proportionality constant is 1 and can be omitted from the equation. Therefore,

$$a = \frac{F}{m}.$$

We usually rearrange this equation to read

$$F = ma;$$

that is, *the force is the product of mass times acceleration.* This relationship is **Newton's second law of motion.** This second law describes the fundamental relationship between forces and the resulting accelerations. It actually includes Newton's first law, since zero force gives zero acceleration.

The second law of motion is the basis for the definition of the SI unit of force. This unit, called the *newton (N),* is the amount of force that gives a mass of 1 kg an acceleration of 1 m/s². As we have said, it is equivalent to 0.225 lb.

Notice that the values of the quantities in Fig. 2.5 are in agreement with Newton's second law. For example, to give the 7.0-kg bowling ball an acceleration of 10 m/s² you need a force given by

$$F = ma$$

$$= 7.0 \text{ kg} \times 10 \text{ m/s}^2$$

$$= 70 \text{ N}.$$

You can verify that the values also check for the softball.

The mass of an object has no direct relationship to its size. A soccer ball and a bowling ball are about the same size, but a bowling ball has much more mass. Thus, it needs much more force for a given acceleration. Have you tried kicking a bowling ball lately?

In Example 2.2 we considered a car capable of an acceleration of 10 (km/h)/s. In SI units, this acceleration is

$$\frac{10 \text{ (km/h)/s} \times 1000 \text{ m/km}}{3600 \text{ s/h}},$$

or 2.8 m/s². If the mass of this car is 1500 kg, we can, from Newton's second law, find the force needed to accelerate it:

$$F = ma$$

$$= 1500 \text{ kg} \times 2.8 \text{ m/s}^2$$

$$= 4200 \text{ N (940 lb)}.$$

Example 2.4   Suppose the flea of Example 2.3 jumps from a sensitive device that can measure the excess force the flea exerts in jumping. If the flea exerts a force of 0.00056 N, what is its mass?

We can rearrange Newton's second law as

$$m = \frac{F}{a},$$

and substitute in the known quantities:

$$m = \frac{0.00056 \text{ N}}{1250 \text{ m/s}^2} = 0.45 \times 10^{-6} \text{ kg},$$

or 0.45 mg.

Exercise   In the last exercise, you found Don Gullett's pitches in the 1976 World Series to give the ball an acceleration of 420 m/s². If the mass of the baseball is 0.15 kg, what average force did Gullett exert on the ball?

Ans: 63 N (or 14 lb)

You may have noticed that Newton's second law, in the way we have stated it, is not always true. For example, if you push on (exert an outside force on) the wall of your room, it probably doesn't accelerate in proportion to the force you apply. The reason it doesn't is that other objects—the floor, the other walls, the ceiling—exert forces in the opposite direction on the wall and thus prevent its acceleration. No net unbalanced force exists on the wall. The outside force that applies in Newton's second law is the *net,* or the *unbalanced,* force on the object. If Jim pulls on Suzie's right arm with a force of 100 N, and Joe pulls in the opposite direction on Suzie's left arm with a force of 300 N, Suzie accelerates to the left under a net force of 200 N (assuming she offers no resistance or assistance either way). Keep in mind, then, that in applying Newton's second law, you must always use the *net* outside force on the object under consideration.

You probably associate the concept of mass with that of weight. A relationship does exist between the two but, as we'll see in the next section, mass and weight are very different quantities.

2.7   Weight versus Mass

Everything near the earth experiences a force directed toward the center of the earth. That "everything" includes you, me, this book, whatever you're sitting or lying on, an apple hanging on a tree, or an airplane flying overhead. We know that force exists because, if whatever holds the object in place is removed, it falls toward the center of the earth. This *gravitational* force that pulls every object toward the earth is defined as the **weight** of the object. We'll now contrast weight and mass, in order to make a clear distinction.

Suppose that you hold this book about waist high, and then suddenly release it. It will, of course, fall to the floor because of the pull of the earth on it—its weight. But, because this book has mass, it also has inertia, and therefore resists that downward acceleration just as it would resist acceleration from any other force. We can apply Newton's second law to this situation. The force that we are considering is the weight, to which we'll give the symbol W. Then $F = ma$ becomes

$$W = ma.$$

If you measured the acceleration of this book after you dropped it, you would find a value of 9.8 m/s². No matter where you dropped it near the surface of the earth, you would find the acceleration approximately the same. In fact, it doesn't matter what the mass of the book is. A book—or rock, or marble, or elephant—of any mass will have the same acceleration right after it is dropped. The acceleration doesn't change with mass because the weight is proportional to the mass. Doubling the mass, for example, doubles the weight so that the acceleration stays the same. This acceleration that a freely falling object experiences is called the *acceleration due to gravity,* and we give it the special symbol g. Substituting this symbol for a in the last equation, we get

$$W = mg.$$

This equation represents the basic relationship between the weight and the mass of an object. Remember: *Weight is a force,* the particular force due to gravity; *mass is the measure of the body's inertia.* Since neither mass nor weight is influenced by whether or not the body is actually accelerating, this relationship is true whether or not the object is moving.

We express mass in units of kilograms. When we refer to a 10-kg rock, we are giving its **mass.** Since weight is a *force,* we express it in *force units* (newtons in the SI system). The weight of the 10-kg rock is given by

$$W = mg$$
$$= 10 \text{ kg} \times 9.8 \text{ m/s}^2,$$

or 98 N.

You may protest that you *know* the last paragraph is wrong—that only last week you bought a jar of peanut butter clearly marked "net weight 1.1 kilograms." We get by with saying that the

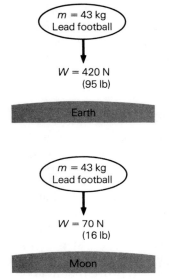

Figure 2.6 The mass of an object is the same on the moon (or anywhere else) as on earth. The weight on the moon is only 1/6 that on earth.

"weight" is 1.1 kg or 1100 g in everyday use because the weight is directly related to the mass through $W = mg$. Knowing the mass, we in effect know the weight. When your bathroom scale registers 52 kg, what it really measures is a force (your weight) proportional to your mass of 52 kg. Even though for simplicity we may say your "weight" is 52 kg, in reality your weight is (52×9.8)N or 510 N.

It's sometimes handy to notice that 9.8 is almost 10, a number much easier to work with. For most purposes we can then use

$$W \text{ (in newtons)} = 10 \times m \text{ (in kilograms)}, \textit{approximately.}$$

On the moon, the gravitational force is only 1/6 what it is on earth. That fact makes the acceleration due to the moon's gravity only 1.6 m/s². However, the mass of an object is the same anywhere: The force needed to give it a particular acceleration is the same anywhere in the universe. It would hurt just as much to kick a football made of lead on the moon as it would on earth! Figure 2.6 contrasts the weight and mass of such a football on the earth and on the moon.

In connection with rocket travel, we sometimes hear that a traveler feels a "force" of several g's—say 4 g's. What this expression actually means is that the traveler experiences a force strong enough to give him an acceleration of four times what he would have if he were falling freely on earth. The net force on his body would be four times his weight on earth.

2.8   The Gravitational Force: One of Nature's Fundamental Forces

You probably already know that the gravitational force acting on objects on earth—that is, their weight—is the same kind of force that keeps the earth and other planets in orbit around the sun, and that keeps the moon and various artificial satellites in orbit around the earth. Let's now look for a more general description of this gravitational force.

In his three laws of motion (of which we so far have discussed only two) Newton clearly restated, using precise definitions and mathematical language, ideas already known to Galileo and others. But his theory of gravitation was a major contribution that simultaneously explained both the weights of ordinary objects and the motions of celestial bodies. To verify the correctness of his theory, Newton had to invent the branch of mathematics we call calculus.

Newton's **law of gravitation** may be stated as follows: *There is a*

$$F \propto \frac{m_1 m_2}{d^2}$$

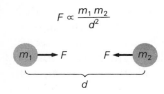

Figure 2.7 Any two masses attract each other with a gravitational force F.

force of attraction between any two masses that is directly proportional to each mass and inversely proportional to the square of the distance between them. Consider the situation presented in Fig. 2.7—two masses m_1 and m_2 separated by a distance d. In mathematical language, this law gives the force F as

$$F = G \frac{m_1 m_2}{d^2}$$

The G is a proportionality constant called the universal gravitational constant, and has the value 6.67×10^{-11} N·m²/kg². Its units make the force come out in newtons when SI units are used for mass and distance.

Newton reportedly started wondering about the gravitational force as he thought about why an apple always falls *downward* from a tree. Let's use the law to calculate the force on a 0.30-kg apple hanging on a tree, as shown in Fig. 2.8 (with the tree and apple somewhat exaggerated in size). This force is, by definition, the *weight* of the apple. Newton needed calculus to prove that if a body is spherical, the force is the same as if all the mass were concentrated at the center. Since the earth is nearly spherical, the relevant distance d in our problem is the earth's radius, 6370 km. We'll arbitrarily take m_1 to be the mass of the apple and m_2 to be that of the earth, 5.98×10^{24} kg. Then

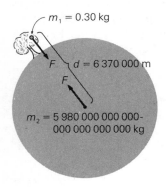

Figure 2.8 A .30-kg apple near the earth's surface is attracted by a force—its weight—of 2.9 N.

$$F = G \frac{m_1 m_2}{d^2}$$

$$= 6.67 \times 10^{-11} \text{ N·m}^2/\text{kg}^2 \times \frac{0.30 \text{ kg} \times 5.98 \times 10^{24} \text{ kg}}{(6.37 \times 10^6 \text{ m})^2}$$

$$= 2.9 \text{ N (or 0.65 lb)},$$

a good weight for a plump juicy apple.

Notice the law says that the two masses attract *each other*. This means that the apple exerts an equal force on the earth. You could turn the emphasis around somewhat and say that, near the surface of a 0.30-kg apple, the earth weighs 2.9 N.

Exercise   Calculate the gravitational force the earth (mass = 6.0×10^{24} kg) exerts on the moon (mass = 7.4×10^{22} kg). How does that compare with the force the moon exerts on the earth? The average distance from earth to moon is 3.8×10^8 m. What would this force be if the earth and moon were touching each other? (radius of earth = 6.4×10^6 m; radius of moon = 1.7×10^6 m).

Ans: 2.0×10^{20} N; same; 4.5×10^{23} N

In Section 2.2 we mentioned that only one force exists that you experience directly that is not electrical in nature. That force is due to gravity—your weight. The gravitational force is one of the four fundamental forces known. The others, which we'll consider in more detail later, are the electrical (or electromagnetic) force and two types of forces related to the nuclei of atoms. Every force we know about fits into one of these four categories.

The gravitational force is very weak between objects of ordinary size. The force becomes significant only when at least one of the objects is extremely massive, like the earth.

Exercise   Calculate the gravitational force between two 1 kg books ½ m apart, and show that the gravitational attraction between them is negligible for practical purposes.

**BLACK HOLE
REFUSES AIRLINE'S JUNK**

CONSTELLATION CYGNUS, (December 21, 2068)—Consolidated Airlines announced today that its attempt to discard its ancient Boeing 747s was not going according to plan. This undertaking, part of their centennial commemoration of man's first moon orbital flight, was to have rid the company of its stockpile of these relics purchased a century ago.

Black holes form during the death of very massive stars. When a star exhausts its nuclear fuel, its component particles are helpless to resist gravitational attraction. It collapses until the gravitational field is so strong that nothing inside—no matter, light, energy in any form—can get out. Anything getting within a certain distance—the so-called "event horizon"—of the black hole, is sucked in with no possibility of getting out or communicating with the outside. It seemed a perfect garbage can for old useless airplanes.

The technician who nudged the first 747 toward the black hole could not believe what she saw. As the plane picked up speed in falling, its color turned from silver to yellow to orange to dull red, and then it faded away. Yet her instruments detected radiation coming from the plane, indicating that the plane moved slower and slower as it approached the event horizon, *never* actually getting there. Airline officials were embarrassed about not having warned their technicians to expect just that, since it was predicted by Albert Einstein in his theory of relativity over 150 years ago.

Inverse-Square Laws

The gravitational force varies inversely as the square of the distance between the masses. Suppose, for example, an astronaut weighs 800 N at the surface of the earth, a distance r from the

$W = 32$ N

$W = 50$ N

$W = 89$ N

$W = 200$ N

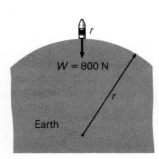

$W = 800$ N

Earth

Figure 2.9 An astronaut experiences a gravitational force that decreases as the square of the distance from earth. The spaceship is shown at distances of one, two, three, four, and five times the radius of the earth.

center of the earth in Fig. 2.9. His weight is shown at various other heights as he recedes from the earth.

The law of gravitation is one of many "inverse-square" laws found in nature. We'll consider others as we explore different topics in this book.

Weights of Objects on Earth

From the law of gravitation we found the weight of a 0.30-kg apple to be 2.9 N. However, in the previous section we found the weight of a body to be related to its mass by $W = mg$, where g is the acceleration due to gravity (9.8 m/s²). That equation also gives the apple's weight as 2.9 N. Some relationship must exist between these two equations.

Let's recast the law of gravitation, pulling m_1 out in front and grouping the other quantities together:

$$F = m_1 \left[G \frac{m_2}{d^2} \right].$$

We can now evaluate the quantity in brackets, taking m_2 as the mass of the earth and d as the distance from the earth's center to the surface—its radius. The value of the quantity in brackets is 9.8 m/s², just the value of g. Therefore, we get the same results as we do when using $W = mg$.

The quantity $[G\, m_2/d^2]$ gives the acceleration due to gravity at any point, not just on the *surface* of the earth, if you use the appropriate value of d. This quantity shows why g is very nearly constant everywhere on the earth's surface: A variation of several kilometers from the earth's center as you move from point to point on the earth's surface, makes little difference out of a total distance of over 6000 km. The bracketed quantity also gives the acceleration due to gravity on other celestial bodies, such as the moon, provided you use those values of m_2 and d that apply.

Center of Gravity

If, using one hand, you pick up a barbell with equal weights on each end, you will have to lift at the center if the barbell is to stay balanced (Fig. 2.10-a). If there is more weight on one end than the other (Fig. 2.10-b), you will need to lift closer to the heavier end for balance. In either case there is one point, called the **center of gravity,** through which all the weight may be considered to act. When

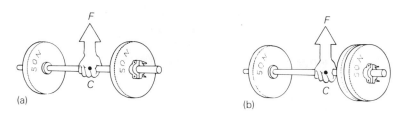

Figure 2.10   (a) The center of gravity of a symmetrical body is at its geometric center. (b) For an unsymmetrical body, the center of gravity is closer to the heavier part.

you lift at that point, the object is balanced, whatever its orientation, and will not tend to rotate when you lift.

The center of gravity of a symmetric object—a sphere, a cube, a barbell with identical weights at each end—is always at the geometric center. For a nonsymmetric body—a chair, an axe, an unbalanced barbell—the center of gravity is near the heavier parts.

We may also define a *center of mass* for a body as the point at which all the mass can be considered concentrated. If you apply a force to an irregularly shaped body, parts of it may follow a wiggly path, but the center of mass will move according to Newton's second law of motion. Except for very tall objects, the center of mass and center of gravity are at the same point. Why should these two points ever be different?

2.9   **Falling Bodies**

A body falling freely as it is acted on only by its own weight experiences a constant acceleration of g near the earth. Yet some bodies, (feathers, for example) do not fall with that acceleration. If all did, it wouldn't be safe to be hit on the head by a raindrop, and it wouldn't help skydivers to carry a parachute. In this section we'll investigate falling bodies, and how their speeds and positions change with time.

Free Fall

The reason all bodies do not fall with the same acceleration is that they experience resistance from the air as they pass through it. For compact bodies that haven't yet reached high speeds, this resistance is small and can be neglected for most purposes. We say that such a body, acted on only by its weight, is in *free fall*.

Let's study free fall by means of a thought experiment. Imagine that you drop a rock out of a window of the 87th floor of the Empire State Building. From windows below, observers precisely measure the time at which the rock passes, and its speed as it passes. (An observer might determine the speed by knowing the height of a window and measuring the time it takes for the rock to pass.)

Figure 2.11 shows the velocities and distances from the point of release as measured at several 1-s intervals from the time of release. These values are the same as those measured by other people in actual free-fall experiments. Time, velocity, and distance are all *zero* at the instant of release.

Since the acceleration is g (9.8 m/s²), the velocity changes by 9.8 m/s each second. For example, after 4 s, the velocity is downward at 4×9.8 m/s, or 39.2 m/s.

The distance of fall is more complicated. Because the speed of the falling object is increasing, the distance traveled in each second gets progressively longer. The distance traveled in 2 s is four times that traveled in 1 s; the distance traveled in 3 s is nine times that traveled in 1 s; and so on. In other words, the distance fallen is proportional to the *square* of the time. Notice that you can get these distances if you multiply the square of each time by 4.9, or ½ the value of the acceleration due to gravity. All these distances d at each time t are consistent with the equation

$$d = \frac{1}{2} gt^2.$$

If you measured the distance for other times and for other objects, you would find the results of any free-fall experiment where the object is dropped from rest to be described by this equation.*

Example 2.5   If you neglect air resistance, how long would it take you to hit the water after diving from a 5-m-high diving tower? Substituting $d = 5$ m, and approximating g as 10 m/s² in our distance equation, we get

$$5 \text{ m} = \frac{1}{2} (10 \text{ m/s}^2)t^2.$$

We can solve this for t to get 1.0 s.

---

*You can derive this equation algebraically by noticing that the average velocity from $t = 0$ to any time t is ½ the velocity at time t. Then, $\bar{d} = (\text{average } v) \times t = (v/2)t$. From the definition of acceleration, $v = gt$. Therefore, $d = (gt/2)t = \frac{1}{2}gt^2$.

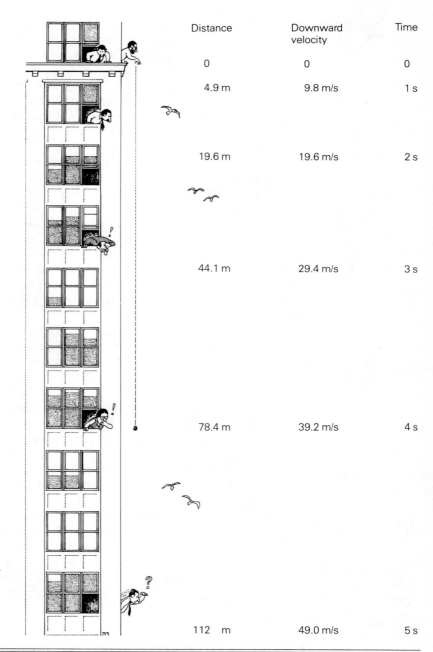

Distance	Downward velocity	Time
0	0	0
4.9 m	9.8 m/s	1 s
19.6 m	19.6 m/s	2 s
44.1 m	29.4 m/s	3 s
78.4 m	39.2 m/s	4 s
112 m	49.0 m/s	5 s

Figure 2.11   The velocities and distances of fall are shown at various times after the rock is dropped.

Exercise   If there were no air resistance, how far would a sky diver fall in 10 s after jumping?   **Ans:** 500 m

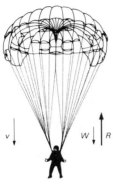

(a) Parachute closed

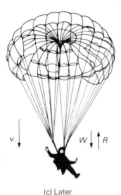

(b) Immediately after
opening parachute

(c) Later

Figure 2.12 (a) A
skydiver reaches terminal
velocity when the upward
resisting force R exactly
equals her weight. (b)
When her parachute
opens, the higher
resistance slows her until
(c) R again equals W.

Effects of Air Resistance

As we have said, if you drop this book, it accelerates with acceleration g. However, if you should (and please don't think about such an unconscionable act!) tear a page from the book, hold the page horizontally and drop it, it would drift down slowly. If you wad the page up, it would fall somewhat faster.

The three fall times differ because of air resistance. The acceleration depends upon the *net* force, and these objects are not falling freely under the action of their weight *only*. The air resistance gives a force in the opposite direction to the motion, and this force retards the acceleration. For the book, the air resistance is small *compared with its weight,* and you notice no retardation for short heights. The force from air resistance on the open page almost equals its weight, so it falls slowly. The wadded paper experiences less air resistance because its shape is more compact.

If there were no air resistance, as in a vacuum, all objects would fall with the same acceleration. You may have seen the film that shows astronauts on the moon dropping a feather and a wrench at the same time. Since no atmosphere exists, these two objects fall at the same speed.

You may have noticed in riding a bicycle or motorcycle, or from sticking your arm out the window of a moving car, that air resistance increases with an increase in speed through the air. As an object picks up speed in falling, the air resistance increases. If it falls far enough, the force of the air pushing up equals the weight of the object pulling down. It then no longer accelerates, and we say it has reached its *terminal velocity.*

Spread-out objects, such as feathers and sheets of paper, reach their terminal velocity quickly. A skydiver (Fig. 2.12) falling through the air without a parachute, has a terminal velocity of about 200 km/h (125 mi/h). She reaches about 80% of this speed while falling 160 m, and gradually approaches the terminal velocity as she continues to fall. When she opens her parachute, the larger size greatly increases the air resistance. The net upward force slows her until the retarding force again exactly balances her weight. She then falls at a lower terminal velocity. The diver descends more slowly with an open parachute for the same reason that the open sheet of paper falls more slowly than the piece wadded up.

Exercise Estimate the speed with which a raindrop, formed a few kilometers above, would hit your head if it experienced no air resistance.

2.10   Other Constant Accelerations

A freely falling body is just *one* example of constant acceleration. The acceleration of such a body is constant because the force on it—its weight—is constant. According to Newton's second law ($F = ma$), any body acted on by a constant net force will undergo a constant acceleration. This means that, for any body starting from rest and having constant acceleration, we can use our equation for distance of travel. We need to use only the actual acceleration of the body, rather than g. In other words,

$$d = \frac{1}{2}at^2,$$

where a is the constant acceleration of the body.

Exercise   How far does a car capable of accelerating at 2.2 m/s² travel in the 10 s needed to reach 80 km/h?   **Ans:** 110 m

---

Exercise   How long would it take Betty Cuthbert to run this distance, if she ran at the speed of her still unbeaten 1960 record of 60 m in 7.2 s?   **Ans:** 13.2 s

---

2.11   The Action-Reaction Law

Force that you exert on the chair

Force that the chair exerts on you

Figure 2.13   Whatever force you exert on the chair, the chair exerts an equal force back on you in the opposite direction.

We need to describe one other important characteristic of forces: You cannot have one single force; forces always come *in pairs*.

If you're sitting in a chair, your body pushes down on the chair. The chair in turn pushes up on your body. You can be sure of this because, if the chair were not there, you would fall to the floor. The force with which the chair pushes on you is exactly equal to the force with which you push on the chair. The force from the chair, however, is applied in the opposite direction of the force from your body (Fig. 2.13). When you hit a tennis ball with a racket, the racket exerts a force on the ball. The ball exerts a force of the same strength on the racket, but in the opposite direction. The bulging of the racket in Fig. 2.14 results from the force of the ball on the racket.

What causes a bicycle to start forward when it is at rest? In order for the bicycle to begin moving, it must experience a force *in the direction of motion*. When you push on the pedal, the back wheel tries to turn, causing the tire to exert a backward force on the road (Fig. 2.15). The road in turn pushes with an equal force in the

Figure 2.14 When the racket pushes on the tennis ball, the ball pushes back on the racket with an equal force.

forward direction on the tire. Since the tire is attached to the bike, this forward force causes it to start to move. (We haven't shown the upward push of the road on the tire that results from the downward push of the tire on the road.) Why do *you* get tired when the road exerts the force to make you go?

Force of road
on the tire

Force of tire
on the road

Figure 2.15 The bike accelerates forward because the road pushes forward on the tire as the tire pushes backward on the road.

These examples illustrate a general law of nature: *For every force exerted by one object on another, the second object exerts an equal force in the opposite direction on the first.* This statement is **Newton's third law of motion,** even though it does not necessarily involve motion. Newton's wording was, "To every action there is always opposed an equal reaction." For this reason the two forces are sometimes called an action-reaction pair.

Exercise   Recall the ways that forces have been exerted in your presence during the past 24 hours. What was each force exerted by and on? What was each reaction force exerted by and on?

Note that the action and reaction forces *always* act on different bodies: When your body pushes on the chair, the chair pushes on you; when the racket exerts a force on the ball, the ball exerts a force on the racket. These forces have nothing to do with whether or not there is a *net* force on any *one* body.

Key Concepts

Newton's first law of motion: A body at rest stays at rest unless acted on by a net outside force; a body in motion stays in motion in a straight line with constant speed unless acted on by a net outside force.

Speed is the rate of change of position. **Velocity** is the rate of change of position with the direction of motion specified or understood. **Acceleration** is the rate of change of velocity; acceleration may involve a change in speed, a change in direction, or both.

When you use any equation that states a definition or a physical law, all quantities in the equation must be expressed in the same **system of units.**

Newton's second law of motion: The acceleration of a body is directly proportional to the net force acting on the body and inversely proportional to the mass of the body. In symbols,

$$F = ma.$$

The **mass** of a body is the quantitative measure of its inertia. The **weight** of a body is the force acting on it due to gravity. The relationship between the weight on earth and the mass is

$$W = mg,$$

where g is the acceleration of an object in free fall near the earth. This acceleration is the same, whatever the mass of the body.

The law of gravitation: There is a force of attraction between any two masses that is directly proportional to each mass and inversely proportional to the square of the distance between them. This law is an example of an inverse-square law. The **center of gravity** is the point on a body through which its weight can be considered to act.

The distance d traveled in time t by a body that starts from rest and has constant acceleration a is

$$d = \frac{1}{2} at^2.$$

Newton's third law: For every force exerted by one object on another, the second object exerts an equal force in the opposite direction on the first.

Questions

Inertia

1. If you're riding a bicycle and hit a boulder in your path, why do you fly over the handlebars?

2. When you go around a curve in an automobile, why does the book on the seat beside you slide off the side of the seat?

3. Some people claim the talent of being able to yank the tablecloth from a set table without disturbing the place settings. If this is possible, why?

4. Why are football linemen usually heavier than backs? Is it their weight or their mass that is important?

5. Why is a breakaway aluminum lamp pole less damaging to an automobile than a solid wooden one (Fig. 2.16)?

Figure 2.16   A collision with a breakaway lamp pole.

6. If a pinball gets stuck, why do you hit the pinball machine in the opposite direction from the way you want the ball to go?

Changes in the State of Motion

7. Is a runner's ability to accelerate usually a major factor in a 200-m race?

8. In what ways would bowling on the moon differ from bowling on earth? For example, could you throw the ball any faster?

9. Why is a person with a weak heart likely to feel dizzy while accelerating upward in an elevator?

10. Why can't you throw a shot-put with the same speed you can throw a softball?

11. Why is a basketball easier to throw than a bowling ball, even though it's bigger?

12. In terms of Newton's second law of motion, explain why inflatable airbags protect the passengers of an automobile during a collision.

Weight versus Mass

13. When you stand on scales, you measure your weight. Which of the following statements is correct?

    (1) "The force exerted on the scales *is* your weight."

    (2) "The force exerted on the scales *is equal to* your weight."

14. What is your mass in kilograms and your weight in newtons?

15. Why can't we express mass and weight in the same units?

16. Explain why, for everyday purposes, we can say that a package of hot dogs "weighs" 454 grams.

17. Was astronaut Neil Armstrong's mass the same on the moon as on earth? How about his weight?

18. Does it mean anything to talk about the *weight* of the earth? If so, what does it mean?

19. Why do a 10-kg rock and a 20-kg rock have the same acceleration when dropped?

The Law of Gravitation

20. The sun is 330 000 times more massive than the earth. On the surface of the earth, you are only 23 000 times farther from

the sun than from the center of the earth. Why doesn't the sun pull you off the earth into itself?

21. Would you expect there to be a place between the earth and sun where an object would experience no net gravitational force?

22. What are some reasons why the acceleration due to gravity *g* might differ a little from point to point on the surface of the earth?

23. Does the law of gravitation discovered by Newton tell us *why* we have weight?

24. Some high-jumpers claim that, as they go over the bar, they bend their bodies in such a way that their centers of mass go *under* the bar while they go over the bar. Do you believe this is possible? (See Fig. 2.17.)

Figure 2.17 Joni Huntley as she clears the bar in a high jump.

Falling Bodies

25. If you go over Niagara Falls in a barrel, do you fall as far during the third second of time as during the second second after you leave the top?

26. Could an astronaut bail out of a spacecraft above the moon and parachute safely to its surface?

27. Discuss the relative merits of skydiving on the earth and on the moon.

28. Why does snow fall more slowly than rain?

29. Legend says that Galileo dropped two stones, a large one and a small one, from the Leaning Tower of Pisa to demonstrate that they both fall with the same acceleration. Would the result of such an experiment be different if one stone were round and smooth, and the other large and flat?

The Action-Reaction Law

30. When you are sitting in a chair, the earth exerts a downward gravitational force on you. Is the corresponding reaction force the upward push of the chair on your body?

31. For a body to have an acceleration, there must be a net force on it *in the direction of the acceleration.* What exerts that force on a car to accelerate it when you push down on the accelerator?

32. The action-reaction law says that if there is a force in one direction, there must be another equal force in the opposite direction. Why doesn't this mean that you can never have a *net* force on *any* body?

33. When a baseball batter ducks to avoid a "bean ball," why does he throw his arms up? Why do you also throw your arms up when you're ducking something coming rapidly at your head?

Problems

1. In 1920, the thoroughbred racehorse Man O'War ran 1⅛ miles (1.8 km) in the Preakness Stakes in 1 min:51 s. What was his speed in m/s and km/h?   **Ans:** 16 m/s, 58 km/h

2. The world's longest recorded human race of 3665 miles from New York City to Los Angeles in 1929 was won in 79 days by Johnny Salo with a running time of 525 h:57 min:20 s. What was his average speed in m/s while running?   **Ans:** 3.114 m/s

3. If the proverbial race between the tortoise and the hare took place over a 2.0 km course, how long was the hare's nap? Take

the tortoise's and hare's speeds to be ⅓ km/h and 60 km/h. respectively. **Ans:** at least 5 h:58 min

4. Suppose Tracy Austin serves a tennis ball that travels at a speed of 34 m/s (76 mi/h) and hits the service line of her opponent's court 19 m away. How long after she hits the ball does it touch the court? (Neglect the slight downward slope and curvature of the ball's path.) **Ans:** 0.56 s

5. A cheetah can run at a speed of 105 km/h. At that speed, how far could it run in 3 h:45 min? **Ans:** 394 km

6. A swimmer can swim 1.5 km in 17 min. How far can she swim in 2 h averaging half this speed? **Ans:** 5.3 km

7. A car can reach a speed of 70 km/h in 7 s. What is its acceleration in m/s²? **Ans:** 2.8 m/s²

8. What speed could the car of Problem 7 reach in 10 s at the same acceleration? **Ans:** 100 km/h

9. How long would it take the car of Problem 8 to reach 120 km/h? **Ans:** 12 s

10. Jack Nicklaus hits a golf ball that leaves his club at a speed of 100 m/s. If the ball and club are in contact for 0.01 s, what is the average acceleration of the ball while it is being hit?

    Ans: 10 000 m/s²

11. If the mass of the golf ball in Problem 10 is 0.1 kg, what average force does the club exert on the ball?

    Ans: 1000 N (220 lb)

12. Calculate the gravitational force acting between two apples of mass 0.25 kg hanging 2 m apart on a tree.

    Ans: 1.0×10^{-12} N

13. Captain Kirk of the Starship *Enterprise* has mass 70 kg and weight 700 N at the surface of Planet Cantaloupius. What would be his mass and weight when he is at a height equal to the planet's radius above its surface? **Ans:** 70 kg, 175 N

14. A swimmer hits the water 1.0 s after diving from a diving platform. With what speed does she hit the water?

    Ans: 9.8 m/s (35 km/h)

15. How high is the diving platform in Problem 14? **Ans:** 4.9 m

16. Wonder Woman jumps from the ground to a tenth story win-

dow, 30 m high. How long is it from the time she leaves the ground until she gets to the window if she jumps straight up? (The time to go up is the same as the time it would take her to come down if she jumped from the window to the ground.) With what speed does she leave the ground?

Ans: 2.47 s; 24 m/s

17. Runner Steve Williams, running at a speed of 10 m/s, passes a parked Cadillac. At the instant he passes, the Cadillac starts to accelerate with a constant acceleration of 8 (km/h)/s (2.2 m/s²). How long does it take for the car to catch Williams? How far does it travel before it catches him?　　　　**Ans:** 9.1 s; 91 m

18. A locust jumps by pushing with its hind legs over a distance of about 4 cm (0.04 m). If it starts from rest and pushes for a period of 0.02 s while moving this distance, what is its acceleration during this time?　　　　**Ans:** 200 m/s² (20 times g)

19. If the mass of the locust in Problem 18 is 3 g (0.003 kg), with what average force does it push during the jump? (Notice that the locust pushes on the ground. The ground pushes back with a reaction force that accelerates the locust.)

Ans: 0.6 N (20 times its weight)

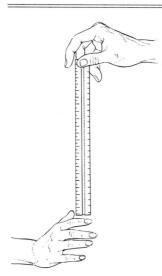

Figure 2.18 A way to estimate reaction time.

Home Experiments

1. Hook a strong rubber band to the handle of a toy wagon or other vehicle you can pull. On a smooth level surface (like the hallway of the physics building) give a friend a ride on the wagon by pulling on the rubber band so that it always stretches the same amount. The constant force will provide a constant acceleration and you will find yourself unable to keep up indefinitely. (You have to pull with a little more force than "just enough" to overcome friction.)

2. Tie the string of a helium-filled balloon to the floor of a car so that the balloon is held just below the ceiling. Does the balloon move in the direction you expect when you accelerate, brake, or round a curve? Explain.

3. Estimate your reaction time by having a friend hold a ruler just above your curved hand and then release it (Fig. 2.18). The distance it falls before you can grab it will tell you, after a simple calculation, your reaction time.

References for Further Reading

Connolly, W. C., "The Physics of Sport Activities, *The Physics Teacher,* Sept. 1978, p. 392. Uses stroboscopic and open-shutter photographs to illustrate the physics of various sporting events.

McMahon, T. A. and P. R. Greene, "Fast Running Tracks," *Scientific American,* Dec. 1978, p. 148. The Harvard University track was designed for faster running by analyzing the mechanics of human running.

"Those Baffling Black Holes," *Time,* Sept. 4, 1978, p. 50. Informative description for the nonscientist of black holes.

3

Mechanical Energy and Momentum

It takes a tremendous amount of energy to be a college student! You need the discipline to direct that energy to the right places: to studying, to keeping your body in good physical condition, to maintaining an alert spirit. All these activities require real energy, just as surely as a train needs energy to pull freight up a mountain. If you aren't convinced, try any activity without eating for awhile, that is, without restocking your body with fuel. Energy is absolutely necessary to carry out the chemical and electrical processes in your brain, as well as to do the work of continually contracting your muscles as you exercise.

Sometimes this energy needs to be channeled toward giving mechanical energy to your body as a whole. When your history professor keeps you late and you don't want to miss that first physics demonstration, you've got to give your body some kinetic energy. As you go up and down the hills or stairways of campus, your body gains and loses potential energy. It takes work to give *your* body kinetic or potential energy just as it does for *any* body, living or inanimate.

In this chapter we will concentrate on mechanical—kinetic and potential—energy, and will develop precise quantitative definitions for these concepts. We can do so only after we first give a precise definition to work. We will, in addition, apply the energy conservation principle in analyzing systems in which mechanical energy changes. Finally, we'll consider the reason why, if you're dashing down the sidewalk and are about to pass the physics building, you can't make your body stop or turn right *instantaneously*. That is, we'll define and use the concept of *momentum,* a quantity for which there is also a conservation principle.

3.1 Work

In Chapter 1 we stated that we do **work** when we cause a force to move something through a distance. Now, let's develop a more quantitative definition of work. To help us to do this, let's consider a situation where the force acts in the *same direction* as (or parallel to) the motion. An example would be the work done by the force that a towrope exerts on a water-skier when both ends of the rope are at the same height above water (Fig. 3.1). The force of the rope then does work in moving the skier along the water.

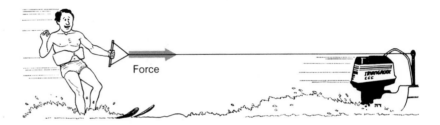

Force

Figure 3.1 The force on the towrope does work in moving the skier.

Work is defined as the product of the force parallel to the motion and the distance through which the force acts:

work = (force parallel to motion) × (distance the force acts).

We can write this definition symbolically as

$$\text{work} = F_p \times d,$$

where the symbols stand for the quantities in parentheses in our word equation. We put a subscript "p" on the force symbol as a reminder that we're considering a case where the force is *parallel* to the direction of motion.

If the boat pulls our skier 200 m in a straight line by exerting a continuous force of 100 N, the work done by this force is then

$$\text{work} = F_p \times d$$
$$= 100\,\text{N} \times 200\,\text{m}$$
$$= 20\,000\,\text{N}\cdot\text{m}$$

This calculation shows the unit of work to be the newton-meter, the result of multiplying force in newtons by distance in meters. This unit of work is defined to be the **joule** (J), the SI unit of work and energy. In other words, the joule is the amount of work done, or energy used, when a force of 1 N acts through a distance of 1 m in the direction of the force.*

For many jobs, the force that does the work is *not* exerted parallel to the motion. For example, if you push a lawn mower, the direction of the force you apply is parallel to the handle, *not* parallel to the motion (Fig. 3.2). Thus, only a fraction of the force is effective in doing work. The smaller the angle between force and motion, the larger the fraction of the force that is effective. This description probably agrees with your experience. With thin short grass, you can push the mower easily at almost any angle. But with tall tough grass, you keep the handle low to apply the force as nearly as possible in the direction of motion. If you haven't had the fun of mowing grass this way, you might have pulled a kid brother in his little red wagon. With only your brother in the wagon, the angle of pull doesn't matter much. When four friends get in with him, however, you may need to decrease the angle to make the same force do more work.

Figure 3.2 If the force and motion are in different directions, only a fraction of the force *is* effective in doing work.

Direction of force

Direction of motion

If the force and the motion are perpendicular, the force does *no* work. For example, when you ride in a car, your weight causes you to push downward on the seat. Since the motion is perpendicular to this direction, the force does *no* work.

For most of the situations we'll consider, the force will be either parallel or perpendicular to the motion. For other angles, you can estimate the effect.

---

*In the British system, the unit of work and energy is the foot-pound, the work done when a force of 1 lb acts through a distance of 1 ft.

Question   Assume that the earth is in a circular orbit around the sun. How much work does the gravitational force on the earth do in one revolution? (See Question 6 also.)

Exercise   If it takes a force of 160 N to slide your dresser across the floor, how much work do you do in sliding it 3.0 m?   **Ans:** 580 J

You may be thinking that there must be something different about "biological" work. You *know* that you get tired holding a stack of books perfectly still, or even just standing still without holding anything. Why do the people in Figs. 1.3 and 1.13 get tired if they aren't doing any work? Notice that for these examples we talked about the work done by the force *on the car* or *on the barbell,* but *not* the work done by the muscles.

Muscles do not "lock in" and become rigid. As the muscle pulls, its many fibers are constantly contracting, then relaxing slightly, then contracting again. The different fibers contract at different times so that the muscle as a whole seems to stay rigid. Each time a fiber contracts, it does real mechanical work—force × distance—and energy is needed for this work.

3.2   Mechanical Energy

In Chapter 1 we described mechanical energy—potential and kinetic energy—in a qualitative way. Now that we have defined work precisely (see Sec. 3.1), considered the effect of forces on motion (Sec. 2.6), and carefully defined the special force we call weight (Sec. 2.7), we can give a quantitative definition to these forms of energy. In considering interchanges between forms of mechanical energy, we can use the principle of energy conservation to understand the motion of a body as it changes position.

Kinetic Energy

Imagine you are throwing a javelin. As Fig. 3.3 shows, you throw it by exerting a continuous force in the direction you want it to go, and it leaves your hand with some velocity. We'll consider a *horizontal* throw in order not to cause any potential energy change while it's in your hand. (Normally, the javelin is thrown at an upward initial angle in order to give it maximum horizontal travel.) We then have a simple case of work being done—force acting through a distance—and all this work going into kinetic energy of an object. Focusing on this example, we can equate the kinetic energy gained to the work done. If we assume the force, and therefore the acceleration, are constant, we can combine Newton's second law with the work-energy equation to get an expression for the **kinetic energy** of a moving object.* The result is

$$\text{kinetic energy } = \frac{1}{2} \times \text{mass} \times (\text{velocity})^2,$$

or

$$\text{KE } = \frac{1}{2}\, mv^2.$$

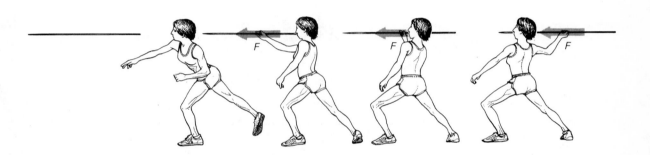

Figure 3.3   To throw the javelin horizontally, you exert a continuous horizontal force as you move it.

In winning the 1976 Olympic Gold Medal, Ruth Fuchs of East Germany threw the 0.60-kg javelin with an initial speed of 26 m/s. The kinetic energy she gave the javelin was then ½(0.60 kg) (26 m/s)², or 200 J.

This equation, KE = ½mv^2, applies not only to bodies initially at rest, but specifies the kinetic energy of any moving body. It says, for example, that if two bodies have the same speed, but one has twice the mass of the other, the more massive one has twice the kinetic energy. However, doubling the speed of a given body qua-

---

*Kinetic energy = work done, or KE = $F \times d = (ma) \times (\frac{1}{2}at^2) = \frac{1}{2}m(at)^2$. For a body starting from rest, the definition of acceleration tells us that $v = at$. Therefore, KE = ½mv^2.

druples its kinetic energy; tripling the speed multiplies the kinetic energy by nine.

Example 3.1   Suppose you're in a head-on automobile collision while traveling at a velocity of 40 km/h (11 m/s). Compare the average force on your head if it is stopped by the car dashboard in a distance of 1.5 cm with the average force on your head if it is stopped by an inflatable air bag in a distance of 25 cm.

 We know that the work done by the force that stops your head must equal the *change* in your head's kinetic energy. Since *all* the kinetic energy is lost,

$$F \times d = \frac{1}{2} mv^2.$$

Taking the mass of the head to be 2.0 kg, we calculate the force F_d of the dashboard from

$$F_d \times 0.015 \text{ m} = \frac{1}{2}(2.0 \text{ kg})(11 \text{ m/s})^2,$$

which gives a force of 8100 N or 1800 lb. We get the force F_{ab} of the air bag from

$$F_{ab} \times 0.25 \text{ m} = \frac{1}{2}(2.0 \text{ kg})(11 \text{ m/s})^2,$$

which gives 480 N or 108 lb.

 What happens to the kinetic energy of your head in either case?

Question   Why does the average braking distance for an automobile go from 10 to 40 m as its speed goes from 40 to 80 km/h?

Exercise   Find the kinetic energy of Don Gullett's 42 m/s fastball (mass = 0.15 kg).

Ans: 130 J

Gravitational Potential Energy

The weights used to operate Thomas Jefferson's seven-day clock (see Sec. 1.2) got their **potential energy** by being lifted to the ceiling. The potential energy gained as they were lifted must equal the work done in lifting. Figure 3.4 shows a weight being lifted to a height h. Lifting it at constant speed—with no acceleration—takes a force equal to its weight ($W = mg$). This force acts through a distance h. Then,

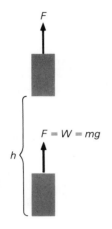

potential energy gained = work done;

$$PE = F \times d = W \times h;$$

or

$$PE = mgh.$$

This equation is a general expression for the potential energy of a body.

Notice that the value of the potential energy depends upon the level from which it is measured. If, as in Fig. 3.5, a 9-kg mass is 2 m above the first floor, the potential energy relative to the first floor is 9 kg $\times$ 10 m/s² $\times$ 2 m, or 180 J. (We've used our approximate value for g.) But, if the floors are 3 m apart, the potential energy relative to the basement is 450 J and, relative to the second floor, is -90 J.

Figure 3.4 To lift the mass at constant speed, a force f equal to the weight must be applied.

Figure 3.5 The value of the potential energy has to be specified relative to some chosen level.

The Ramp Imagine that you're moving across town, and that you've packed all your books into a single box and rolled the box out to the street balanced on your roommate's skateboard. But somehow you're not especially inspired to lift the box onto the back

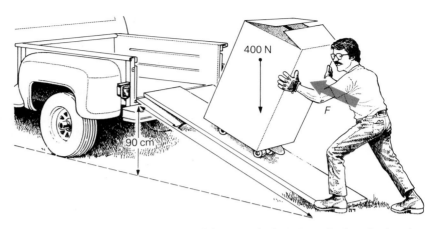

Figure 3.6   The change in potential energy is the same whether the box is lifted directly onto the truck or rolled up the ramp. If friction is neglected, the work done is the same either way, but the force F needed to push the box up the plank is only 3/10 as much.

of that rented pickup truck. An alternative approach, one that would require much less force, would be to lay a plank from the street to the truck and roll both box and skateboard up the plank (Fig. 3.6).

Suppose the truck floor is 90 cm from the street and the box weighs 400 N (mass 40 kg). The work you need to do equals the change in potential energy, *mgh* or *Wh*. This is 400 N × 0.90 m = 360 J. The change in potential energy of the box is the same whether you lift it directly or roll it up the ramp. If we neglect the weight of the skateboard and the friction opposing its rolling, the work of pushing the box up the ramp is 360 J. Therefore,

$$\text{work} = F \times 3\,\text{m} = 360\,\text{J}.$$

We can solve this equation for F and find the force needed is 120 N, only 3/10 that needed to lift the box directly. In reality, the force would have to be stronger than this by the amount of frictional resistance plus the force needed to raise the skateboard itself.

Question   Why are roads up steep mountains made to wind around rather than to go straight up the mountain?

Figure 3.7 shows some devices that use the same principle as the ramp. The difference is that these do their work against the forces holding together the object being penetrated, rather than against the gravitational force.

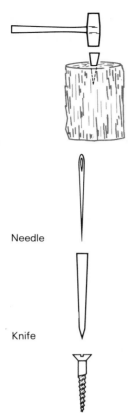

Figure 3.7 The wedge, the needle, the knife, and the screw are all ramps in disguise.

Interchanges of Mechanical Energy

We mentioned in Chapter 1 that, in some situations, energy is interchanged between kinetic and potential energy, with other forms of energy remaining unchanged. When that is true,

kinetic energy + potential energy = constant.

Now that we have quantitative definitions of potential and kinetic energy, we can use this conservation principle to study the variations in such a system. We'll illustrate with two examples.

First, let's find the speed with which a diver hits the water, after diving from a 5-m-high platform (Fig. 3.8). We'll neglect any air resistance, so that total mechanical energy stays constant. That means the diver's

total energy at platform = total energy at water,

or   (KE + PE) at platform = (KE + PE) at water.

The KE is zero at the platform, and we can choose the PE to be zero at the water. Then

$$mgh = \frac{1}{2}mv^2,$$

where the left-hand side applies to the diver at the platform, the right-hand side at the water. We can divide both sides of the equation by m, and get

$$v = \sqrt{2gh} = \sqrt{2(10 \text{ m/s}^2)(5 \text{ m})},$$

or 10 m/s.

We could have solved this problem by the methods we previously used in studying free fall. The real advantage of the energy method appears in solving a problem where the direction of motion changes.

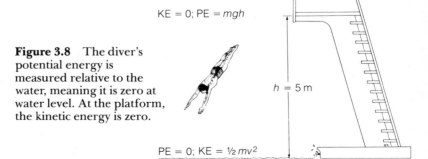

Figure 3.8 The diver's potential energy is measured relative to the water, meaning it is zero at water level. At the platform, the kinetic energy is zero.

KE = 0; PE = mgh

h = 5 m

PE = 0; KE = ½ mv²

As our second example, consider a roller coaster traversing the path of Fig. 3.9. If it starts at rest at point *A*, what is its speed when it gets to point *B*, assuming there is *no* friction? All we need to know is that energy is conserved: Total energy at point *A* equals total energy at point *B*. We don't care what happens in the middle. Since KE = 0 at point *A* and we can choose PE = 0 at point *B*, we have

$$mgh = \frac{1}{2} mv^2.$$

Again, we get $v = \sqrt{2gh} = \sqrt{2(10 \text{ m/s}^2)(7.2 \text{ m})} = 12$ m/s. Using the energy method, we don't need to know any forces, heights, or directions at intermediate points.

KE = 0;
PE = *mgh*

A

7.2 m

PE = 0
KE = ½ *mv²*

B

Figure 3.9   When there is no friction, the roller coaster's *total* energy at point *B* equals its *total* energy at point *A*. When there *is* friction, its total energy at *B* is less than that at *A* by the amount of heat generated as work is done against friction.

Notice that if we had chosen the zero of PE at any other point, we would have gotten the same answer. For instance, if we had taken the zero at the bottom of the pool in the first example, it would have added the same amount of potential energy to *both* sides of our equation, which would not have changed the result.

Frictional Effects

In the roller coaster problem we assumed there was no friction. Any system with moving parts has *some* friction, and we can modify our approach somewhat to include its effect. Total energy is still conserved, but some of the mechanical energy does work against the frictional forces. This loss in *mechanical* energy shows up as heat. Our energy equation then reads

total energy at one instant

= total energy at later instant + work done against friction.

We can apply this equation to the roller coaster problem (Fig. 3.9). Suppose the coaster car and its occupants have a total mass of 1000 kg; the total path length from A to B is 50 m; and a continuous frictional force of 400 N opposes the motion at every point. The energy conservation equation becomes

$$\text{PE at } A = \text{KE at } B + \text{work against friction,}$$

or

$$mgh = \frac{1}{2}mv^2 + (\text{frictional force} \times \text{distance}),$$

or

$$(1000 \text{ kg})(10 \text{ m/s}^2)(7.2 \text{ m}) = \frac{1}{2}(1000 \text{ kg})(v^2) + (400 \text{ N})(50 \text{ m}).$$

We can solve this equation for v to get a velocity of 10 m/s. As you would expect, this speed is a little lower than if there were no friction.

The energy conservation techniques that we've applied to these specific and rather frivolous examples have general applicability, and are used routinely by scientists and engineers to solve problems of quite practical significance.

3.3 Momentum

When a fleet-footed running back pauses briefly, then sidesteps as the fast-moving linebacker approaches, the would-be tackler passes helplessly on by, unable to change his course. Or, when you're driving your car along a gradual uphill slope, and the engine suddenly dies, the car doesn't stop immediately—it coasts a distance on up the hill. In either of these cases, is there a force acting on the body to keep it going? No. These bodies keep moving because they have what we call **momentum.** A body can lose or change the direction of its momentum only when external forces act on it.

Newton's first law of motion tells us that a body tends to continue in motion in the direction it is going. Momentum measures this tendency quantitatively by taking into account the two proper-

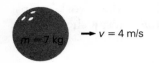

Momentum = mv = 28 kg·m/s

m = 0.15 kg

v = 187 m/s

Momentum = mv = 28 kg·m/s

Figure 3.10 The lower mass has a higher velocity so that the momentum of the baseball equals that of the bowling ball.

ties that determine it: mass and velocity. *Momentum is defined as the product of mass and velocity.* Thus,

$$\text{momentum} = \text{mass} \times \text{velocity}$$
$$= mv.$$

A 7-kg bowling ball moving at a speed of 4 m/s has a momentum mv of 7 kg × 4 m/s, or 28 kg·m/s. But a 0.15-kg baseball moving at a speed of 187 m/s (83 mi/h) has the same momentum (Fig. 3.10).

Impulse

Momentum tells us how much effort it takes to stop a moving body, or to give it a certain velocity if it is at rest. That "effort" depends on two factors: How much force you apply, and how long you apply it. Here's an example.

Suppose your car stalls on Main Street and you have to get out and push it down this level street to a parking area (Fig. 3.11-a). If you can push with a continuous 200-N force, you might get the car up to a speed of 10 km/h in a time of 30 s. On the other hand, if two of your friends come to your rescue (Fig. 3.11-b), the three of you each pushing with a continuous 200-N force could get the car to the same speed in 10 s, or ⅓ the time. The final momentum of the car is the same in both cases.

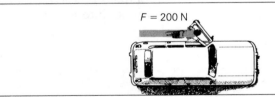

(a) 10 km/h in 30 seconds

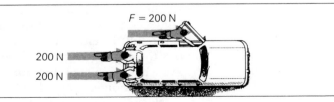

(b) 10 km/h in 10 seconds

Figure 3.11   (a) If only one person pushes, it takes three times as long to reach some speed as (b) if three people push with the same force.

The product of the force applied and the time of application is what we call *the impulse.* Thus,

$$\text{impulse} = \text{force} \times \text{time}$$

$$= Ft.$$

Our example illustrates that the impulse applied is proportional to the change in momentum. In fact, the two are exactly equal:*

$$\text{impulse} = \text{change in momentum,}$$

or $$Ft = \text{change in } mv.$$

The analysis of a problem from the impulse-momentum viewpoint offers an alternative to applying Newton's second law. It is often easier to use the momentum approach for problems in which the force is applied for short periods of time, such as a ball being hit with a bat, two vehicles colliding, or a bullet being fired from a rifle.

In any case, you can achieve a given change in momentum with a smaller force by increasing the length of time it acts, and vice versa. You can increase the momentum change by increasing either the force, the time, or both. Here are a few everyday situations that you can analyze on the basis of impulse and momentum. You might at the same time analyze them from a work-energy viewpoint, and see if you reach the same conclusion.

1. Why are you less likely to do serious injury to yourself and your car if you crash into a collapsible guardrail rather than a concrete bridge support?

2. Whether you crash into the guardrail or concrete support, why are your survival chances better if you are wearing seat belts or if your car has inflatable air bags?

3. Why is a hard right to a boxer's jaw less damaging if he's moving away from it at the time?

4. Why do some trapeze performers use a safety net underneath themselves?

5. If you jump from a high wall, why do you bend your knees when you land rather than hitting with your legs stiff?

6. Why can you safely jump from a much higher point into water than onto dry land?

---

*We can show this experimentally proven fact from Newton's second law of motion, $F = ma$. Since $a = (\text{change in } v)/t$, $F = m \times (\text{change in } v)/t$. Multiplying both sides by t, we get $Ft = \text{change in } mv$.

Example 3.2   A 70-kg diver hits the water with a downward velocity of 10 m/s, after diving from a 5-m-high platform. If his downward motion stops 0.7 s after hitting the water, what is the average net upward force on him as he is being stopped?

$$\text{Impulse} = \text{change in momentum}$$

or

$$Ft = \text{change in } mv.$$

Since the diver's velocity goes to zero,

$$\text{change in momentum} = \text{initial momentum}$$

$$= mv$$

$$= (70 \text{ kg})(10 \text{ m/s}) = 700 \text{ kg} \cdot \text{m/s}.$$

Therefore,

$$F \times 0.7 \text{ s} = 700 \text{ kg} \cdot \text{m/s},$$

or

$$F = 1000 \text{ N}.$$

Conservation of Momentum

Another important property of momentum very useful in understanding physical processes is the observed fact that *the total momentum of an isolated system remains constant* (an isolated system here being one with no net *force* on it from outside). This is a statement of the principle of **conservation of momentum.** Even though it is similar in wording to our statement of the principle of conservation of energy, the two principles are not the same.

The conservation of momentum principle is often applied in comparing the state of motion of some system before and after a particular event in that system. Such an event might be a collision, an explosion, or a rapid expansion or contraction of the parts of the system. We can then restate the momentum conservation principle for such a situation: *The total momentum of the system after the event is equal to the total momentum before the event.*

We will consider a few examples of how the principle of conservation of momentum can be used to understand the motion of the parts of a system after some event occurs within it.

Billiard Ball Collisions   Suppose a moving billiard ball makes a direct head-on collision with another ball at rest on the table. If there is no english on the ball—that is, no special spin other than from rolling—what is the resulting motion of the balls? If you're a billiard player, you know that the first ball stops completely and the second one moves off with the same speed the first one had. This situation is easy to understand from the conservation of momentum principle.

Figure 3.12 shows the cue and eight balls before and after a head-on collision. If the mass of each ball is m, the momentum before the collision is mv. Therefore, the momentum after the collision must be mv. This can happen if the cue ball stops, and the eight ball moves off with speed v, since the mass m of both balls is the same. If the ball has any spin other than from its normal rolling across the table, then other forces act during the collision, making the analysis more complicated.

Before collision   After collision

Figure 3.12 A head-on collision between the cue ball with no "english" and the eight ball at rest.

You might have seen an apparatus such as the one in Fig. 3.13 used either as a toy or as a physics classroom demonstration. You can understand it in the same way you can understand the billiard ball collisions. If you pull one ball aside and release it, another ball will fly off from the other end after the collision. If you pull two aside and release them, two others will fly off after the collision. The effect can be spectacular if you construct the apparatus of bowling balls and hang them from the ceiling.

Both billiard balls and the hanging balls of this toy deform slightly during collision, but then spring back to their original shapes. As a result, any kinetic energy used to deform the balls when they come together is regained as they separate. Except for small frictional losses, the total kinetic energy after the collision equals the total kinetic energy before the collision. A collision that conserves kinetic energy, as well as momentum, is called an **elastic collision.**

Many types of bodies can have *nearly* elastic collisions: billiard balls, bowling balls, marbles, springs. Only atoms and subatomic particles can have truly elastic collisions. Our next example discusses an obviously *inelastic* collision.

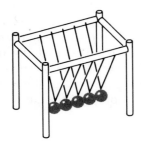

Figure 3.13 A demonstration of the conservation of momentum.

Automobile Collisions Suppose a 1000-kg car moving at 80 km/h (22 m/s) collides on slick ice with a stationary 10 000-kg truck and the two stick together. With what speed do they slide across the ice together after the collision (Fig. 3.14)? Friction is involved with the crunching of the vehicles' metal. Therefore, the total kinetic energy

Figure 3.14 An inelastic collision conserves momentum but not *kinetic* energy.

does not stay the same: The collision is inelastic. In fact, most of the kinetic energy is converted to heat during the collision. Since momentum *is* conserved:

momentum before collision = momentum after collision,

or $1000 \text{ kg} \times 22 \text{ m/s} = (1000 \text{ kg} + 10\ 000 \text{ kg}) \times v,$

where v is the final velocity of the stuck-together vehicles. We solve this equation and find this velocity to be 2.0 m/s (7.2 km/h). Calculate and compare the kinetic energy of the wreckage immediately after the collision with that of the car before.

The Firing of a Gun Consider the cannon shown in Fig. 3.15. Before firing, the entire system—cannon, cannonball, and powder—are not moving and therefore have zero momentum. From conservation of momentum, we know that the total momentum after the firing must also be zero. That may seem strange, since you know the cannonball has momentum. The total momentum can still be zero, because momentum has a direction associated with it. We associate a *positive* sign with momentum in one direction. Then momentum in the other direction is *negative*. Since the cannon recoils in the opposite direction to the motion of the cannonball, we must have

(momentum of ball) − (momentum of cannon) = 0

or

(mass of ball) × (ball velocity) − (mass of cannon) × (recoil velocity) = 0.

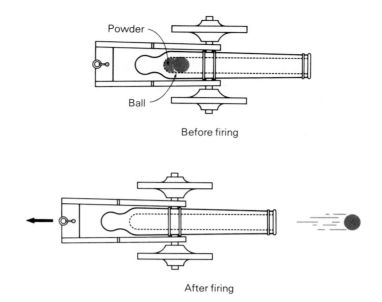

Before firing

After firing

Figure 3.15 The total momentum is zero both before and after the cannon is fired.

The mass of the cannon is greater than the mass of the ball; therefore, the speed of the ball is greater than the recoil speed of the cannon.

The recoil effect just described for a cannon applies to the firing of any gun or rifle. For example, if you fire a high-powered rifle, you will feel a strong "kick" against your shoulder. This kick happens because the rifle must move backward with the same momentum the projectile has moving forward. The rifle, of course, has a much lower speed than the projectile because the rifle has much greater mass.

We can better understand this situation by applying Newton's action-reaction law to it. The cannon and burned powder push forward on the ball, so the ball must push with the same force in the opposite direction on the cannon and burned powder. Since the force on each acts for the same period of time, the impulse-momentum relation tells us that the same amount of momentum is given to each, but in opposite directions.

Rocket Propulsion   Look again at the cannon of Fig. 3.15. Suppose the cannon has a large supply of cannonballs stored inside it, and that it is capable of firing those balls in rapid succession, one immediately after the other. Each time a ball is fired to the right with a certain momentum, the cannon must gain that same amount of momentum to the left. If the balls are fired often enough and with enough velocity, the cannon must acquire a high velocity in the opposite direction. The cannon in effect becomes a rocket!

This action is exactly what happens in a real rocket. Figure 3.16 shows a simplified rocket diagram. Fuel is burned in the combustion chamber producing gases at extremely high pressure. These gases escape out the nozzle in the rear, so shaped that the gas particles exit with exceedingly high velocities. The momentum of the gas particles shooting out the rear (the equivalent of cannon balls) must be matched exactly by the forward momentum of the rocket in the opposite direction. In this way, the rocket is propelled forward.

Figure 3.16   The momentum of the rocket moving forward equals the momentum of the gases moving backward.

This process is also what happens when you blow up a balloon and turn it loose to fly around the room. The backward momentum of the air shooting out the rear must be matched by the gain in forward momentum of the balloon.

You may doubt that something as light as gas molecules could have enough momentum to balance the forward momentum of a massive rocket. Remember, though, that billions of billions of billions of gas molecules come out in a short time at speeds much higher than the speed of sound.

Key Concepts

Work is defined as the product of the force parallel to the motion and the distance through which the force acts. As the angle between a force and the direction of motion increases, the work decreases, becoming zero when the two are perpendicular. A nonmoving muscle does work as its fibers continually contract and relax.

Kinetic energy = ½ × mass × (velocity)² = $\frac{1}{2}mv^2$. For a mass m, a height h above the level at which the potential energy is chosen as zero,

$$\textbf{potential energy} = mgh = Wh,$$

where g is the acceleration due to gravity and W is the weight of the mass.

For a system in which the total mechanical energy does not change,

kinetic energy + potential energy = constant.

If some mechanical energy does work against friction, that amount of energy is converted to heat. Then,

total energy at one instant

= total energy at later instant + work done against friction.

Momentum = mass × velocity = mv.

Impulse = change in momentum:

Ft = change in mv.

For a system with no net force on it from outside, the total momentum remains constant. In an **elastic collision,** both kinetic energy and momentum are conserved.

Questions

Work

1. On June 3, 1921 German strongman Herman Görner carried a 6423-N (1444-lb) piano a horizontal distance of 16 m (52½ ft). How much work did he do on the piano in carrying it this distance?

2. How much mechanical work is involved in:

   (a) lifting a 2-kg brick a height of 1 m;

   (b) holding the brick 1 m off the floor;

   (c) slowly moving the brick a *horizontal* distance of 10 m;

   (d) letting the brick fall a distance of 1 m?

Mechanical Energy

3. We found that pitcher Don Gullett could give a baseball a kinetic energy of 130 J and that javelin-thrower Ruth Fuchs

could give a javelin a kinetic energy of 200 J. Do these facts mean that Ruth Fuchs is stronger than Don Gullett?

4. Why does pulling the bow back farther cause an arrow to have a higher speed when released?

5. Figure 3.17 is taken from a state driver's manual. Explain why reaction distance is proportional to speed, whereas braking distance increases much more significantly with increase in speed. Mathematically, how does braking distance vary with speed?

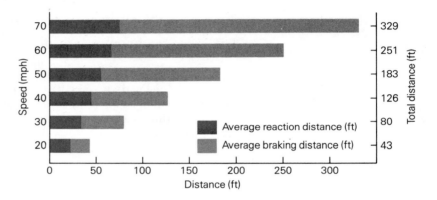

Figure 3.17   Reaction and braking distances in driving an automobile. Reaction distance is the distance traveled from the time you start to move your foot until you actually apply the brakes. Braking distance is the distance needed to stop once the brakes are applied.

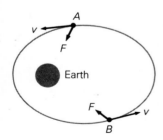

Figure 3.18   The arrows *v* and *F* show the velocity of, and gravitational force on, a satellite at two different points in an elliptical orbit.

6. Many satellites move in elliptical rather than circular orbits. When the satellite is at a position such as point *A* in Fig. 3.18, the gravitational force is acting in a direction to speed it up. When the satellite is at point *B*, the gravitational force slows it down. Show that these changes are consistent with the principle of conservation of energy, since the potential energy decreases as the satellite gets closer to earth.

7. You can easily catch a 2-kg brick dropped from a height of ½ m. However, if the back of your hand is resting on a tabletop when the brick first contacts it, your hand may be injured. Why does catching the brick with your hand on a table increase the force on your hand so much?

8. As you catch a fast-moving baseball, why is it wise to move your hand in the direction of the ball's motion?

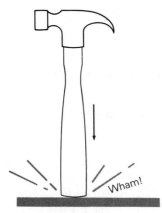

Figure 3.19   The way to put a loose handle back into a hammer.

9. If you're waiting for a stoplight to change, and you're about to be rammed from the rear, you'd probably prefer that it be by a Honda than by a loaded Mack truck. From an energy standpoint, why do you have this preference?

Momentum and Impulse

10. When the handle is coming out of your hammer, why should you hold it vertically as shown in Fig. 3.19 and slam the handle on the floor rather than turning the handle up and slamming the handle into the hammer?

11. Why does a long-jumper build up all the speed he can get before his jump, whereas a high-jumper runs fairly slowly?

12. Why is follow-through more important when hitting a softball than when hitting a baseball?

13. Some people claim that, in an accident, they are less likely to be injured if they are intoxicated than if they are sober. Can you think of a reason why this might be true?

14. Why does a firefighter need great strength to hold the nozzle of a hose squirting a high-speed stream of water?

Conservation of Momentum

15. If you drop a bowling ball, what happens to the momentum it had just before hitting the ground? Answer the same question for a Ping-Pong ball.

16. Charlie Brown is abandoned by his friends on perfectly slick ice (absolutely no friction). Is there any way for him to get off the ice?

17. Convince your roommate that having air behind a rocket or jet airplane for the expelled gases to push against has nothing to do with the propulsion. (For example, would the kick of a rifle be any different if it were fired in vacuum?)

18. In 1854 an Englishman, J. Howard, long-jumped 29 ft, 7 in with the help of two 5-lb weights. He sprinted down the track with the weights held high, and threw them to the ground just as he launched himself. Explain what the weights did for him. (It took another 114 years before Bob Beamon broke the 29-ft mark without the aid of any weights.)

Problems

1. Suppose that in throwing a ball you exert an average force of 50 N on it for a distance of 2 m. How much work do you do on the ball?   **Ans:** 100 J

2. If you exert this force for a time of 0.10 s, what is the power used on the ball? What is this in horsepower?

   Ans: 1.0 kW; 1.3 hp

3. Suppose you use a 200-N force to slide a large box 5 m across the floor in 20 s. You then use a 100-N force to slide a small box 20 m across the floor in 40 s. Which of these efforts takes more work?

   Ans: moving the smaller box (takes twice as much work)

4. Which of the jobs in Problem 3 takes more power?

   Ans: both jobs take the same amount

5. Based on the data in Fig. 3.17, how much frictional force exists between the tires and the road when a 1000-kg car moving at a speed of 65 km/h (40 mi/h) is stopped in the shortest possible distance? (Note: 1 m = 3.3 ft.)   **Ans:** 6000 N (1300 lb)

6. Riding in a car at 80 km/h (22 m/s), you collide head-on with an identical car moving in the opposite direction at the same speed. What was the kinetic energy of your body just before the collision? If it stops within a distance of 1 m during the collision, what average force will your body feel?

   Ans: (for a 60-kg body): 14 000 J; 14 000 N (3200 lb)

7. If your car traveling 50 km/h will stop in 15 m after you hit the brakes, what distance will it need to stop at 150 km/h?

   Ans: 135 m

8. Henri La Mothe holds the record for the highest shallow dive. He dived 12 m (40 ft) from the Flatiron Building in New York City into 0.32 m (12½ in) of water. What was his kinetic energy when he hit the water? What average net force was exerted on him as he stopped? (Assume his mass to be 70 kg.)

   Ans: 8400 J; 26 000 N (5900 lb)

9. A 50-kg stewardess slides down the emergency chute from a flaming 747. If the top of the chute is 4 m above ground, and friction is negligible, with what speed does she hit the ground?

Would her speed be different if she "weighed" 60 kg?

Ans: 8.9 m/s

10. If a sprinter running at a speed of 10 m/s could convert all his kinetic energy into upward motion, how high could he jump?

   Ans: 5.1 m

11. From what height would a car have to be dropped to have the same kinetic energy it would have if it were moving horizontally at 95 km/h (59 mi/h)?   **Ans:** 35 m

12. Lucy Van Pelt slides down a sliding board whose top is 2 m off the ground. If her mass is 20 kg and she hits the ground at a speed of 5 m/s, how much heat is produced during her slide?

   Ans: 140 J

13. If Lucy's sliding board (Problem 12) is 3 m long, what is the average frictional force slowing her motion?   **Ans:** 47 N

14. If the frictional force opposing the rolling of the box up the ramp of Fig. 3.6 is 20 N, what is the efficiency of the ramp and rollers? Assume the 400 N includes the weight of the roller system, and recall that we found F to be 120 N with no friction.

   Ans: 86%

15. A billiard ball collides head-on with another that is at rest. Momentum would be conserved if both balls moved off with ½ the velocity of the incoming ball and in the same direction. Show that kinetic energy would *not* be conserved in this situation. You can guess the mass and assume a velocity, or you can use symbols.

16. A chunk of putty with mass 1 kg slides with speed 33 m/s across a frictionless surface and hits a 10-kg block of steel at rest. If the putty sticks to the steel, with what speed do they both slide across the surface?   **Ans:** 3 m/s

17. Snoopy, whose mass is 5 kg, runs at a speed of 7 m/s and jumps onto his 2-kg skateboard. How fast do he and the skateboard move?   **Ans:** 5 m/s

18. Springs and other elastic bodies return to their original shape when the distorting force is removed. The distance x that a spring stretches (or compresses), is proportional to the applied force F. Thus, $F = kx$, where k is called the spring constant. The *average* force needed to stretch (or compress) the spring a distance x is then ½ the force *at* distance x. Show that the elastic potential energy of the spring, which equals the work done in deforming it, is $\frac{1}{2}kx^2$.

19. Use the result of Problem 18. A spring with spring constant 20 000 N/m is compressed a distance of 15 cm. How much work is done during this compression?   **Ans:** 220 J

20. Suppose a mass of 1 kg is placed on the spring of Problem 19. If the spring is compressed vertically 10 cm (0.10 m) by an outside force and then released quickly, how far will the 1 kg mass fly above the point of release? (Hint: Use conservation of energy.)   **Ans:** 10 m

Home Experiments

1. Have two friends hold a bed sheet in an approximately vertical plane by gripping the four corners. Stand in front of the sheet and throw a raw egg at it as hard as you can. In terms of work-energy and in terms of impulse-momentum, explain why the egg does not break when it hits the sheet. (Have your friends hold the bottom corners up somewhat to form a valley in the sheet to catch the egg. Otherwise, the egg will fall to the floor and break.)

2. Make a vertical stack of several identical coins on a smooth table. Lay another such coin on the table several centimeters away. Hit this extra coin with your finger so that it slides across and hits the bottom of the stack. If your aim is good, the extra coin will exactly replace the bottom one in the stack. Explain in terms of conservation of momentum.

3. The next time you go to a football game, study the effects of energy conservation and momentum conservation on the players and the ball. Keep in mind that most of the collisions between players are very inelastic.

References for Further Reading

Knight, P., "The Physics of Traffic Accidents," *Physics Education,* Jan. 1975, p. 30.

Lin, H., "Newtonian Mechanics and the Human Body: Some Estimates of Performance," *American Journal of Physics,* Jan., 1978, p. 15. On the basis of energy considerations and other basic physics, the author estimates the maximum human performance for field and track events, as well as certain everyday tasks. He compares these estimates with actual performance.

4

Rotational Effects

We all spend our time going around in circles. We rotate with the earth once every 24 hours. This motion is superimposed on the much larger path of the earth's revolution around the sun. The influences of rotation on our lives also appear in spinning phonograph records, turning automobile engines and wheels, rotating fan blades, churning food processors, rotating flywheels, and cranked ice-cream freezers, to name a few.

In this chapter we will look at rotational effects in three main categories: (1) in machines, both simple tools as well as more complex devices; (2) in rigid bodies that can rotate freely; and (3) in bodies that move as a whole in curved paths.

4.1 Simple Machines

In today's language the word *machine* can mean anything from a garden hoe to a digital computer. We will talk in this chapter only about mechanical machines, but even these can be as simple as a screwdriver or as complex as an automobile. Most mechanical devices are based on the idea that a small force moving through a large distance can do the same amount of work as a large force moving through a small distance. In this section we'll see how this application of the principle of conservation of energy makes some kinds of work much easier.

Did you ever have to move a piano? If so, you probably needed to lift it first, in order to put casters under its legs. If you couldn't get help, you might have raised one end at a time by using a plank as a lever (Fig. 4.1). As you exerted a fairly small force downward on the plank, it exerted a much larger force upward on the piano. Thus, the plank served as a "force magnifier." This might seem to violate the principle of conservation of energy but, as we'll see, it does not.

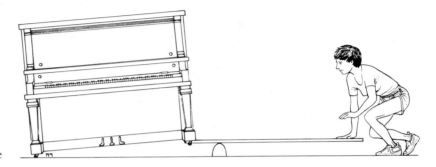

Figure 4.1 The work done by the plank on the piano equals the work done by the person on the plank.

Let's look at the situation more closely with the help of Fig. 4.2. You exert a force f downward at the right-hand end, which moves a distance D. The force F upward at the left-hand end moves only a distance d, much shorter than D. Since the lever is not a source of energy, the work done by f on the right must equal the work done by F on the left. (We are neglecting the small amount of friction between the lever and the objects it touches.) Thus,

$$\text{work on left} = \text{work on right.}$$

From the definition of work—force parallel to the motion times the distance through which it acts—we have

$$F \times d = f \times D.$$

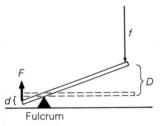

Figure 4.2 Forces on, and distances moved by, each end of a lever.

Since d is much less than D, F must be much greater than f for this equation to hold. Therefore, the "gain" in force results from a "loss" in distance—the larger force moves through a much smaller distance. The closer you move the fulcrum, or pivot point, to the left-hand end, the larger the force F you can exert for a given applied force f; however, the distance d decreases accordingly.

For example, if the piano weighs 2000 N (200-kg mass), you need to exert a force of 1000 N to lift one end. Suppose your plank is 240 cm long and you place the pivot 40 cm from the piano. The length of plank to the right of the pivot is five times that to the left; therefore, for each centimeter moved by the left-hand end, the right-hand end moves 5 cm. By balancing the work on the two ends, we obtain

$$1000\,\text{N} \times 0.01\,\text{m} = f \times 0.05\,\text{m.}$$

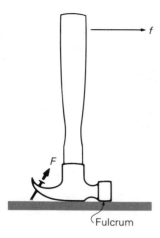

Figure 4.3 The claw hammer is just a lever when it's used to pull a nail.

Solving for f, you find that the downward force you need is 200 N (a force equal to the weight of 20 kg), only 1/5 the weight of one end of the piano.

In summary, the plank functions in two ways. First, it serves as a "transfer agent" that transfers work from your hands to the piano. Second, it gives you an advantage, letting you raise the piano with a much smaller force than you would need if you had to lift the piano directly. You pay for this smaller force by having to move it a larger distance. If you don't need to move a piano, try lifting some small objects using a meter stick or other short stick. Check out the effect of different locations and different heights for the support point that serves as a fulcrum.

In this discussion we have used the symbol f for the weaker force and F for the stronger force, d for the shorter distance and D for the longer distance. *Throughout this chapter,* wherever we contrast smaller and larger values of the same physical property, we will use capital letters for the larger values and lowercase letters for the smaller values.

You've probably used a claw hammer to pull a nail. The head serves as a built-in fulcrum (Fig. 4.3). A small force f from your hand causes the hammer to pull with a much larger force F on the nail. You, of course, have to move your hand much farther than the nail moves. Figure 4.4 shows several other examples of ordinary tools that magnify force at the expense of a larger distance. For each one, f represents the smaller force applied by the hand, whereas F represents the larger force exerted by the device. The fulcrum, or pivot point, which is shown for each device, is often—(a), (e), and (f)—at the end of the "lever." The force F is always closer to the fulcrum than f, so that the distance F moves is less than the distance f moves. The nutcracker has a fulcrum for each arm.

Analyze each of these devices in terms of the applied and resulting forces, and the work done by each. The wrench of Fig. 4.4(c) may seem different from the other machines. In order to tighten the nut, you need to apply forces F to turn it as shown. These forces move in a circular path with a radius approximately equal to the "average radius" of the nut. Since the hand pushing with force f moves in a much larger circle in doing the same amount of work, the hand's force can be much smaller than that required at the nut itself.

In all our examples so far, the device supplies a resulting force larger than the force the user applies to it. Sometimes, however, the need is not so much for a *larger* force but one that can move a *longer distance*. If that's true, we may be willing to apply more force to the device than we get out of it in return for being able to move the weaker force over a longer distance. Examples of this are certain

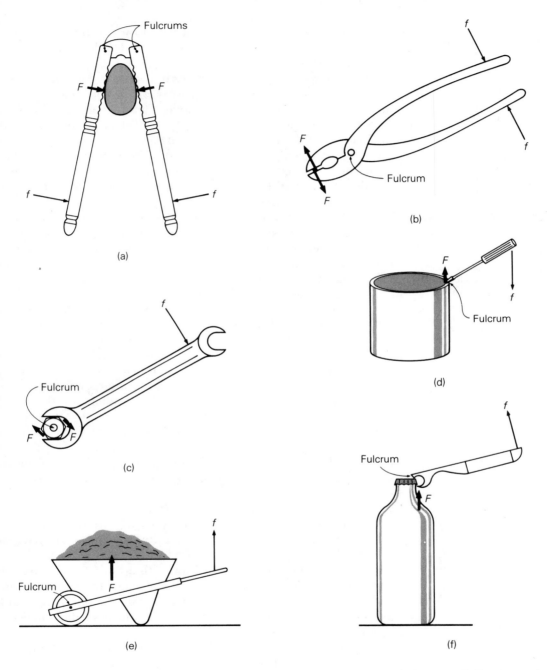

Figure 4.4   Devices that exert a large force F over a small distance as the result of applying a small force f over a long distance: (a) nutcracker; (b) pliers; (c) wrench; (d) screwdriver; (e) wheelbarrow; (f) bottle opener.

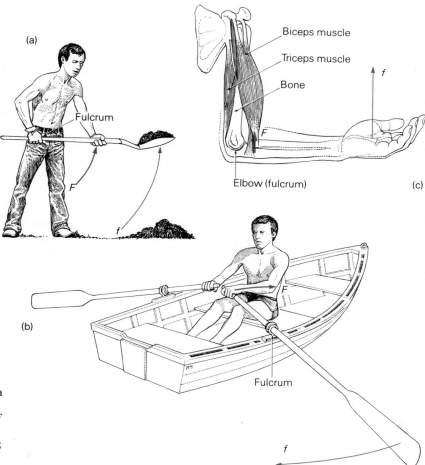

Biceps muscle

Triceps muscle

Bone

(c)

Elbow (fulcrum)

(a)

Fulcrum

F

f

(b)

Fulcrum

f

Figure 4.5 Devices that exert a small force *f* over a long distance as the result of applying a large force *F* over a short distance: (a) shovel; (b) oar of rowboat; (c) human forearm.

uses of the shovel, and the oar of a rowboat. (See Figs. 4.5-a and 4.5-b.)

A person shoveling dirt from a loose pile often holds the end of the shovel handle almost stationary while exerting a force about midway up the handle. This grasp pivots the shovel about the end of the handle. The force on the dirt is less than the force you exert on the handle, but acts over a larger distance, thereby moving the dirt farther and faster than by hand. The work done on the dirt by the force *f* is equal to the work done by the shoveler's hand exerting force *F* on the handle. The rower of a boat wants the end of the oar in the water to move farther and faster than he can move his hand. This process can happen only if the force *f* on the water is less than

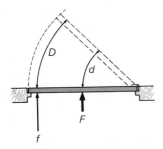

Figure 4.6 If you push on the door in the middle to open it, you have to push twice as hard as if you push on the edge opposite the hinge.

the force F with which the rower pulls on the oar. Thus, more oar must be outside the boat than inside.

An example even closer to home is your own forearm (Fig. 4.5-c). The biceps muscle is attached typically about 5 cm from the elbow. If the palm of the hand is 35 cm from the elbow, the biceps muscle must apply a force of 154 N (35 lb) to lift a 22-N (5-lb) bag of sugar.

We often intuitively do things in a way that uses as little force as possible. We pay for this small force by a longer distance of movement. For example, when you open a heavy department store door, you normally would exert a force such as f in Fig. 4.6 near the edge *opposite* the hinge. The door could also be opened by pushing with a force F closer to the hinge. The same amount of work is done (or energy expended) by either f or F, but F does this work in moving the shorter distance d; thus, F must be larger than f. Specifically, if f is exerted near the edge, but F near the midpoint, then F must be twice as large as f.

4.2 Torque

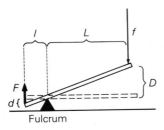

Figure 4.7 The lever of Fig. 4.2 showing the lever arm for each force.

Another way of describing the action of the simple lever-type machines we have talked about is to use a quantity called **torque.** We can use the lever shown in Figs. 4.1 and 4.2 to define this term. We'll redraw this diagram (as in Fig. 4.7) to indicate the perpendicular distances from the fulcrum to the lines along which the forces act—their *lines of action.*

The symbols l and L represent the smaller and larger of these distances, respectively. Each such perpendicular distance from the line of action of the force to the fulcrum, or pivot point, is defined as the **lever arm** for that force. (This distance will actually vary somewhat as the lever tilts. However, if the lever doesn't rotate far from a horizontal position, this change is relatively small and negligible for our purposes.)

We can now define torque. *Torque is the product of a force and its lever arm.* That is,

$$\text{torque} = \text{force} \times (\text{lever arm})$$

$$= \text{force} \times (\text{perpendicular distance from fulcrum to line of action}).$$

For the two forces shown in Fig. 4.7,

$$\text{torque on left} = F \times l$$

and

$$\text{torque on right} = f \times L.$$

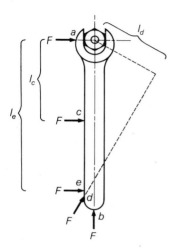

Figure 4.8 The force F applied at point e has the largest turning effect because it has the longest lever arm. The lever arms for points a and b are zero. Those for c, d, and e are shown as l_c, l_d, and l_e.

You may ask: Why do we need to define a quantity such as torque anyway? Let's answer this question. Suppose that you're using a wrench to loosen a very tight nut, as in Fig. 4.8. You certainly must apply a force to the wrench. But there are many ways of applying it, a few of which are shown. If you apply it as shown at a and b, you will certainly not make any headway. Applying the same force at c or d is somewhat better, but position e is the most likely to be successful. Why? The force is the same every time—the lever arm is what makes the difference. The lever arm is zero for positions a and b, but for position e, its value l_e is larger than for any other points and lines of force shown. For position d, the force is applied farther down the wrench than for e, but because of the angle of the line of force, its lever arm is the same as for c.

The ability to turn the nut depends equally on the force and the lever arm. That is, it depends on the *torque*. Torque, then, is a measure of the **turning effect** of the applied force. The greater the torque, the greater the turning effect.

To start anything rotating, you need to apply a torque to it: to turn a doorknob, you apply a torque about its axis; to dial a telephone, you apply a torque around the center of the dial; to spin a plate on his finger, a juggler applies a torque about his finger supporting the plate by exerting a force at the rim. (See Fig. 4.9.)

Now let's go back to the torques on the lever of Fig. 4.7. Is there some relationship between the torque on the left and that on the right? We pointed out earlier that the work done by the two forces F and f must be the same, or

$$\text{work} = F \times d = f \times D.$$

Figure 4.9 A juggler starts a plate rotating by applying a torque on it about his finger.

But a definite relationship exists between the lever arms and the distances the ends of the lever move. (We used this relationship in finding the force necessary to lift the piano.) If you remember some geometry, you can prove that the ratio of the distance moved by the large force to its lever arm is the same as that of the distance moved by the small force to its lever arm. That is,

$$\frac{\text{distance large force acts}}{\text{its lever arm}} = \frac{\text{distance small force acts}}{\text{its lever arm}},$$

or

$$\frac{d}{l} = \frac{D}{L}.$$

(If you don't remember any geometry, you can test the equation by measuring the four distances on the figure.)

By using this relationship in our equation for the work on each end of the lever, we get, after a few algebraic manipulations,*

$$F \times l = f \times L.$$

In words, this equation is:

(large force) × (its lever arm)

= (small force) × (its lever arm).

But from our definition of torque,

torque = force × (lever arm),

which means that

torque on left = torque on right.

This relationship, which comes from the equality of work done by each force, is sometimes easier to use than the work relationship. The reason is that the lever arm is known from the shape and size of the device, but the distances moved depend on the way the device is used.

All the devices discussed in the last section (Figs. 4.1 through 4.6) may be described as well in terms of torques as in terms of work. As you analyze each device, keep in mind that the lever arm for each force is the perpendicular distance from the line of action of that force to the pivot point. Then the torque that the user applies to the device must equal the torque of the resulting force. Our piano example can be solved directly in terms of torques by equating the torques on the two ends of the plank:

$$1000 \text{ N} \times 0.40 \text{ m} = f \times 2.0 \text{ m}.$$

When we solve this equation for *f*, we find, as before, that a 200-N force is needed at the right end of the plank.

Balanced Torques Some devices, for example the juggler's plate (Fig. 4.9), rotate freely when a torque is applied, and rotate faster and faster as long as the torque acts. In Section 4.5 we'll look more carefully at bodies that can rotate freely. For the lever devices we've been considering, there are balancing torques so that the amount of actual rotation is either limited or nonexistent. When you lift the piano and hold it still (Fig. 4.1), you apply a clockwise torque to the

---

*Divide $F \times d$ by d/l and divide $f \times D$ by D/L (which is equal to d/l). Then *d* cancels on the left and *D* cancels on the right; *l* and *L* come to the numerator to give the desired equation.

Figure 4.10 For proper seesawing, the torque on both ends should be the same size, but in the opposite sense.

lever. The piano in turn applies an equal torque *to the lever* in a counterclockwise sense. As a result, there is no net rotation of the lever. There is a slight unbalance in the torques as the motion starts and stops, but this unbalance is negligible in analyzing the forces and torques.

If you use a wheelbarrow (Fig. 4.4-e), the counterclockwise torque you apply is balanced by one in the clockwise sense due to the weight of the load. For any device of the type we've been discussing, the sum of the counterclockwise torques on the device must equal the sum of the clockwise torques.

For smooth seesawing, the torques from the weights of the two seesawers must balance. If you want to seesaw with someone twice your weight, you need to be twice as far from the pivot point as that heavier person in order for the torques on the two ends to balance (Fig. 4.10).

| Exercise | If one seesawer 2 m (6.5 ft) from the pivot point "weighs" 60 kg (132 lb) and the other "weighs" 80 kg (176 lb), how far should the heavier person sit from the pivot? What can you say about the work done by each in moving up and down? What about their changes in potential energy? Why is the word "weighs" in quotation marks?   **Ans:** 1.5 m |

Torques Due to Gravity   Several of the torques we've described depend on the force of gravity: the weight of the piano, the weight of the dirt in the wheelbarrow, the weight of the seesawers. When the weight of a body is the force providing a torque, the line of action of the force is downward through the **center of gravity.** The center of gravity is the point through which all the weight of a body can be considered to act, and is nearer the more massive parts of an irregularly shaped object (Sec. 2.7). Sometimes the weight of the lever itself provides enough torque to influence the other forces needed.

The location of the center of gravity relative to where a body is supported determines whether or not a body can rest stably without toppling over. In Fig. 4.11(a), the center of gravity of the chair is located so that the weight acts downward between the points of support. There is no net torque tending to rotate the chair. For the high-backed chair of Fig. 4.11(b), the center of gravity is located so that the weight acts outside the points of support. The weight gives a net torque that rotates the chair about point A. When the chair back hits the floor, as in Fig. 4.11(c), the force from the floor at point B provides a torque that prevents further rotation.

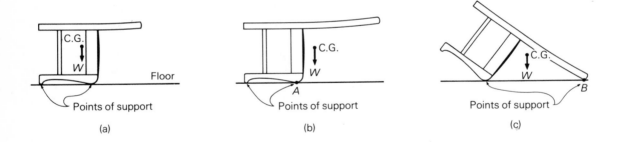

Figure 4.11   The chairs are stable as shown in parts (a) and (c), but not as in (b). C.G. is the center of gravity and W is the weight of the chair.

Question   If you stand with your heels against a vertical wall, why can you not touch your toes, no matter how athletic you are? Try it!

Figure 4.12 shows a couple of devices you might have used (maybe in chemistry lab) that depend for their operation on equal torques in opposite senses. Platform scales used by many doctors are basically the same as part *b* of the figure. The interconnecting linkages, however, are somewhat more complicated.

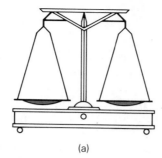
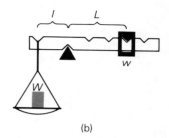

Figure 4.12 (a) The equal arm analytical balance; (b) the unequal arm balance.

(a)

(b)

4.3 Physics versus Technology

In reading this kind of book, you may find it hard to distinguish the *physics* from the *technology*. We use many technological devices to illustrate and clarify the principles of physics. Let's talk about the difference.

So far, we have discussed several laws of physics. Among them are the law of conservation of energy, the law of gravitation, Newton's laws of motion (that describe the relationship between forces and motion), and the law of momentum conservation. These laws are basic principles that apply to objects or to systems of objects in general. The basic principles describe what will happen to these objects under certain conditions.

The term "technology" encompasses the body of methods and materials used to achieve practical objectives. The engineer's job is to understand and to use known physical laws in order to design devices and systems for practical purposes.

When we refer to a claw hammer or a bicycle, we are talking about part of our technology. These devices operate according to certain physical laws that we are trying to explain in this book. We hope that you will not only understand how these laws apply to the specific gadgets we point to in our examples, but will also understand the laws in general and can thus apply them to other devices or events.

**4.4 Applications to Complex Machines*

The principles in Secs. 4.1 and 4.2 lay the foundation for all machines that transmit mechanical motion from one location to

---

*Sections with the ** symbol describe applications only, and contain no *new* physics.

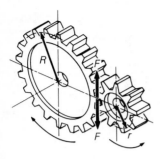

Figure 4.13 Interlocking gears. The teeth of the right-hand gear exert a force F downward on the teeth of the left-hand gear. The left-hand teeth exert an equal force upward on the right-hand ones.

another, and that transform this motion into its most useful form. Examples of some complex machines are: the chain-and-sprocket system of a bicycle, the can opener, the egg beater, the door latch, and the automobile transmission. The principles we have described for the single-component machines apply to each component of a complex machine.

Many such devices use *gears* to transmit work. A gear is a wheel bearing teeth designed to mesh with the teeth of another gear. (See Fig. 4.13.) A gear is in reality nothing more than an odd-shaped lever. If the gear on the right in the figure turns counterclockwise, it exerts a force F on the tooth of the left gear. That force, acting a distance R from the axis of the left-hand gear, provides a torque that tends to rotate the gear. Since the force F on each gear is the same strength, the torques on each are proportional to the lever arms—the radii of the gears.

If you're a bicycle rider and not a mechanic, you might be more familiar with gears that use wrap-around chains rather than those that mesh directly.

Question   For a ten-speed bicycle, explain on the basis of forces and torques why it's easier to climb a steep hill with the chain on the smaller front chain wheel. Why is hill-climbing easier with the chain on the largest rear sprocket wheel?

4.5   Rotating Rigid Bodies

In Chapters 2 and 3 we studied the factors that influence motion in a straight line. There, *force* was the important consideration. A body continues in its state of motion—at rest or with constant velocity—unless a net outside force acts on it. When a net force *does* act on the body, that body accelerates in proportion to the force.

A similar principle holds for rotating bodies. However, *torque does for rotation what force does for linear motion.* Torque is what influences the state of rotation. If a body is not rotating, or *is* rotating at a given rate, it continues to do so unless there is a net outside torque on it. Compare this with Newton's first law of motion.

For each of the quantities that describe and influence some aspect of linear motion, there is a corresponding quantity for rotational motion. Table 4.1 summarizes and compares these quantities. We'll comment on each of the entries as we compare them with the linear motion that you already know about.

Table 4.1   Comparison of Linear and Rotational Motion

Linear Motion	Rotational Motion
distance	angle of rotation
velocity	angular velocity: number of revolutions per second
acceleration	angular acceleration: rate of change of angular velocity
force	torque
mass	rotational inertia

Newton's second law:
    force = mass × acceleration
	Newton's second law for rotation:
	    torque = mass × (rotational inertia)

translational kinetic energy:
    ½ × mass × (velocity)²
	rotational kinetic energy:
	    ½ × (rotational inertia) × (angular velocity)²

momentum:
    mass × velocity
	angular momentum:
	    (rotational inertia) × (angular velocity)

The first three quantities describe rotational motion just as speed, velocity, and acceleration describe linear motion. The angle of rotation corresponds to the distance; the angular velocity (the number of revolutions per second) replaces the velocity; and the angular acceleration tells us the rate of change of angular velocity. The other entries in Table 4.1 need somewhat more detailed comment.

Newton's Second Law for Rotation

Just as Newton's second law—force = mass × acceleration—describes how forces influence linear motion, a variation of this law describes how torques influence rotational motion. This variation has exactly the same form as the second law:

torque = (rotational inertia) × (angular acceleration).

As you would expect, a greater torque on a particular body gives it a greater angular acceleration. But the quantity that measures the resistance to change in the rotational state is not mass, but rather something we call **rotational inertia.** This term is usually called *moment of inertia,* but "rotational inertia" describes its role better.

Rotational inertia depends on mass: It's certainly easier to spin a wooden block on a table than a lead brick of the same size. But rotational inertia also depends on how that mass is distributed relative to the axis of rotation. Let's illustrate. If you let a baseball bat hang in your hand, as in Fig. 4.14(a), you can easily spin it back and forth about a vertical axis. But hold it as in Fig. 4.14(b), and it's much harder—takes more torque—to spin it back and forth. The mass is the same both ways, but in Fig. 4.14(b) it is distributed farther from the axis of rotation. The farther the mass is distributed from the axis, the greater the rotational inertia.

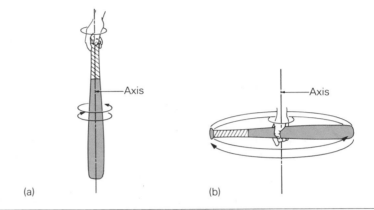

Figure 4.14 The rotational inertia of a body depends on how the mass is distributed relative to the axis.

(a)   (b)

Question   (1) Why do children just learning to walk hold their hands above their heads as they walk (Fig. 4.15-a)? (2) Why does a tightrope walker carry a long skinny pole rather than a short fat one (Fig. 4.15-b)?

Figure 4.15 Maximizing rotational inertia.

Rotational Kinetic Energy and Angular Momentum

Based on our analogy with linear motion, the definitions of **rotational kinetic energy** and **angular momentum** in Table 4.1 should be reasonable to you. Both depend on rotational inertia in the same way that translational kinetic energy and linear momentum depend on mass.

Question   Why are flywheels for storing rotational kinetic energy often made with most of their mass near the outer radius of the wheel rather than near the axis?

Rotational kinetic energy, like any other energy form, must comply with the principle of conservation of energy. Anytime the rotational kinetic energy of a body increases, that increase must be at the expense of a decrease in some other form or forms of energy, and vice versa.

A conservation principle also exists for angular momentum: The total angular momentum of a body, or system of bodies, remains constant if there is no torque on the body or bodies from outside.

Have you ever wondered how figure-skaters can be rotating slowly and then suddenly, as if by invisible forces, begin to spin rapidly? How they pull this off can illustrate both conservation of energy and conservation of angular momentum in a rotating body.

Imagine a very skinny skater with fat hands rotating slowly with arms extended, as in Fig. 4.16(a). What happens if he pulls his hands in close to his body, as in Fig. 4.16(b)? Because much of his mass is then closer to the axis of rotation, his rotational inertia is *less* than before. Let's see what the two conservation principles, energy and angular momentum, tell us about what *should* happen.

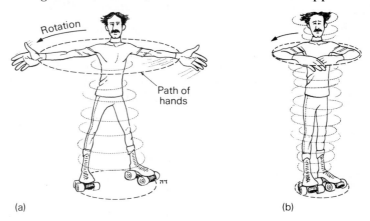

(a)   (b)

Figure 4.16   A skinny skater with fat hands can easily reduce his rotational inertia by pulling in his hands.

Figure 4.17 A skater uses the principles of energy and momentum conservation to increase rotational speed.

Conservation of energy says that the total energy after the skater pulls in his hands must be the same as before, since no energy enters or leaves the skater. (We'll neglect any friction in the air or the skates.) Since rotational kinetic energy = ½ × rotational inertia × (angular velocity)², his angular velocity must increase to compensate for a decrease in rotational inertia. In addition, chemical energy stored in his body lets his muscles do the work of pulling in his hands, thereby further increasing the rotational kinetic energy and adding to his rotational speed. Conclusion: The skater spins much faster with his hands pulled in. A real skater with normal body and hands accomplishes the same result by pulling in arms and hands, plus maybe an extended leg (Fig. 4.17).

If you find the idea of rotational inertia somewhat hard to swallow, we also can understand what happens in terms of translational kinetic energy. The skater's hands have translational kinetic energy associated with their motion along a circular path. The direction of motion is continually changing, but that doesn't change the kinetic energy. When the skater pulls in his hands, the radius of the circular path decreases, which decreases the circumference of the path. Since a hand then goes a shorter distance in one revolution, it must make more revolutions per second to keep the same speed and therefore the same kinetic energy. As before, the kinetic energy also increases by the amount of work done in pulling in the hands. Therefore, pulling in his hands increases his rotational speed.

We can reach the same conclusion on the basis of angular momentum conservation. If the small torques from friction in the skates and the air are neglected, the angular momentum after pulling in the arms must equal that before:

(rotational inertia × angular velocity) after
            = (rotational inertia × angular velocity) before.

When the rotational inertia decreases by pulling in the arms, the angular velocity must increase to compensate. Reextending the arms reverses the process and slows the rotation.

4.6   **Bodies in Circular Motion**

Another aspect of rotational motion deals with the movement of bodies as a whole along a curved path, rather than with their rotation about an axis through the body. An important example of this type of motion is the revolution of planets around the sun, and of satellites around the earth. Our emphasis in this section will be on bodies moving along a particular kind of curved path—a circle—with constant speed.

Centripetal Acceleration and Centripetal Force

We know from Newton's first law that the natural tendency of matter in motion is to move in a straight line with a constant speed. For it to do otherwise, there must be a force on it. If you've ever tried to slam on brakes or turn your car sharply on an icy road, you know this well enough. There is not enough frictional force from the ice to stop you or change your direction, at least not as fast as you'd probably like.

Any deviation from a constant straight-line velocity involves an acceleration. From Newton's second law, any acceleration requires a force. A body moving in a circle is changing direction continually and, therefore, is accelerating continually. Since we are talking about a change in direction and not a change in speed, the force that accelerates the body is perpendicular to its motion and is directed toward the center of the circle. As we said in connection with Fig. 2.3, a rock twirling on a string moves in a circular path only as long as the string exerts a force on it toward the center of the circle (Fig. 4.18).

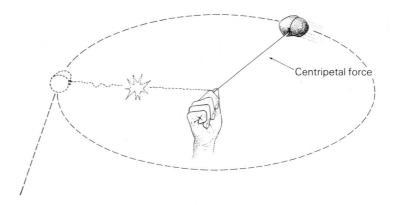

Centripetal force

Figure 4.18   The string provides the centripetal force that keeps the rock moving in a circle. When the string breaks, there is no force and the rock moves in a straight line.

The force acting on a body that keeps it moving in a circle is called **centripetal force.** The name itself describes the force— "centri" meaning "center," and "-petal" meaning "to go toward" from the Latin *petere.* The centripetal force always acts *toward* the *center* of curvature. This term is just a name given to whatever force

Figure 4.19 The nearly vertical walls of the track provide this skateboarder with a centripetal force.

acts on the body to keep it in circular motion. For the rock on the string, the centripetal force is the pull of the string. For the stunt-riding skateboarder on the nearly vertical walls of a bowl-shaped track (Fig. 4.19), the walls exert the centripetal force. For the communications satellite in circular orbit around the earth, the gravitational force is the centripetal force.

The acceleration that accompanies the force is called a **centripetal acceleration** because it, too, is directed toward the center of curvature.

Question   An automobile rounding a curve is moving in a circular path. What provides the centripetal force for this motion? (Hint: See Fig. 4.20.) Why do highway builders usually "bank" sharp curves—that is, slope them downward toward the inside of the curve?

Figure 4.20   The orientation of the front tires during the rounding of a curve.

The centripetal force needed to keep a body in a circular path increases with increase in its speed, and decreases as the radius of its path increases. That's why you need to slow down for a sharp curve (short radius of curvature) in the road. Otherwise, even dry pavement can't provide enough frictional force for the needed centripetal force.

Centrifugal Force

You have certainly had the experience of riding in a curved path and having to lean toward the center to keep from falling over or falling off. This may have been on a merry-go-round, a car moving around a curve, a bicycle, or a skateboard, for example. We will choose the merry-go-round to illustrate our point, but if you identify better with one of the other machines, the ideas still work. As the person in Fig. 4.21 rides on the merry-go-round, he is thinking about staying on the horse. (We'll assume the horse only moves around in a circle and does not move up and down.) But the rider feels what appears to be a force F_{away} pulling him off the horse.

In reality, is there such a force? If you were watching from outside, you would certainly answer "No." The rider's body, having inertia, tends to move in a straight line. At the instant shown he is moving to the right. Without an outside force on him, he would continue in a straight line. But the horse he's on is moving in

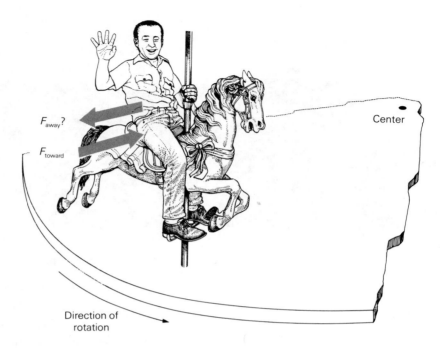

Figure 4.21   A person on a merry-go-round feels a force F_{away} pulling him away from the center of his motion. To an outside observer, there is only a centripetal force F_{toward} pulling him toward the center.

a circle and so curves to the left of that straight line. That is, the horse tries to move from under the rider. As it does, it exerts a force F_{toward} on the rider's leg. From your point of view, that is the only horizontal force on the rider. It provides the centripetal force needed for the rider to continue to move in the circle.

Then what is this force the rider feels trying to pull him off the horse? The rider considers himself to be part of a rotating system; physicists would say that he is in the *reference frame* of the merry-go-round—a rotating reference frame. Because of that, he feels a very real force away from the center. We call this a **centrifugal** (center-fleeing) **force.** To you, an outside observer, the force doesn't exist: It is fictitious. The effect you observe is due entirely to the rider's tendency to move in a straight line.

The next time you get slammed against the car door because the driver went around a curve too fast, remember that the force pushing you there is purely fictitious from the reference frame of those of us sitting still in our chairs. It is no more real than the force that seems to throw you against the dash when the driver slams on the brakes. Both effects occur because your body obeys Newton's first law of motion.

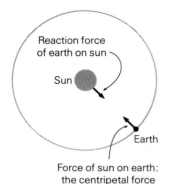

Reaction force
of earth on sun

Sun

Earth

Force of sun on earth:
the centripetal force
on the earth

Figure 4.22 The centripetal force that keeps the earth in orbit around the sun is the gravitational attraction.

Planets and Satellites

A planet moves in a curved path around the sun. Satellites move in curved paths around the earth. Keeping a satellite in a curved path requires a force toward the center of curvature. The reason we have said "curved" rather than "circular" is that most planets and satellites move in elliptical rather than circular orbits. Since the same basic ideas apply in either case, we can assume circular orbits.

The centripetal force needed for the circular motion comes from the gravitational attraction between the satellite and the earth, or between the planet and the sun (Fig. 4.22). The sun exerts a gravitational force on the earth, and this force provides the centripetal force. The earth exerts an equal reaction force on the sun. Comparing this with the rock on a string, you could say that the gravitational force acts as the "string" that transmits force between the sun and earth.

Weightlessness   In this age of space travel we hear lots of talk about weightlessness. Those of us who have not had the privilege of traveling in space have probably seen films of weightless astronauts floating and frolicking around in their capsules. What does it mean to be weightless?

The weight of an object is defined as the gravitational force on the object from the earth or whatever body it's on. An astronaut and his spaceship in orbit around the earth are still acted on by the gravitational force of the earth. Otherwise, his spaceship would not stay in orbit. Then how can he be weightless? If he were on the way to the moon he would pass a point where his weight due to the moon would exactly balance his weight due to the earth. (Would this point be halfway?) He would be weightless there, but beyond that point would have weight directed toward the moon. Only if he were far out in free space, a great distance from any other body, would his weight be nearly zero. Then why do we say that a space traveler near the earth or moon is weightless?

We'll answer that question by considering the way a satellite is placed in orbit. Consider first a special satellite launched from a special launchpad—a rock launched from your hand.

Figure 4.23 shows the paths taken by identical rocks thrown horizontally at different speeds on level ground. As you throw harder, the rock travels farther horizontally before hitting the ground. But the earth pulls downward with the same force on each rock, so that the time required for the rock to fall to the ground is the same for each throw. It goes farther when you throw harder because its faster horizontal speed covers a larger horizontal distance in the same amount of time.

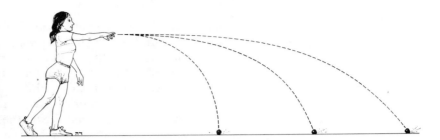

Figure 4.23 The harder you throw the rock, the farther it goes before hitting the ground.

But suppose we look at you from much farther away as you throw the rock. We'll retreat far enough to see the entire earth, and draw you large enough that we can see you (Fig. 4.24). As you throw harder and harder, the rock travels farther and farther around the earth as it falls to the ground. When you throw the rock fast enough (path *O*), its curved path as it falls is exactly parallel to the surface of the earth. The rock continues to fall completely around the earth to the starting point. It would then keep on going indefinitely. For the rock to follow this path, the gravitational force between the earth and the rock must provide the *exact* centripetal force necessary for the circular motion. The centripetal force needed depends on the speed, and would equal the gravitational force when the speed is about 28 000 km/h (18 000 mi/h). To

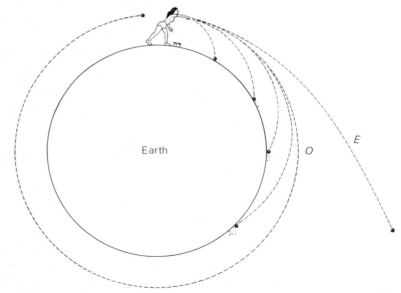

Figure 4.24 If you throw hard enough, the rock falls completely around the earth.

completely escape the gravitational field of the earth (path *E*), the rock must have a speed of at least 40 000 km/h. (Underlying this discussion is the assumption that you have a very good pitching arm!)

The same ideas apply to the launching and orbiting of any space satellite. The satellite in orbit is continually falling, but moves with the proper horizontal speed to stay in orbit.

Now let's go back to our question concerning the meaning of weightlessness. Suppose Senator John Glenn, a former astronaut, reverts to space travel and is in a satellite orbiting the earth. Both the satellite and Glenn are continuously falling around the earth, as we just said. If he were to stand on scales inside his satellite, he would exert no force on them because the satellite and scales are falling along with him. He *appears* to be weightless because he exerts no force on the scales or anything else in the satellite. The effect inside the spacecraft is the same as if there were no gravitational force. But viewed from the earth, each object has a weight and is continually falling because of that weight.

You would experience the same effect if you were inside an elevator and the supporting cable broke. The elevator and all its contents, including you, would be in free fall and therefore "weightless" until you hit the bottom. In the meantime, you could "have a ball" floating weightlessly about the elevator.

A spacecraft on the way to the moon usually gains enough speed early in its flight to coast the rest of the way, except for minor corrections. During this coasting period the travelers are again weightless because the spacecraft is falling freely. Weightlessness ends when the retrorockets are fired for slowing the spacecraft down as it nears the moon.

Satellite Descent   Our discussion so far has implied that, once a satellite is in orbit, it stays there forever, continually "falling" around the earth. Yet you know from news reports that satellites sometimes fall *to* earth. In 1978 a Soviet Union *Cosmos* satellite with a nuclear reactor aboard created worldwide publicity and concern when it fell to earth, spilling radioactive materials over parts of Canada. Another example: In 1978 certain specialists from the National Aeronautics and Space Administration worked intensively but unsuccessfully by remote methods to stabilize the orbit of *Skylab,* the space laboratory that had been idle since 1974. They hoped to prevent its plunge to earth before the Space Shuttle was available to boost it to a higher orbit.

Orbiting artificial satellites eventually return to earth because, at their altitudes above earth, there is *some* atmosphere that supplies a frictional "drag." Work done against this friction comes from the kinetic energy of the satellite, slowing it enough that it loses altitude. The *Skylab* story has some interesting, though costly, surprises. When the astronauts returned from it in 1974, they left it in an orbit that was expected to keep it safely until the mid-1980s. But unexpected activity on the sun spewed out excess energy that heated the earth's atmosphere slightly. The resulting expansion pushed more atmosphere into the region of *Skylab,* thereby increasing the drag on it. The mid-1978 maneuvering was an attempt to orient the satellite for minimum drag. Its plunge to earth in 1979 speaks for the failure of the effort.

Key Concepts

If only two forces act on a lever, and we neglect friction, both forces do the same amount of work. This fact is the basis of mechanical machines, both simple and complex. The use of a machine can reduce the external *force* needed to do a certain amount of work, but the distance the force acts must increase correspondingly.

Torque = force × lever arm,

where the **lever arm** is the perpendicular distance from the force to its line of action. Torque is a **turning effect.** When a weight is the force providing a torque, its line of action is downward through the **center of gravity.**

A body that is either not rotating or rotating at a given rate, continues to do so unless there is a net outside torque on it. The rate of change of angular velocity, the angular acceleration, is directly proportional to the net applied torque and inversely proportional to the body's **rotational inertia.** Rotational inertia depends on mass and on how that mass is distributed relative to the axis of rotation; rotational inertia is greater the farther the mass is distributed from the axis.

Rotational kinetic energy and *angular momentum* depend on rotational inertia and angular velocity in the same way that translational kinetic energy and linear momentum depend on mass and linear velocity, respectively. The angular momentum of a system on which there are no external torques remains constant.

A body in circular motion must be acted on by a **centripetal force** directed toward the center of curvature. The corresponding acceleration is the **centripetal acceleration.** A **centrifugal force** is experienced by a body in a rotating reference system, but is fictitious from the viewpoint of an outside observer.

The centripetal force that keeps planets and satellites in orbit is the gravitational force on them. Persons in orbit appear *weightless* because they, as well as everything in their spacecraft, are falling at the same rate.

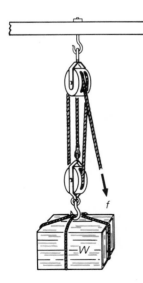

Figure 4.25   The block and tackle.

Questions

Simple Machines and Torque

1. How many machines that serve as force magnifiers (such as those of Fig. 4.4) have you used in the past week? How about those that exert a weaker-than-applied force, but over a larger distance (as in Fig. 4.5)?

2. In Fig. 4.5(b), what exerts the force that makes the boat go forward? Why does the boat go in a circle if you row from only one side?

3. Suppose you use a wrench to loosen a spark plug on your car. If you do 10 J of work on your end of the wrench, about how much work does the other end of the wrench do on the spark plug?

4. Why do the block-and-tackle systems of Fig. 4.25 result in more load W being lifted than force F exerted? Estimate the ratio W/f for each system. (Hint: Consider the relative distance of travel by W and f.)

5. When a threaded pipe is especially hard to unscrew, plumbers sometimes put a piece of pipe over the handle of their pipe-wrench to extend the length of the handle. How does this help?

6. In sweeping the floor, you usually hold the top of the broom handle with one hand. Compare the relative merits of different positions along the broom handle for the other hand.

7. For aesthetic reasons, doorknobs are sometimes put in the center of a door. What are the disadvantages of this arrangement?

8. Where should a doorstop be located relative to the door in order to minimize the force on the hinge screws when the door gets stopped?

9. It's easy for the coattail of a coat hanging in a closet to get caught between the closet door and the door casing on the hinged side of the door. Why will this easily damage the hinges, even though you don't push very hard to close the door?

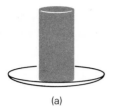

(a)

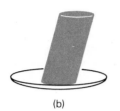

(b)

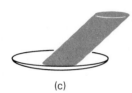

(c)

Figure 4.26   When does the Jell-O tower start to topple?

10. Where does the torque come from that causes the arms of a windmill to rotate?

11. What torques act on a diving board to keep it from rotating when you stand on the end of the board? What are the relative senses (clockwise or counterclockwise) of these torques?

12. Archimedes, the noted Greek mathematician of the third century B.C., reportedly said that if he had a suitable fulcrum on which to rest his lever, he could lift the world. In what ways is his boast meaningless? (See Chapter 2, Question 18.)

13. Suppose you make a cylindrically shaped hunk of Jell-O and put it on a plate as in Fig. 4.26(a). If it gradually distorts as shown from (a) to (b) to (c), draw its configuration when it first starts to topple over.

14. Freddie the Wino had four pallbearers at his funeral. If the front two hold the casket ¼ its length from the end, and the rear two hold it at the end, compare the relative loads carried by the pallbearers.

15. The curvy piece of wood in the hand of Fig. 4.27(a), sometimes called a "sky hook," would obviously not stay in a horizontal position if you supported it only at the tip. Yet, when a belt is inserted in it as in Fig. 4.27(b), you *can* support it that way. Why?

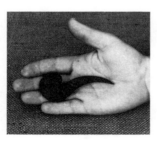

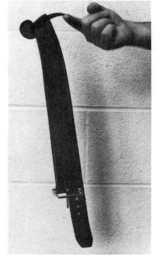

Figure 4.27   The "sky hook": (a) shown; (b) in action.

16. For maximum ease in pedaling uphill, should the front chain wheel of a bicycle be large or small? What about the rear one?

Rotating Rigid Bodies

17. Why does choking the bat (grasping it a short distance from the end, rather than at the end) make it easier for a baseball batter to swing?

18. In terms of rotational inertia, explain why: (a) short-legged people step more quickly than long-legged ones; (b) it's hard to walk without bending your knees; (c) you hold your arms straight out to the side when you're walking on a log across a creek; and (d) you can swing a short bat faster than a long one of the same weight.

19. Why does a helicopter (Fig. 4.28) need a second propeller on its tail?

Figure 4.28 A helicopter needs two propellers.

Bodies in Circular Motion

20. Why does a bicycle rider lean toward the inside in going around a curve?

21. Is a bug riding on the rim of a 33⅓ RPM phonograph record accelerating?

22. If Santa Claus became the Arctic ambassador to Ecuador, would his accurate bathroom scales read the same when he stood on them in the embassy in Quito (latitude 0°) as they did at his home?

23. During the spin cycle of a washing machine's operation, water is separated from the clothes as the tub rotates (Fig. 4.29). Would you say that water is thrown from the clothes, or that the clothes are pulled away from the water? Your answer may depend upon whether you consider yourself to be outside the machine or inside it, spinning with the clothes.

24. Why does mud on the tires of a truck fly off as the truck picks up speed? What is the direction of motion of a hunk of mud relative to the tire?

25. Why do riders not fall out as they go through the loop-the-loop of the "Loch Ness Monster" (Fig. 4.30)?

Figure 4.29 A spinning washing machine tub.

Figure 4.30

26. In the future, large numbers of people may live in space colonies—large self-sufficient space stations that produce "artificial gravity" by slow rotation of the entire station. Why would this rotation produce the same sensation as gravity to a person living on the *inner* surface of a shell surrounding the station?

27. Would you weigh more if the earth were not rotating?

28. Is it correct to say that the earth is continuously falling around the sun?

29. If you jump from a 2-m-high wall, are you weightless until you hit the ground?

30. Is a skydiver weightless after jumping from an airplane, but before opening the parachute?

Problems

1. With what force must you push down on the plank of Fig. 4.1 if the plank is 1-m long, if the piano weighs 1600 N, and if the fulcrum is placed 20 cm from the piano?   **Ans:** 200 N

2. In Fig. 4.31 what force F is exerted by the biceps muscle on the forearm as the ball of mass 15kg is held in the position shown? Remember to distinguish between mass and weight.)

   Ans: 1300 N (290 lb)

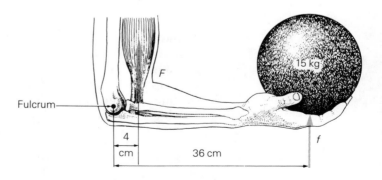

Figure 4.31   Some typical dimensions for the forearm.

3. Suppose there is a pecan 2 cm from the fulcrums in the nut-cracker of Fig. 4.4(a). If you squeeze with a 40-N force 8 cm beyond the nut, with what force do the jaws push on the nut?

   Ans: 200 N

4. If the wrench of Fig. 4.4(c) is 20-cm long, what torque can you exert on the nut with a force f of 150 N?   **Ans:** 30 N·m

5. On the wheelbarrow of Fig. 4.4(e) the distance from the wheel axle to the center of the body is 2 ft, and the distance from the wheel axle to the end of the handles is 5 ft. About how much force is needed on the handles to lift 200 lb of dirt? (Neglect the weight of the wheelbarrow.)   **Ans:** 80 lb

6. Superman is shoveling coal with a supershovel (Fig. 4.5-a). He has 900 kg (mass) of coal in the shovel as he holds it in a horizontal position. If his left hand is halfway between the coal and his right hand, what force does his left hand exert? If we consider his left hand as the fulcrum, what is the direction and strength of the force exerted by his right hand?

   Ans: 18 000 N; 9000 N downward

7. The beautiful princess is rowing toward the charming prince in a boat with oars 2-m long (Fig. 4.5-b). Where must the oar's fulcrum be if she wants the end in the water to move three times as far as the end in the boat? If she pulls with a force of 200 N on her end of the oar, what force does the other end exert on the water?   **Ans:** 50 cm from inside end; 67 N

8. If you are seesawing with a friend and you weigh only ⅔ as much as your friend, where should he sit if you sit on one end of the seesaw? Assume the pivot point to be at the center.

   Ans: ⅔ as far from the pivot

9. In the unequal arm balance of Fig. 4.12(b), suppose the empty pan exactly balances the empty beam with the movable weight w removed. The distance l is 3 cm, the distance from the fulcrum to the first notch is 2 cm, and the notches are 2 cm apart. Putting w in the first notch balances a 10-N weight in the pan. Moving w each additional notch to the right balances an additional 10-N weight. What is the weight of w?   **Ans:** 15 N

10. A continuous torque of 50 N·m brings a flywheel from rest to an angular velocity of 6000 revolutions/min in 30 s. How long will it take a continuous torque of 150 N·m to bring it to the same angular velocity?   **Ans:** 10 s

Home Experiments

1. With a meter stick or similar bar as a lever, try lifting objects of different weights. Compare the effect of different positions and different heights for the fulcrum.

2. Tie a rope around your waist and throw the free end over the limb of a tree. Pull yourself up to the limb by pulling down on the free end. Compare the strength needed in your arms this way with that needed to climb straight up a single length of rope tied to the limb.

3. Analyze the motion of a rocking chair. When you rock backwards and lift your feet, what provides the torque that rocks you forward again? What are the energy transformations as you rock all the way back, and then stay still and let the oscillations die out?

4. If you don't like shaking down your clinical thermometer, get one in a plastic case, drill two holes in the top of the case, and loop a string through it as shown in Fig. 4.32. With the thermometer inside, take a loop in each hand and twirl the case to wind up the string. As you pull outward the rapid spinning will "shake down" the thermometer. You probably will find that a string smaller than about 1-mm diameter will not work. Why?

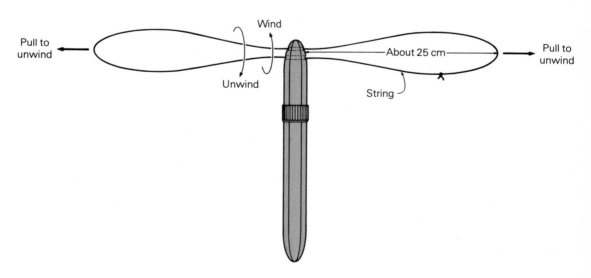

Figure 4.32   The no-wrist-action way to shake down a clinical thermometer inside its case.

References for Further Reading

Edelson, E., "Saving Skylab—the Untold Story," *Popular Science,* Jan. 1979, p. 64. The story of the mid-1978 earth-bound efforts to revive the sleeping *Skylab* and save it in orbit until the arrival of the Space Shuttle.

Koff, R. M., *How Does It Work?* (Signet, New York, 1961). Simple but clear discussion of the workings of many everyday gadgets.

The Way Things Work, An Illustrated Encyclopedia of Technology, V. I and V. II (Simon and Schuster, New York, 1967). Detailed mechanisms as well as some of the physics of hundreds of everyday and industrial devices.

5

Fluids

You, I, and all other forms of life live immersed in a sea of fluid. For fish, that sea is the water in their ocean, river, lake, or stream. For you and me, and for most of the life forms around us, that "sea" is the air we breathe. Fluids influence our lives in countless ways: we travel on them in airplanes and boats; we burn them as fuel in our furnaces and ranges; we take baths in them; we drink them; and we depend on them to carry nutrients through arteries to all parts of our bodies.

Even though we often take the word "fluid" to mean liquid, a **fluid** is a substance that *can flow.* That category includes both *liquids* and *gases.* The molecules of a liquid can move around, letting the liquid take on the shape of its container. Yet these molecules stay about the same distance apart, making the liquid essentially *incompressible.* That means a given mass of liquid keeps the same volume, aside from minute changes. For a gas, however, the molecules bounce around randomly, filling up the complete volume of the container. A gas is, therefore, *compressible,* and behaves differently than a liquid in certain ways. In spite of the differences, many properties of a fluid are independent of whether that fluid is a liquid or a gas.

In this chapter we'll examine some of the many aspects of the behavior of fluids. Before we can begin, however, we need to know the meaning of two important concepts that apply to solids as well as to fluids: namely, density and pressure. We can then discuss atmospheric pressure, and the way pressure varies throughout any fluid. This pressure variation within a fluid is what accounts for the buoyant forces that make objects float on, or seem lighter in, fluids.

Practical use of fluids often depends upon our ability to move them—to pump them—from one place to another. After looking at the physical bases for pumping, we'll consider a few properties of fluids in motion.

We can explain our final topics, *surface tension* and *capillary action* in liquids, only on the basis of the interactions between individual molecules of the liquid.

5.1   Density and Specific Gravity

Suppose you have three bricks, all of the same shape and size, but made of different materials: one wood, one concrete, and one lead. You know already that each of these has a different weight, and therefore a different mass. Substances with the same volume but different mass, differ in their *densities*.

Density is the mass of a unit of volume of a substance. Thus,

$$\text{density} = \frac{\text{mass}}{\text{volume}}.$$

Density may vary widely from substance to substance. For a gas, which is compressible, the density can vary over a wide range, depending on how much the given gas is compressed.

If we have a block of aluminum 20 cm $\times$ 10 cm $\times$ 5.0 cm and we measure its mass to be 2.7 kg, we can calculate its density by

$$\text{density} = \frac{\text{mass}}{\text{volume}} = \frac{2.7 \text{ kg}}{0.20 \text{ m} \times 0.10 \text{ m} \times 0.050 \text{ m}},$$

or 2700 kg/m³.

Table 5.1 gives the densities of a few ordinary substances. Rather than using the SI units kg/m³ as in Table 5.1, it's sometimes easier to work with densities in the smaller metric unit g/cm³. You can get the density in g/cm³ by dividing the values in this table by 1000. Why?

Table 5.1   Densities of Selected Substances

Substance	Density (kg/m³)	Substance	Density (kg/m³)
air (sea level)	1.29	concrete	2 800
pine wood	500	diamond	3 300
butter	860	steel	7 830
paper	920	lead	11 300
ice	920	mercury	13 600
water	1000	gold	19 300

Exercise   What is the mass of the mercury that fills a 1-liter bucket? (1 liter = 1000 cm³ = 0.001 m³)   **Ans:** 13.6 kg

Are you surprised that water has such a convenient density: 1000 kg/m³? This number is not an accident of nature, but comes from the original definition of the gram as the mass of 1 cm³ of water.

Since water is such an omnipresent substance, the densities of other substances are sometimes given as the ratio of *their* densities to that of water. This "relative density," called **specific gravity,** is defined as:

$$\text{specific gravity} = \frac{\text{density of the substance}}{\text{density of water}}.$$

Being a ratio, specific gravity has no units.

Question   Why does the specific gravity of a substance have the same numerical value as its density in g/cm³, or 1/1000 the numerical value of its density in kg/m³?

We sometimes use the quantity **weight density:**

$$\text{weight density} = \frac{\text{weight}}{\text{volume}},$$

with SI units N/m³. (The British units are lb/ft³.) Since the weight of an object equals its mass multiplied by the acceleration due to gravity ($W = mg$, Chapter 2),

$$\text{weight density} = \text{mass density} \times \text{acceleration due to gravity}$$

$$= \text{mass density} \times g.$$

Since g (9.8 m/s²) is almost 10 m/s², we can get the weight density closely enough for most purposes by adding a zero to the value of the mass density (in SI units only). For water, the weight density is 9800 N/m³ (62.4 lb/ft³), or roughly 10 000 N/m³. When we use only the word "density," we'll mean *mass* density.

Knowing the density of a substance, we can get the mass of a certain volume of it from

$$\text{mass} = \text{density} \times \text{volume}.$$

Using symbols for the quantities in this equation, we have

$$m = DV.$$

Similarly, the weight is given by

$$\text{weight} = (\text{weight density}) \times \text{volume},$$

$$W = D_wV,$$

where we use the subscript "w" to distinguish weight density from mass density. From $W = mg$, we get

$$W = DgV,$$

or

$$\text{weight} = \text{density} \times (\text{acceleration due to gravity}) \times \text{volume}.$$

5.2 Pressure

Hold a sharp pencil between your fingers and push as shown in Fig. 5.1. You can push hard and the eraser end won't hurt. But the finger touching the pointed end hurts! You're applying the same force, but the point concentrates it in a much smaller area. The difference is in the pressure. **Pressure** is defined as the force per unit of area:

$$\text{pressure} = \frac{\text{force}}{\text{area}}.$$

Figure 5.1   The same force on both fingers gives a much higher pressure on the right than on the left because the area of contact is much smaller on the right.

If the force you exert with the pencil is 25 N and it is distributed over the eraser of area 32 mm² or 3.2×10^{-5} m², then

$$\text{pressure} = \frac{25\,\text{N}}{3.2 \times 10^{-5}\,\text{m}^2} = 7.8 \times 10^5\,\text{N/m}^2.$$

Estimate the pressure you expect from the same force applied to the point. If the point area is 0.20 mm², you can calculate the pressure to be 1.25×10^8 N/m², or 160 times as much as with the eraser.

Many footballs are marked "inflate to 13 lb." But I doubt seriously that the one Sonny Jurgensen passed for a 99-yard Redskin touchdown play in 1968 *weighed* 13 lb. It was, however, probably inflated to 13 *lb/in²* of pressure—a force of 13 lb on each square inch of area inside the ball. Similarly, inflating your tires to 29 "pounds" means 29 lb/in² of pressure.

Question   Why is a period typed with a typewriter harder to erase than an m?

Exercise   Convert a pressure of 29 lb/in² to newtons/meter².

Ans: 200 kN/m² (k = 1000)

Atmospheric Pressure

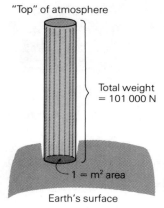

"Top" of atmosphere

Total weight = 101 000 N

1 = m² area

Earth's surface

Figure 5.2 Atmospheric pressure is caused by the weight of the air above the area. A column of air of 1 m² in cross-sectional area from sea level up weighs 101 000 N.

The sea of air that we live in—the atmosphere—exerts a pressure on us and everything else submerged in it. This pressure comes from the weight of the air itself as it pushes down on the earth and its occupants.

Figure 5.2 illustrates this point. Consider a square meter of area on the earth and the cylinder of air above it 1 m² in cross section extending to the "top" of the atmosphere. If the surface area is at sea level, the amount of air in that cylinder weighs 101 000 N. This means the **atmospheric pressure** at sea level is 101 000 N/m² (101 kN/m²), or 14.7 lb/in². At higher altitudes there is not as much air above a given area, and therefore the atmospheric pressure is lower. That's why your ears "pop" as you go up or down a tall mountain.

Atmospheric pressure not only pushes down on the top of horizontal surfaces, but upward on their bottoms and sideways on their sides. It pushes perpendicular to *any* surface, whatever its orientation. Since the air is free to flow, it exerts equal pressure in all directions at a given location. This condition holds true for any fluid under pressure. When you dive deep into water, you feel the pressure in all directions on your body.

Exercise   Estimate the force with which the atmosphere pushes on your chest.

Ans: (for 0.06-m² chest): 6100 N (1400 lb)

We don't feel the roughly 10 N on each square centimeter of our bodies (15 lb on each square inch) because we are used to it, and our bodies push outward with an equal but opposite pressure all over.

Atmospheric pressure exerts a force of about 3700 N (820 lb) on the side of the empty can shown in Fig. 5.3(a). Why doesn't it crush the can? You know the answer is that air is also inside the can pushing out. But if we connect a hose from the can to a vacuum pump, and pump out the air inside, atmospheric pressure quickly crushes the can into the distorted mess shown in Fig. 5.3(b). It's almost as entertaining to see the can come back nearly to its original

 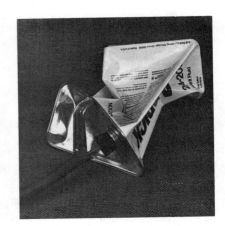

Figure 5.3 When the air is pumped from an empty can, the outside atmospheric pressure easily crushes it.

shape when you pump in air at a pressure a little higher than atmospheric.

We often use sea level atmospheric pressure as a *unit* of pressure. For example, we say that a pressure of 250 kN/m² is 2.5 atmospheres, since this pressure is 2.5 times "standard" atmospheric pressure.

The 90 kN/m² (13 lb/in²) of pressure in your football and 200 kN/m² (29 lb/in²) in your car tires are actually the difference in pressure between the inside of the given inflatable object and the atmosphere. The usual way of measuring pressure is just that: the difference between the internal pressure and the external atmospheric pressure. The total air pressure inside the tire is, therefore, 301 kN/m².

Pressure in Liquids

When we spoke earlier of the "top" of the atmosphere, we put the word "top" in quotes because the atmosphere doesn't actually have a top, but gradually thins out from about 6×10^{20} molecules/cm³ at sea level to about 1 molecule/cm³ in interstellar space. The density at any particular point depends upon the pressure at that point.

Liquids, on the other hand, are relatively incompressible: Their densities stay almost the same for large variations in pressure. For that reason, we can easily find the pressure at a given depth caused by the liquid above it.

Consider the square area on top of the fish in the tank of Fig. 5.4. The downward force on this area is caused by the weight of the liquid in the column directly above: its density $\times$ g $\times$ its

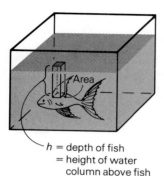

h = depth of fish
= height of water
column above fish

Figure 5.4 The pressure due to the water on the fish at depth h below the surface is determined by the weight of the column per unit of area. This weight depends only on the density of the liquid and the depth.

volume. Since the volume of the column is its area × its height, we can calculate the pressure from

$$\text{pressure} = \frac{\text{force}}{\text{area}} = \frac{\text{weight}}{\text{area}} = \frac{\text{density} \times g \times \text{area} \times \text{height}}{\text{area}}$$

$$= \text{density} \times g \times \text{height}.$$

In symbols, this equation is

$$p = Dgh.$$

Notice that the height of the water column is the *depth* under water of the point where we're finding the pressure.

The pressure is directly proportional to the depth in the liquid, and is independent of volume or shape of the container holding the liquid. You've probably seen water storage tanks shaped like the one in Fig. 5.5(a). The pressure at the bottom is the same for both tanks (a) and (b). For (a), the tank walls push up on the lower sloping walls of water to partly support its weight.

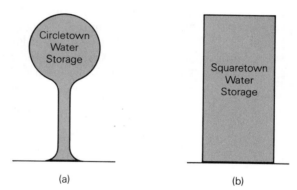

Circletown
Water
Storage

Squaretown
Water
Storage

(a)   (b)

Figure 5.5 The pressure at the bottom of both tanks is the same.

If, as usual, there is atmospheric pressure on the surface of a liquid, the total pressure at some depth would be the atmospheric pressure *plus* that due to the liquid. We'll illustrate with an example.

Example 5.1   Compare the pressure 50 cm below the water surface with that 3.0 m down. The pressure due to water only is given by

$$p = Dgh.$$

At 0.50 m,

$$p = 1000 \text{ kg/m}^3 \times 9.8 \text{ m/s}^2 \times 0.50 \text{ m} = 4900 \text{ N/m}^2.$$

Thus, if p_a is atmospheric pressure, then

total pressure $= p + p_a = 4.9 \text{ kN/m}^2 + 101 \text{ kN/m}^2 = 106 \text{ kN/m}^2.$

At $h = 3.0$ m, a similar calculation gives the pressure of the water only to be 29 kN/m², and a total pressure of 130 kN/m²: 23% more than near the surface. Does that explain why your ears may hurt when you go deep into the water after a high dive?

Question   Did the little Dutch boy who held back the whole North Sea with one finger have superhuman strength?

Measuring Atmospheric Pressure

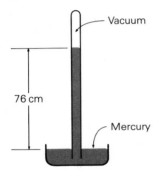

Imagine that you take a glass tube about a meter long that is closed at one end, hold it with the open end up, and fill it with mercury. Then carefully invert it in a bowl so that the open end stays immersed in the mercury that runs out. You would find that not all the mercury would run out, but only enough to leave about a 76-cm-high column of mercury in the tube, as shown in Fig. 5.6.

To find out why, let's calculate the pressure at a depth of 76.0 cm below the surface of mercury, which has a density of 13 600 kg/m³. Since there is vacuum (nothing) in the tube above the mercury, the pressure at the bottom is due entirely to the mercury:

$$p = Dgh = 13\ 600 \text{ kg/m}^3 \times 9.80 \text{ m/s}^2 \times 0.760 \text{ m}$$

$$= 101\ 000 \text{ N/m}^2.$$

Figure 5.6 Sea level atmospheric pressure pushing down on the open surface of mercury holds the mercury 76 cm up in the tube. This principle is the basis of the mercury barometer.

But this value is exactly sea level atmospheric pressure. The atmosphere pushing down on the open surface of the mercury pushes it up into the tube until the pressure at the surface level caused by the weight of the mercury column exactly balances the atmospheric pressure.

This principle is used in making a mercury *barometer*—a device for measuring atmospheric pressure. Atmospheric (*barometric*) pressure varies somewhat from place to place, and from time to time at a given place. The height of the mercury column supported in this way is a measure of the atmospheric pressure. In weather reports, this pressure is usually reported directly as the height of the mercury column—for example "760 mm" or "29.92 inches."

Question   Could you make a barometer using water instead of mercury? How tall would the water column need to be?   **Ans:** 10.3 m (33.8 ft)

5.3   **Pressure Distribution in a Fluid**

Blaise Pascal, a seventeenth-century French theologian and scientist, discovered that if you apply a pressure to one part of an enclosed fluid, the pressure *increases* by that same amount everywhere in the fluid. We call this effect **Pascal's principle.**

---

*We can derive Pascal's principle from the energy conservation principle. The apparatus shown in Fig. 5.7 contains an incompressible liquid. The weight on the left pushes down with force f on the left and moves that piston down a distance D; the one on the right must move up some distance—call it d. From the principle of conservation of energy, the work done against the force F must equal the work done by the force f, neglecting friction. That is,

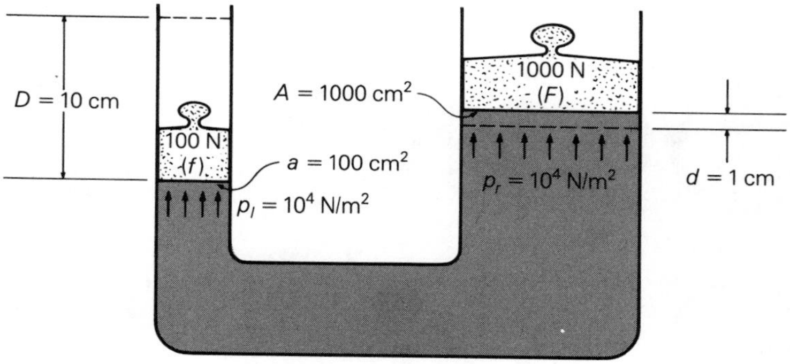

Figure 5.7   In a hydraulic press, the pressure is the same on both pistons. The force is therefore proportional to the area. The smaller piston must move farther in the same proportion.

work by f = work on F,

or $f \times D = F \times d.$

If pressure p_l acts over area a on the left and pressure p_r over area A on the right, then $f = p_l a$ and $F = p_r A$. Putting this into the work equation, we get

$$(p_l a) \times D = (p_r A) \times d,$$

that we can regroup to get

$$p_l \times (aD) = p_r \times (Ad).$$

---

*Paragraphs marked with these special brackets are somewhat more mathematical than most, and are not essential in understanding later material.

The quantities in parentheses here—area × distance—are the volume changes on each side. Since the fluid is incompressible, these volume changes are the same. That means $p_l = p_r$: The pressure is the same on both sides, just as Pascal's principle says. You can put in the actual values in Fig. 5.7 and show that they agree with the equations here.

In the "hydraulic press" of Fig. 5.7, the force F must be as much greater than f as the area A is larger than a, since the pressure is the same. In other words, the press acts as a force magnifier in the same manner as the levers of Chapter 4. Just as in the case of the levers, the smaller force must act over a proportionately longer distance. This principle is used in many hydraulic devices, including jacks and car lifts used by service stations, dentist and barber chairs, and brakes in cars and trucks (Fig. 5.8).

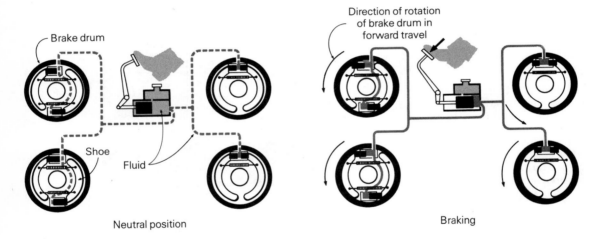

Figure 5.8   A hydraulic brake for automobiles. When you push on the brake pedal, the increased pressure is transmitted throughout the brake fluid, thereby pushing the brake shoe against the drum.

5.4   Buoyancy

You probably have found that it's easy to lift a person who is almost completely underwater. Also, when you've been in the water for a long time and climb out, you feel unusually heavy as you get out because you've gotten used to being partly held up by the water. Evidently, as these two examples hint, there is an upward force on objects immersed in a liquid. That force is strong enough that some

objects—boats, rafts, persons with life preservers, beach balls—
float on top of liquids. This effect is not just confined to liquids.
Helium-filled balloons and the Goodyear blimp float in air. Any
object totally or partially submerged in a fluid experiences a *buoyant
force* that is responsible for all the effects we've mentioned.

The buoyant force exists simply because the pressure increases
with depth in the fluid. Consider the fish relaxing in the sun in Fig.
5.9. Because its bottom is a little lower than its top, the pressure is a
little greater on its bottom. The water, therefore, pushes up on it
somewhat harder than it pushes down. The difference is just
enough to hold the fish up so that it doesn't sink. The lead weight

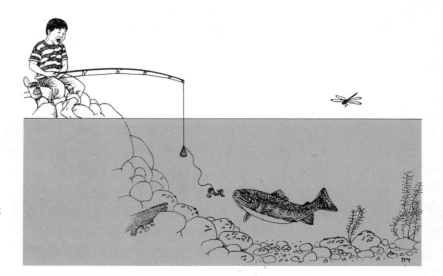

Figure 5.9 The buoyant
force on the fish or the
weight comes from the
pressure being higher
below than above it.

on the fishing line, being denser than the water, would sink if dropped. But the weight feels lighter to us under water than out because the pressure on its bottom is a little greater than on its top.

In the third century B.C., Archimedes observed a simple but useful fact: *The buoyant force pushing up on a submerged object equals the weight of the fluid it displaces.* We call this statement **Archimedes' principle.** You can understand the reason for this principle by looking at the fish or the weight in Fig. 5.9. If the object were not there, water would be in its place. The force pushing up on the water would hold it up, and therefore would have to equal its weight.

When you're underwater, you displace a volume of water equal to your body's volume. Since your body is mostly water, its weight is about the same as that of an equal volume of water: Its density is about the same as that of water. The buoyant force is therefore about equal to your weight so that it takes little or no additional force to hold you up. If your head is out of the water, you displace less water and some force is needed. The farther you get out of the water, the less water is displaced, and the more upward force you must supply to keep yourself up.

A floating object, like a boat, sinks far enough to displace a weight of water equal to that of the boat and its contents. If a 700-N person (mass about 70 kg) steps into the boat, it sinks a little bit—far enough to displace 700 N. We'll illustrate the use of Archimedes' principle with one numerical example.

Example 5.2   How many 70-kg people could get into a 15-ft canoe in still water without sinking it, if everybody stands perfectly still? To make calculation easier, assume that the volume of the parts of the canoe lower than point M (the first point to go under), is the same as that of the rectangle shown in dashed lines in Fig. 5.10—0.61 m³. When the canoe is about to go under,

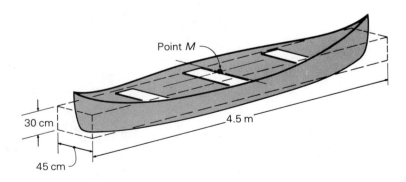

Point M

30 cm

45 cm

4.5 m

Figure 5.10   Estimate the volume of the canoe by that of the dashed rectangle.

weight of water displaced = density of water $\times g \times$ volume

$$= 100 \text{ kg/m}^3 \times 9.8 \text{ m/s}^2 \times 0.61 \text{ m}^3 = 6000 \text{ N}.$$

The buoyant force can support this total weight, which allows for a 340-N (76-lb) canoe and eight 700-N (157-lb) people.

Question   In terms of Archimedes' principle, explain why an upright steel ship will float but a capsized one will not.

Exercise   In 1895 Josephine Blatt raised 15 850 N (3564 lb) in a "hip and harness" lift. In 1957 Paul Anderson raised 27 890 N (6270 lb) in a back lift. If Mr. Anderson's weights occupied a volume of 0.37 m³, could Ms. Blatt have lifted his weights underwater?   **Ans:** No; they would weigh 24 260 N underwater.

Some liquids will float or sink in other liquids, if the two do not combine chemically when mixed. The less-dense liquid is pushed to the top by the denser one. That's why oil spills stay on the top of water. Did you ever drink nonhomogenized milk and have to shake it to mix the cream and milk? If not, then maybe you've had to shake the oil-and-vinegar dressing before putting it on your salad.

**5.5   Pumps*

Fluids that we must use do not do us any good until we get them to the place where we need them: blood must be pumped to our brains, water to our bathtubs, natural gas to our furnaces. In this section we'll look at how some of the principles we've discussed apply to this practical job of pumping. You won't find any new physical principles here, but you'll see how several of the old ones work together.

Breathing

A pump extremely important to your life is the one that pumps air into and out of your lungs many times a minute. This pump is

---

*Sections marked with the ** symbol are for illustration only and contain no new principles of physics.

simple in principle. As your diaphragm (the muscle that separates your chest and abdominal cavities) moves down, the chest cavity becomes larger. This decreases the pressure inside your chest, making the pressure in your lungs lower than atmospheric. The atmosphere then pushes air into your lungs. When your diaphragm moves upward, the higher pressure inside the lungs pushes air back outside.

Even when you open your mouth and pull in a breath quickly, you're not really *pulling* it in. You're merely lowering the diaphragm quickly and *letting* the air rush in faster.

Drinking from a Straw

You probably don't think of a straw as a pump, but you and the straw together do make up one. When you start to suck on the straw, you lower the pressure in the top of the straw and in your mouth, letting atmospheric pressure *push* the drink into your mouth. You don't *pull* it in at all!

The Siphon

A siphon is not a pump in the usual sense, but it depends upon some of the same principles. Figure 5.11 shows two possible situations that might exist if you try siphoning gasoline from your car tank to use in your lawn mower. In part (a) the gasoline level outside is higher than that in the tank; in part (b) it's lower. Based on what we've already said, *you* explain why in (a) the gas will run back into the tank but in (b) it will run out. Three hints: (1) the

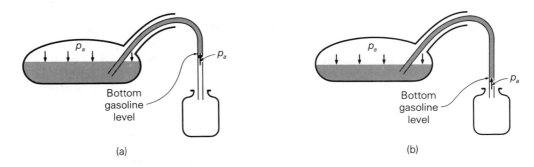

(a)   (b)

Figure 5.11   (a) The outside level of the liquid in the siphon is higher than that in the tank. (b) Here, the outside level of the liquid is lower. Why does the liquid run out only in (b)?

same atmospheric pressure acts down on the gas in the tank and up on the exposed gas surface at the other end; (2) the pressure difference between two heights in a liquid column is proportional to the height difference; and (3) at the highest point, look at the pressure trying to push liquid to the left and the pressure trying to push it to the right.

The Force Pump

The pump shown schematically in Fig. 5.12 may seem more appropriate as a machine of your grandmother's youth. However, this "force pump" serves as a good introduction to the pump we'll discuss next—one that certainly will not be outdated in the near future.

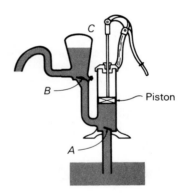

Figure 5.12   A force pump.

The main body of the force pump is a cylinder containing a piston that moves up and down as the handle is moved. A pipe extends from this cylinder down into the water. As the piston moves up, atmospheric pressure pushing down on the water in the well *pushes* water up into the additional space in the cylinder. As the piston moves down, the water pressure from above closes check valve *A*, forcing the water to go up through check valve *B* into chamber *C*. Some of this water compresses the air in the top of the chamber; some goes out the spout. When the piston stops and starts to move up again, the pressure in chamber *C* closes valve *B*, and the expanding air in the chamber continues to push water out the spout. The cycle continues in this way, with the compressing and expanding of the air in the chamber providing a fairly steady flow of water.

Question   Explain why such a pump cannot operate on the moon, and why there is a limit on earth to how high the pump cylinder can be above the water in the well? Hint: What role does the atmosphere play?

The Heart

Your heart is basically a force pump that pushes blood through the arteries, capillaries, and veins in the different parts of your body. Rather than having a piston that moves up and down, the heart has muscular walls that contract and relax, changing the volume of the heart's chambers as it "beats."

Figure 5.13 shows the main steps in the pumping action. Notice the action of the valves that do the same job as the check valves in the force pump. The arteries are very elastic—a feature that smooths the blood flow just as the air chamber in the pump smooths the water flow. When the heart beats and pumps blood into the arteries, they expand. Their elastic contraction keeps the blood flowing as the heart relaxes for the next beat. The two sides of the heart are separate pumps. The right side pumps blood to the lungs and back to the left side, where it is pumped to the rest of the body.

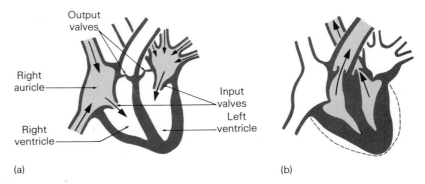

(a)

(b)

Figure 5.13   Pumping by the human heart. (a) Low pressure in the veins forces blood into the relaxed auricles and ventricles. The output valves are closed by the pressure in the arteries. (b) Contraction of the ventricles—the heart beat—pushes blood into the arteries. Pressure within the ventricles closes the input valves between the auricles and ventricles.

Question   Blood pressures are reported as two numbers—for example, 120/80, which reads "120 over 80"—that correspond to the highest (systolic) and lowest (diastolic) pressures in the artery of the arm during each pumping cycle. The units are millimeters of mercury. Why does the pressure vary up and down during each beat cycle? How would you expect this ratio to change when a person has arteriosclerosis (or hardening of the arteries, a condition where artery walls become rigid and inflexible)?

Notice that in our discussion of pumping we have made *no* mention (which is often made) of creating a partial vacuum that "pulls" the fluid into some area of the pump. Pumps do create a partial vacuum—a pressure lower than atmospheric. But *a vacuum can do no pulling!* It merely lets the higher outside pressure *push* the fluid in.

5.6   **Fluids in Motion—Flying Airplanes and Curving Baseballs**

When fluids are moving, the rules are different than when they're at rest. In looking at some of these different rules, we'll start by considering a fluid confined inside a pipe, where we can account for all of it. We can use what we learn about that situation to describe what happens when unconfined fluids move past solid objects.

Suppose Uncle Charlie plants his petunias so far from the faucet that his hose won't reach to water them. We'll assume for now that the connection between his hose and the smaller one he borrows from his neighbor to reach his petunia patch is very smooth, as shown in Fig. 5.14. Because the water is incompressible, the volume of a given mass doesn't change with pressure. For a steady rate of flow out the end, the same volume of water must pass both points *B* and *A* each second. However, the area is smaller at point *B*, so that the water must move faster at that point to get the same amount through per second. You can probably convince yourself that the speed must be exactly inversely proportional to the cross-sectional area of the pipe. That is, if the area is ½ as large, the speed is doubled.

Compare this with the situation where two lanes of traffic merge into one to cross a bridge. If you're caught in traffic that is backed up and moving slowly in the two-lane section, then you move about twice as fast once you get into the one-lane part. Assuming the fluid is incompressible is like assuming the cars stay the same distance apart in both one- and two-lane sections.

If the fluid is compressible, as air or some other gas would be, the inverse proportionality doesn't hold exactly, but the speed does increase with decreasing pipe size.

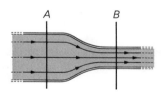

Figure 5.14   Since the same volume of liquid passes *B* and *A* in a given time interval, the speed is larger at *B*. The energy conservation principle shows that the pressure is lower at *B* where the speed is higher.

The lines with arrowheads, called *streamlines,* show the direction of flow at any point, and help in visualizing the flow pattern. Since the streamlines are continuous, they are *closer together* in a region of *higher speed.* If the junction between the large and small regions of hose is abrupt rather than smooth, there may be some turbulence in the junction region instead of the smooth lines of flow we've shown.

Question   With a garden hose, why can you shoot a stream of water much farther if you have a nozzle with a small opening than if you just use the end of the hose? Almost the same *amount* of water comes out per second either way. Does partially covering the end of the hose with your finger make it squirt farther for the same reason?

Next consider what happens to the *pressure* as the speed of the fluid particles changes. Looking again at Fig. 5.14, you might guess that the pressure in the pipe at point *B* is larger than at point *A,* since the fluid is being pushed through a smaller pipe. But if you measured the pressure at both points, you'd find just the opposite case: It's *lower* at *B!*

The reason for the lower pressure at *B* is the higher speed of the fluid particles, as we can see from conservation of energy. For the fluid particles to gain kinetic energy in going from *A* to *B,* work has to be done on them. This work requires a net force in the direction of the increased speed. Therefore, the pressure pushing on the fluid between *A* and *B* from the left of *A* has to be larger than that pushing on it from the right of *B.* This means that the pressure in the fluid at *B* has to be less than that in the fluid at *A.*

If the fluid flow is from the smaller to the larger region, the pressure is still higher in the larger region. The higher pressure then slows the fluid particles to give the lower speed consistent with a larger hose.

Even though we've shown the relationship between pressure and speed for the special case of a liquid in a pipe of varying size, it is generally true that *the pressure in a fluid decreases as the velocity increases,* and increases as the velocity decreases. Scientists call the relationship between the pressure and energy of a fluid in motion **Bernoulli's principle.**

We can demonstrate Bernoulli's principle by some simple experiments in which air flows over solid objects. One way is by holding a piece of paper as shown in Fig. 5.15 and blowing across the top. The lower pressure of the moving air on top lets the higher pressure of the still air under the paper push it up. Or you can

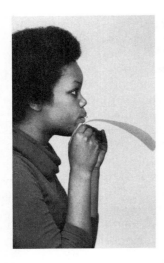

Figure 5.15 You can demonstrate Bernoulli's principle by blowing across the top of a piece of paper that is hanging down.

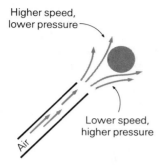

Higher speed, lower pressure

Lower speed, higher pressure

Air

Figure 5.16 The higher pressure in the low-speed region of air pushes the ball toward the centerline of the air coming out the hose.

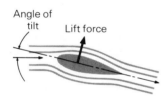

Angle of tilt

Lift force

Figure 5.17 As the air is pushed down, the reaction force pushes up on the airplane wing.

suspend a light ball—Ping-Pong or beach ball—in midair by hooking a hose to the air-output end of a vacuum cleaner, and aiming it up at the ball as in Fig. 5.16. The bombardment by the air pushes the ball upward. As the ball falls below the centerline of air flow, it feels a lower pressure near the centerline because of the higher speed of the air there. The ball is therefore pushed back toward the centerline by the higher pressure of the slower air.

We'll now discuss two additional questions related to the motion of fluids: What holds an airplane up in flight, and what makes a baseball curve? Viewed from the ground, both of these involve the motion of an object *through* the air. But they are both easier to understand as viewed from the reference frame of the object itself. There, the object is stationary and the air rushes by. The force causing these effects is often explained in terms of Bernoulli's principle, but it's much easier to understand in terms of Newton's third law of motion—the *action-reaction law.*

First, let's examine the force, or "lift," that holds the airplane up. If you were to cut off an airplane wing in its middle, the end where you made the cut would have a shape something like that shown in Fig. 5.17. From the reference frame of the airplane, air is passing the wing as shown by the streamlines. The wing is shaped and oriented so that the air, after passing the wing, has a partial *downward* motion. If the wing pushes the air *downward,* the action-reaction law tells us the air must push on the wing upward. This upward reaction force provides the lift. Except for low-speed flight, the wing would not be tilted as much as we've shown. Some wing designs have the lower surface fairly flat and the upper one curved. Air curving over the upper surface leaves the wing with the needed downward motion.

Question In terms of the number of air particles passing the wing, explain why the angle of tilt can be decreased as the speed increases.

Understanding why a baseball curves is a little more complicated. From the ball's frame of reference, air passes a *nonspinning* ball as shown in Fig. 5.18(a). The air cannot follow the sharp curvature of the ball, and separates, leaving a turbulent area behind the ball. Since the flow is the same around both sides, there is no net sideways force. But when Catfish Hunter "turns loose" with a curve ball, Fig. 5.18(b) shows what happens. He puts a fast spin on the ball. As it spins, the seams drag a thin layer of air around with it. The air spinning with the ball on the left side opposes the overall flow there, causing an early separation of air from the ball. On the right side, the spinning air and passing air move in the same general direction, letting the air stick to the ball longer and thus leave in a slight leftward direction. The reaction to this push to the left on the air pushes the ball to the right, making it curve to the right.

Figure 5.18 (a) When a baseball is not spinning, the air flow is the same on both sides and there is no net force. (b) When a "curve" is pitched, the spinning ball drags air around with it, giving the air a slight sideward motion as it passes. The reaction force pushes the ball in the other direction.

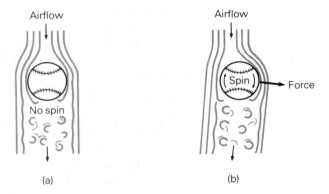

The same ideas apply, of course, to a spinning tennis ball or golf ball that has a surface rough enough to drag air around with it.

Question   In what direction must a ball be spinning when the pitcher throws a "drop"?

5.7 Molecular Forces and Surface Tension in Liquids

So far, we haven't paid any attention to the fact that fluids are made up of molecules. However, to explain an important characteristic of liquids—their surface tension—we have to look at interactions between individual molecules. Strongly dependent on surface tension is the phenomenon called **capillary action,** a process that nature uses in many practical ways.

Surface Tension

You know that a solid chunk of steel won't float in water. To make it float (as a steel ship floats), you have to shape the steel so that it has a hollow region that will displace a weight of water equal to the weight of the steel. This holds for a chunk of any size—large or small. If you drop a needle or razor blade into a glass of water, either will certainly sink.

Now, suppose you don't *drop* these small items, but instead lay them gently on the water surface with their long or flat side parallel to the water. Figure 5.19 shows what happens: They stay on top of the water. The surface molecules of the water are attracted to their neighbors on the surface with a force called **surface tension.** The force is strong enough that the weight of small thin objects cannot break the surface tension and get through to the interior of the liquid. Once a corner or edge breaks through the surface, the object immediately sinks.

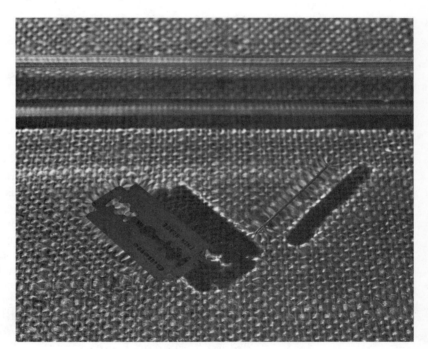

Figure 5.19 Laid carefully on the surface, a needle or razor blade will stay on top of the water.

Let's see why surface tension exists. In Fig. 5.20 the circles represent the centers of the molecules of liquid. Each molecule, such as the one shown in black and labeled *A,* exerts a force on its nearby neighbors. The force on any one molecule acts roughly as it would if all the molecules were rubber balls embedded in bubble gum. There would be a given distance between a molecule and its neighbor at which there would be no force. If the two molecules got closer they would be repelled, with the repulsion increasing drasti-

cally as the distance decreased (the rubber balls would r‹
other as their sides started to touch). As the molecules m‹
from the no-force distance, they are attracted together wi
that first increases, but then decreases with increasing s
(the bubble gum would stretch giving a strong force at firs
a weaker one as it stretched farther and thinned down).

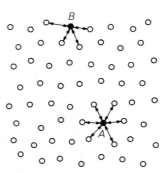

Figure 5.20 A molecule in the interior of a liquid is attracted in all directions by its neighbors. One on the surface is attracted only sideways and downward. This accounts for the surface tension of liquids.

Our rubber ball and bubble gum analogy breaks down, in that the molecules are free to move about throughout the liquid. As molecule *A* moves around, it feels the same forces, whatever its neighbors. Averaged over time, there is no *net* force on it.

But what about molecule *B,* which is on the surface? It has no molecules above it that exert forces on it, so it's pulled and pushed only by its neighbors beside and below it. If something tries to increase the amount of surface, either by penetrating it or stretching it, the net downward and sideways attractive forces resist this increase. This tight binding along the surface makes up the *surface tension.*

Surface tension not only resists the penetration of a liquid surface, but causes it to take on a shape that minimizes the surface area. Drops of liquid that do not stick to a surface (for example, drops of water on a greasy pan) or falling drops of liquid are roughly spherical because a sphere gives the smallest surface area for a given volume. The supporting surface or the air resistance distorts the drop from being perfectly spherical.

Detergents

If you pour a few drops of dishwashing detergent in the water of Fig. 5.19, the needle and razor blade immediately sink. Evidently,

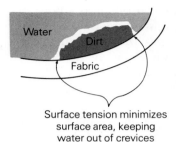

Surface tension minimizes surface area, keeping water out of crevices

Figure 5.21 Dirt is not washed away by pure water because surface tension keeps water from seeping into the cracks between dirt and fabric.

the detergent reduces the surface tension of the water. In fact, reducing surface tension is one of the main jobs of a detergent.

What does reduced surface tension have to do with cleaning? Most "dirt"—either from clothes or dishes—is greasy or oily material that doesn't mix with water. As Fig. 5.21 shows, surface tension keeps a smooth surface on the water, not letting it get under the dirt to wash it away. Detergent reduces the surface tension, thus allowing the water to get between dirt and fabric. Detergent makes the water "wetter."

In hot water, the molecules bounce around faster than in cold. They are, therefore, not bound as tightly and thus hot water has lower surface tension than cold. That's one reason hot water washes better than cold, although many modern detergents are effective even in cold water.

Capillary Action

If you look closely at the water in a clean glass, you will see that the surface curls up at the edges where it meets the glass (Fig. 5.22-a). This curl happens because water molecules are attracted more strongly to glass molecules than to each other. This attraction between unlike molecules is called *adhesion,* whereas the attraction between like molecules is called *cohesion.* Because of the stronger adhesive forces, water is pulled a small distance up the glass surface. The travel stops when the adhesive forces between glass and water, the cohesive forces in the water, and the weight of the small elevated ring of water are in balance.

Figure 5.22 (a) Adhesion between water and glass is stronger than cohesion in the water, causing the contact region to curve up. (b) For mercury, cohesion in the liquid is stronger than adhesion to the glass, causing the contact region to curve down.

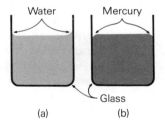

Between mercury and glass, the adhesive forces are much weaker than the cohesive forces within the mercury. In this kind of situation, the surface tension in the liquid causes it to curve downward (see Fig. 5.22-b).

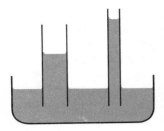

Figure 5.23 In a small glass tube, water will rise above the level of the water in the bowl. The smaller the tube, the farther the water rises.

Suppose one end of a small glass tube is put into a bowl of water, as in Fig. 5.23. The level of water in the tube will be higher than in the bowl. Also (as the figure illustrates) the smaller the tube, the farther the water will rise. In a 0.2-mm diameter tube, water will rise about 15 cm; in a 0.02-mm tube, it will rise about 150 cm. This rising of liquid in such a tube is called **capillary action,** or *capillarity.*

Water rises in the tube because adhesion pulls water up along the glass walls. Surface tension in the liquid then pulls the center up more nearly level with the edges. This allows the edges to move up more, with cohesion then pulling the center up. This upward motion continues until the forces pulling the liquid up balance the weight of the liquid in the tube. Since, for a given height, the weight is less for a smaller tube, the liquid can rise higher in the smaller tube.

Capillarity is responsible for such phenomena as the rising of melted wax in candle wicks and the traveling of water through a towel, even though only an edge of the towel may be in the liquid. But nature uses capillarity in many important ways. These include the bringing of water through the soil to the roots of plants, assisting other mechanisms in raising sap in the plant stems, and transferring blood through capillaries (very small blood vessels in the human body).

Key Concepts

A **fluid** is a substance that can flow—either a liquid or a gas. Liquids are incompressible, gasses compressible.

Density is the mass per unit of volume. **Weight density** is the weight per unit of volume. The **specific gravity** of a substance is the ratio of its density to that of water.

Pressure is the force per unit of area. Sea level *atmospheric pressure* is 101 kN/m², which can support a column of mercury 76 cm high. The pressure at any depth in a liquid caused by the liquid itself is given by density × g × depth.

Pascal's principle states that increasing the pressure at one place in a stationary fluid increases the pressure by the same amount throughout. Hydraulic devices operate on this basis.

Archimedes' principle states that a body partially or totally immersed in a fluid is buoyed up by a force equal to the weight of the fluid it displaces.

Bernoulli's principle expresses conservation of energy for a moving fluid. Regions of lower velocity of fluid particles have higher pressure.

A molecule in a liquid experiences a force from its nearby neighbors. The force repels it if it gets too close, attracts it as it moves away. At the surface of the liquid, molecules are bound together with a **surface tension** that tries to minimize the surface area.

Capillary action causes a liquid to move through small tubes or between the fibers making up a material. It can raise the liquid to a height such that the capillary lift equals the weight of the liquid raised.

Questions

Density and Specific Gravity

1. A mixture of 50% water and 50% antifreeze (enough anti-freeze to lower the freezing point to −30°C) has a specific gravity of 1.070. The acid solution in a fully charged car battery has a specific gravity of about 1.28, and a "dead" battery has a specific gravity of about 1.13. Why do service station attendants use a hydrometer—an instrument for measuring specific gravity—to check your antifreeze and to find out if your battery needs charging? (See Problem 16.)

Figure 5.24 A tire flattens until the total force on the road from the tire pressure equals the weight supported by the tire.

Pressure

2. Why are water beds sometimes recommended for eliminating bed sores on bedridden persons? (Bed sores usually occur at the points of strongest contact between the body and the bed.)

3. Figure 5.24 shows how a tire flattens out on the road under the weight of a vehicle. Why do larger tractor tires require only low air pressure (about 80 kN/m²), whereas bicycle tires require high pressure (around 500 kN/m²)? (See Problem 8.)

4. My son recently received the gift of a soccer ball that is labeled "Inflate to 0.6 kg or 8 lb." What is missing from these units? What is especially bad about this use of the kilogram?

5. The total downward force from the atmosphere on top of a Ping-Pong table is 421 000 N (47 tons). Why do you not have to lift that hard to lift the table?

6. How do suction cups work? (No, vacuum does not hold them on.)

7. With what force does each square centimeter of your body push outward on the atmosphere?

8. Even though it is seemingly not a popular practice, some women have been known to supplement their own anatomies by wearing inflatable bras. Comment on the hazards of this practice when flying in high-altitude airplanes, especially if the cabin suddenly depressurizes.

9. Is the pressure on your body when you are 2 m below the surface of the water in a swimming pool twice what it is when you are 1 m down?

10. Large coffee urns often have a vertical glass tube along the outside to show the coffee level. Why is the level in the tube the same as that in the pot?

11. True or false: When you get into a half-filled swimming pool, the pressure all over the bottom of the pool increases a small amount.

12. True or false: A mercury barometer works because there is a vacuum in the top of the tube that pulls up on the mercury.

13. When you're washing dishes, and you raise a submerged glass out of the water with its *bottom up,* why does no water leave the glass until the glass is completely out of the water?

14. You operate a plunger for unclogging drains by covering the drain with the plunger, pushing down slowly, and then *quickly* pulling up. The *upward* motion is supposed to shake loose the clog, which then gets washed down by the water. Is the clog *pulled* up or *pushed* up when you pull up?

Pascal's Law

15. In a hydraulic lift (Fig. 5.7) you get a larger force at the larger piston, since the pressure is the same at both. Does this mean you get more work done *by* the larger piston than you do *on* the smaller one?

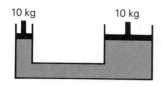

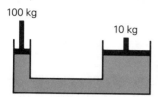

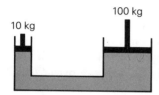

Figure 5.25 The three hydraulic systems contain movable pistons. Which system is in balance?

16. Which of the three systems shown in Fig. 5.25 is balanced? In each case the small piston has an area of 10 cm², the large one 100 cm².

Buoyancy

17. Most people just barely float in fresh water, although some cannot. Estimate the average density of the human body.

18. Which, if either, is easier to float in: deep water or shallow? (Assume you don't touch the bottom either way.) Does the size of the pool make a difference?

19. Crocodiles have an average density somewhat less than 1000 kg/m³. They sometimes swallow stones, and this habit assists them in sneaking up on their prey in water. Why?

20. Sea water has a specific gravity of about 1.025. Why is it easier to float in the Atlantic Ocean than in the Mississippi River?

21. Since ice is less dense than water, it floats in water with a small fraction of the ice sticking out of the water. If a glass is filled to the very top with ice and water, will some water run out as the ice melts?

Pumps

22. Nurse Goodbody tells Grandpa Jones that he has hypertension (high blood pressure). He thinks that means he's unusually healthy because his heart is pumping at a higher pressure. Actually, his circulatory system has tightened up, increasing the resistance to blood flow. How would you expect this condition to influence the load on Grandpa's heart and the likelihood of heart problems?

23. A giraffe has a blood pressure of about 260 mm of mercury when standing, but only about 120 mm when lying down. A dog's blood pressure is about 110 mm either way. Explain why these differences are appropriate.

Fluids in Motion

24. A pipe of 2-cm diameter branches off to four pipes of 1-cm diameter that supply water to four faucets. If *one* faucet is turned on, how does the number of liters of water leaving the faucet per minute compare with the number of liters passing some point in the 2-cm pipe? How does the speed of water in the 1-cm pipe supplying the open faucet compare with that in the 2-cm pipe?

25. Why does a steady stream of water from a faucet get narrower, the farther it gets from the faucet? Does this narrowing process continue, no matter how far it falls?

26. When a convertible travels down a highway, why does the top bulge upward as far as possible?

27. The effect of the Bernoulli principle is one reason a shower curtain pulls in at the bottom when you turn the water on. Explain.

28. When the roof of a house comes off during a tornado, the reason is usually not that the wind blows it off, but rather that the air pressure inside the house pushes it up and then the wind blows it away. Explain. Would it help to open the windows?

Surface Tension and Capillary Action

29. If you had a body of water densely populated with hexopuses all holding hands and pulling on each other, would there be a surface tension around the outer surface of the hexopuses? (Note: A hexopus can be created by surgically removing two tentacles from an octopus.)

30. Why are drops of mercury on a table and drops of water on a greasy pan nearly spherical, except where they are supported by the table or pan?

31. If you dip a dry, soft-bristled paint brush with spread-out bristles into water and remove it, the bristles all stick together. Why?

32. Why is the sand at the beach wet considerably beyond the highest point the water reaches?

33. Where power lines are connected to a house, "drip loops" are provided to keep rainwater dripping outside, rather than running down the entrance line (see Fig. 5.26). Discuss the action of adhesive and cohesive forces in making the water run both ways toward the sag, and drip only near the bottom of the sag.

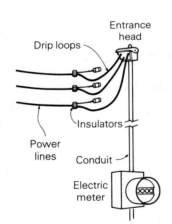

Figure 5.26 "Drip loops" are provided where power lines connect to the side of the house, to prevent water running down the entrance wire.

Problems

1. If you have 4.3 g of a substance that has a volume of 5.0 cm³, and you know that it is one of the substances in Table 5.1, which substance is it?

2. Superman compresses a rock the size of a tennis ball (volume 160 cm³) with a density of 2.5 g/cm³ into the size of a marble (volume 1.1 cm³). What is the new density?   **Ans:** 360 g/cm³

3. What is the specific gravity of each of the substances listed in Table 5.1? What is the weight density of each in N/m³?

4. A freezer rests on four legs, each 4.5 cm in diameter. If the food and freezer together weigh 3500 N (790 lb), what is the pressure under the legs if the load is equally distributed among them?   **Ans:** 550 kN/m² (80 lb/in²)

5. Women's shoe fashions fluctuate from year to year, but heel sizes reached a minimum in the early 1960s. If we *guess* that then–First Lady Jackie Kennedy's mass was 50 kg, and assume that ½ her weight was supported by her two heels, then how much pressure did her 1-cm-diameter heels exert on the White House floor?   **Ans:** 1600 kN/m² (230 lb/in²)

6. Find the pressure on a phonograph record from a needle that has a circular point 0.025 mm in diameter and "tracks at 2 grams."   **Ans:** 40 × 10⁶ N/m² (5800 lb/in²)

7. A queen-size water bed typically has a mass, including the frame, of about 800 kg, and is usually supported on a pedestal shaped as illustrated in Fig. 5.27. A conventional bed typically has a mass of about 90 kg and is supported on four posts. If the water bed pedestal is made of 2-cm-thick lumber and the conventional bed posts are 4 cm in diameter, compare the pressure on the floor for the two beds when each holds two people totaling 140 kg. (Caution: In considering the feasibility of a water bed, you also have to consider whether or not the floor structure can support the *entire* weight.)

   Ans: water bed—39 kN/m²; conventional bed—450 kN/m²

8. Suppose a car and its contents have a mass of 1400 kg, and the air pressure in the tires is 200 kN/m². How much tire area of each tire is in contact with the road? (See Question 3.)

   Ans: 0.017 m² (170 cm²)

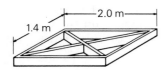

Figure 5.27 A queen-size water bed pedestal.

9. The water pressure in city water lines is typically about 40 lb/in². Express this pressure in N/m² and atmospheres.

Ans: 280 kN/m²; 2.7 atm

10. What is the force from the atmosphere on an eardrum of area 0.60 cm²? A 100-m increase in altitude above sea level changes the air pressure about 1%. If, while ascending this distance in an elevator, the pressure of the fluid inside the ear does not change, what is the net outward force on the eardrum?

Ans: 6.1 N; 0.06 N

11. Swimming pools are sometimes built with observation windows in the side for underwater viewing. What is the total net force on a window 2.0 m by 1.0 m and centered 2.0 m below the surface? Why do you not need to consider atmospheric pressure?

Ans: 39 000 N

12. Pressure gauges on older boilers for hot-water home-heating systems usually show the pressure in the boiler as "feet of water." If the gauge reads "14 ft," what is the pressure in N/m² and atmospheres?

Ans: 42kN/m²; 0.41 atm

13. A hydraulic automobile lift has a 30-cm diameter piston. What pressure is needed in the fluid to lift a 1200-kg car?

Ans: 170 kN/m² (1.6 atm)

14. How many 70-kg people can float on an empty leakless 160-liter (42-gal) barrel without submerging it, if the barrel's mass is 12 kg?

Ans: 2

15. Consider a uniform solid cylinder submerged upright in a liquid. Show, from the difference in pressure on the top and bottom, that the net upward force on the cylinder equals the weight of the water it displaces.

16. A hydrometer for measuring the specific gravity of antifreeze solution or battery acid is made as shown in Fig. 5.28. (See Question 1.) Liquid is drawn into the large tube by squeezing and releasing the bulb at top while the tube is in the liquid. Suppose the float rests as shown, and displaces exactly 100 cm³ of pure water. If a mixture of 50% antifreeze and 50% water with a specific gravity of 1.07 is drawn in, how much higher will the float rest if its stem is 1.0 cm in diameter?

Ans: 8.3 cm

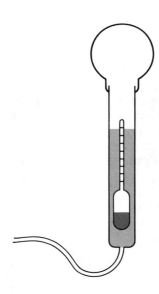

Figure 5.28   A hydrometer for measuring the specific gravity of a liquid.

17. If the atmospheric pressure above a 100 m² roof is reduced 1% during a hurricane (see Question 28), what is the net atmospheric force acting upward on the roof?

Ans: 100 000 N (11 tons)

Home Experiments

1. The next time you go "tube-ing," either down the river or down the hill on snow, blow the inner tube up with your breath. You may be surprised how tight you get it that way. Then estimate the pressure with which you blew as follows: Lay the tube in sand or snow and put a piece of plywood or other stiff board big enough to cover most of the tube on top of it. Stand on the middle of the board so that it stays level. Then get off, measure the approximate diameter and width of the indentation left by the tube, and calculate the pressure from your weight and the indentation area. This value is roughly the pressure with which you blew. (You can do this experiment in your room by sprinkling powder or flour and then measuring the area flattened by the tube. I found I could blow to a pressure of a little over 6 kN/m².)

2. If you can get a hot water bottle, a tube for connection to it that is a little longer than your roommate is tall, and a funnel to fit in the tube, ask your roommate to stand on a book on the water bottle. Hold the tube up and pour in water until your roommate is lifted. Calculate and compare the pressure from your roommate's weight, and that from the water column. Notice that the water column weighs much, much less than your roommate. Does it matter what size the tube is?

3. You can carry out the can crushing of Fig. 5.3 without having a vacuum pump. Get a rectangular can of about 4-liter volume—such as a paint thinner can—with a tight-fitting top. Put about 1 cm of water in the bottom and put the can *with the top off* on the surface element of a range or hotplate. When the water starts to boil, remove the can from the heat, put the top on tightly, and let it cool. Running cold water on it speeds the cooling. The steam in the can condenses, greatly reducing the pressure inside, and letting the atmospheric pressure crush the can.

4. Fill a glass with water, put a piece of paper or cardboard tightly on top, turn the glass upside down while holding the paper in place, then release the paper. Why does the water not run out? You can also do it with a partially filled glass, but you may need to push up a little on the center of the paper before releasing it.

5. Carefully put a fresh egg in a glass of water and notice that it sinks. Remove the egg, stir in as much salt as will dissolve, and put the egg back. It will now float. What can you say about the relative densities of water, saltwater, and fresh egg?

6. You can demonstrate Bernoulli's principle with several simple experiments. (1) Lay a Ping-Pong ball on the edge of a table and put your mouth down where you can blow along the tabletop and beside the ball. A quick puff will make the ball roll toward the blown air, not away. (2) Put a straw in your mouth and aim it upward at a slight angle to the vertical direction. Hold a Ping-Pong ball above the end of the straw and blow with a steady even breath. As you start to blow, release the ball. With practice you can make the ball hang there a few seconds. Experiment with different angles and different starting ball positions. (3) Hang two apples from strings tied to their stems so that they are a few centimeters apart. Blow between them. Are you surprised by the direction of their motion? (4) Blow between the flames of two candles placed a few centimeters apart. Which way does the flame tilt? (5) Hold a straw vertically in a glass of water and blow with a quick gust *across* the top of the straw. With the right technique, water will come up the straw and spray out with your air. (You, the glass, and the straw will have become an atomizer.)

7. You can simulate the curving of a golf ball, tennis ball, or baseball by tying a string around a Ping-Pong ball and hanging it from about one meter of string. Pull it aside and let it swing as a pendulum. It swings in a straight line. Then twist the string about 50 turns and again pull it aside to swing back and forth. The direction of oscillation gradually rotates. Does it curve the right way?

8. Carefully lay a needle, a razor blade, or a piece of wire screen on the surface of a bowl of cold water. (The first two will be easier to place if you rub them with a light coat of oil.) Now touch the surface with a bar of soap, or pour in a little liquid detergent to break the surface tension. Kerplunk!

9. Sprinkle pepper on the surface of a bowl of water. Then touch the center of the water surface with a piece of soap and watch the pepper "split."

s of glass—windowpane or picture ges together, and leave a small gap es. Dip them vertically into a pan of pillary action pulls the water up further where the plates are closer together.

References for Further Reading

Boys, C. V., *Soap Bubbles* (Doubleday & Company, Garden City, New York, 1959). These three lectures given by Boys in 1889 and 1890 describe many interesting effects of surface tension and capillary action. Includes a list of experiments to demonstrate these effects.

Milne, L. J. and M. Milne, "Insects of the Water Surface," *Scientific American,* April 1978, p. 134. Four insects found all over the world depend on the surface tension of water for their existence.

Smith, N. J., "Bernoulli and Newton in Fluid Mechanics," *The Physics Teacher,* Nov. 1972, p. 451. Focuses on the explanation of airplane lift and baseball curves.

Strother, G. K., *Physics with Applications in the Life Sciences* (Houghton Mifflin Co., Boston, 1977). Chapter 11 includes many applications of fluid principles in biological systems.

Walker, J., *The Flying Circus of Physics* (John Wiley & Sons, New York, 1977). See especially Chapter 4. Considers many interesting effects in fluids and gives references to literature that discusses in detail those things that we could only briefly cover here.

Part II

Heat

I . . . am persuaded, that a habit of keeping
the eyes open to everything that is going on
in the ordinary course of the business of life
has oftener led, as it were by accident, or in
the playful excursions of the imagination, . . .
to useful doubts and sensible schemes for
. . . improvement, than all the more intense
meditations of philosophers, in the hours
expressly set apart for study.

Count Rumford

Benjamin Thompson, born in Woburn, now
Concord, Massachusetts, in 1753, was an
opportunist active in political life during the
time of the American Revolution. Even though
he was in general loyal to the British side, he was
careful always to be identified with the winners.
His scientific and technological contributions
included major breakthroughs in
understanding the nature of heat, improvement
in the fireplace, and invention of the drip
coffeemaker. He made many of his discoveries
during the course of his everyday routine.
Among his honors, King George III conferred
knighthood on him and he was made Count of
the Holy Roman Empire, choosing for his title
the name Rumford.

6

Heat Energy

How many roles does heat play in your life? Some of these roles come to mind at once: its effects on climate, personal comfort, cooking and the warming of houses. But heat also influences our lives in subtle ways that we don't usually think about.

Take another look at Fig. 1.1 of Chapter 1, and notice that heat was often the "middleman" between source and use of energy. Almost all electricity is generated from heat: Some form of fuel is burned to release heat that produces steam that turns a turbine that runs a generator that generates electricity. Automobile, train, and airplane engines burn fuel that releases heat to make the engines go. Heat is involved in the manufacture of thousands of useful everyday products: steel and aluminum production requires large amounts of it; it is used to dry paint on bicycles and to cure colors in fabrics; and it's needed to put the "memory" into permanent press fabrics. Your own body depends on heat for proper functioning. Almost all the energy we use is in the form of heat in one stage of its exploitation.

In this chapter and the next three we'll describe how heat performs its many jobs. We'll find that when heat is applied to a substance, it can have either or both of two effects: Heat can change the temperature of the substance or it can change the substance's state of matter. We'll consider the subject of "change of state" in Chapter 7. Here, we will define temperature and heat and describe how they are measured. Then we'll discuss specifically how heat and temperature relate to each other. We'll consider sources of heat and, finally, we will examine an effect that usually accompanies a temperature change, an effect of tremendous practical importance: expansion or contraction. In Chapter 8 we will find out how heat moves from one location to another; that is, how heat is transferred. In Chapter 9 we will discover how we can use heat to do work, and how we can pump heat from one area to another of higher temperature.

6.1 **Temperature versus Heat**

When we begin to discuss any new topic, the first step is to define the terms involved. This topic is no exception. Even though the terms "temperature" and "heat" are both familiar, we must carefully distinguish them.

Temperature

When people talk about heat, they often mean **temperature.** For example, you may say, "Turn up the heat, I'm freezing." But what you are really saying is, "Set the thermostat so that the *temperature* will get higher and I will be warmer." When you are heating water for that 3:00 A.M. cup of coffee while you study for an English test, you know that the water boils when it gets to a certain *temperature.* When it's "104 in the shade," it's the *temperature* that gets to you. When you are sick, your body *temperature* goes up.

Temperature, then, *is a measure of the "hotness" or "coldness" of a body or region.*

Heat

Knowing that we want a certain temperature doesn't tell us what to do to *change* the temperature. If you have a choice of boiling your coffee water in a large or a small pot, you know that it would take longer to bring the large potful of water to the boiling point than it would the small one placed on the same burner. Yet, either one will burn your tongue just as quickly because both boil at the same temperature. For similar reasons it takes a much larger furnace to heat a 15-room house than a neighboring 3-room house constructed of the same materials.

To see the why, consider an analogy. Suppose you have a pint jar and a bathtub side by side, each containing water up to a level of 5 cm from their bottoms. If you then add a cup of water to each, what does this additional amount do to the water levels? In the pint jar, it about doubles the level. In the bathtub, you probably don't see any change. Yet the same amount of water was added to each container. In summary, we can change the water level in either container by adding water to it. But the *amount* the level changes for a given amount of water added depends upon the volume available to absorb the water. Or, to change both levels by the same amount, we must add more water to the larger-volume container.

The relationship between heat and temperature is the same as the relationship between water and water level in a container. To change the temperature of something, you have to add heat to it. The more of the "something" you have, the more heat it takes to change its temperature a certain amount. Figure 6.1 symbolizes the situation for the large and small pots of water. More heat is needed to bring a large pot of water to a boil than a small one, even though both start and stop at the same temperature. Similarly, more heat is needed to change the temperature of the air and contents of the 15-room house a fixed amount than to do the same for the 3-room one.

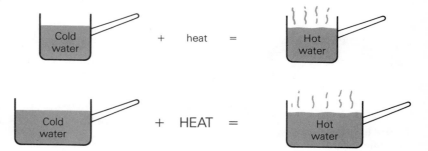

Figure 6.1 The larger letters symbolize the larger quantity of heat needed to change the temperature of the larger pan of water the same amount.

One way of defining heat is to say what it does. *Heat raises the temperature of matter to which it is added, and lowers the temperature of matter from which it is removed,* provided the matter does not change state—such as from solid to liquid or liquid to gas. (As mentioned, we'll consider that possibility in Chapter 7.)

Internal Energy versus Heat

We have said what heat *does,* but not what it *is.* In order to say what it is, we need to consider the fact that all matter is made up of molecules—either single atoms or combinations of atoms. There are 33 million billion billion of them, for example, in a liter of water. These molecules continuously jiggle around at random.

In gases at ordinary (near atmospheric) pressures, the molecules bounce around freely, interacting little except during collisions. In liquids and solids, the molecules continuously push

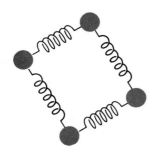

Figure 6.2 The electrical force between molecules in a solid acts as if the molecules were tied together by tiny springs. The circles represent four molecules in a substance.

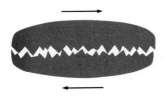

Figure 6.3 The magnified appearance of smooth surfaces shows how frictional effects transfer energy to the molecules as the parts slide over each other.

and pull on each other by means of electrical forces. In the case of solids, these electrical forces hold the molecules in the same average position relative to each other: The molecules behave *as if* they were held to each other by tiny springs (shown in Fig. 6.2 for only four molecules). The atoms *within* a molecule interact with each other through similar forces.

As a molecule jiggles around, its total energy consists of its kinetic energy and its share of the potential energy of the compressed or stretched springlike forces. We call this energy associated with the random molecular motion **thermal energy.** The total energy of all the molecules that make up a substance is what we call the **internal energy** of the substance.

In forming a mental image of what's going on inside a substance, it helps to concentrate on the kinetic energy associated with the jiggling molecules. Keep in mind, however, that potential energy is also present. The molecular energies vary over a wide range within the substance but, at any instant, have some *average* value.

To see what all this motion has to do with heat, we'll look first at why friction warms a body. In Fig. 1.10, we saw how a smooth surface looks when magnified. When one such "smooth" surface slides over another, the jags interlock (see Fig. 6.3) and cause a frictional resistance. As the molecules on the surface bang together, they gain kinetic energy. The work done against the frictional force goes to thermal energy of the molecules and thereby increases the internal energy of the body.

When we add or remove internal energy, we say we have added or removed **heat.** Thus, *heat is internal energy that is added to or removed from a piece of matter.* In other words, if we refer to the thermal energy of the molecules of a body, we are talking about internal energy. But if we mention adding or removing internal energy to or from the body, we are talking about heat. Heat, then, is not really a new form of energy. It is merely mechanical energy associated with the individual molecules and atoms.

The principle of conservation of energy still applies, but now must include changes in the internal energy of the body. When this principle does in fact include the effect of internal energy changes, we usually call it the *first law of thermodynamics.* We'll say more about this special form of the energy conservation principle in Chapter 9.

But why would heat enter or leave a body? Figure 6.4 shows an imaginary slice through an electric heating element and a pan. The molecules of the heater and the pan are jiggling around with a certain amount of thermal energy. Those molecules on the heater surface in contact with the pan collide with the molecules on the pan's surface. Similarly, those of the pan collide with the molecules

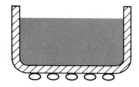

Figure 6.4 Heat is transferred from the range heating element as molecules of the heater give up thermal energy to molecules of the pan.

of the heater. If, on the average, the molecules of the heater have more energy than those of the pan, the heater molecules will give up some of their energy as they collide. That way, the heater will lose some internal energy, and the pan will gain some. This process means that heat has been transferred from the heater to the pan.

We have talked only about the molecules on the surfaces. The surface molecules of the pan will collide harder with the next layer of molecules than they did before, thus giving up some of their newfound energy. The ones in that layer, in turn, will collide harder with the next layer up, passing on some of their energy. This process goes on distributing energy through the pan and into the water.

As long as the heating element is operating, electricity supplies the heater with internal energy that is transferred to the heater's surface by molecular activity. When you turn the heater *off,* molecules on its surface still give up energy to pan molecules, thereby decreasing their average energy. These surface molecules, in turn, can pick up energy from molecules below them, so that the average energy of all the heater molecules decreases.

With the heater off, these processes go on until the *average* energy of the molecules in both heater and pan is the same. But you already know that, with the heater off, both it and the pan eventually come to the same temperature. Anytime a hot body and a cold body are put in contact, the hot one cools and the cold one warms until they both reach the same final temperature. If we combine this fact with what we just said about the heater and pan, we can conclude that when two bodies are at the same temperature, the average thermal energy of the molecules of the bodies is the same. In other words, *temperature is a measure of the average thermal energy* of molecules of the substance. As the *average* molecular energy goes up, the temperature goes up. As the *average* molecular energy goes down, the temperature goes down.

Two bodies are at the same temperature when: (1) the average thermal energy of their molecules is the same; and (2) no heat flows from one to the other if they are placed in contact. Actually (1) and (2) are equivalent—if the average molecular energy of one were higher than that of the other, energy would be transferred. If two bodies in contact are at different temperatures, *heat always flows from the hotter to the colder,* and never the reverse.

Let's summarize the four terms we have defined.

1. *Thermal energy:* energy associated with random molecular motion.

2. *Internal energy:* the *total energy* associated with all the molecules of the substance.

3. *Heat:* internal energy being added to or removed from the substance.

4. *Temperature:* the hotness or coldness of the material, as determined by the *average* thermal energy of the individual molecules.

We often rather loosely use the word "heat" when we really are talking about internal energy. You can usually tell from the context which idea is intended.

To illustrate the different terms, let's suppose the small pan in Fig. 6.1 contains 1×10^{25} molecules and the large pan 3×10^{25} molecules. If both pans are at room temperature, the total internal energy of all the molecules in the large pan is three times as much as that in the small pan. But because they are both at the same temperature, the average thermal energy of the molecules is the same in each. If we put them both on a stove and raise their temperatures to the boiling point, we must add three times as much heat to the large one, since it has three times as many molecules. But once both are at the temperature corresponding to the boiling point, the average molecular energy is the same in each pan. This average is, of course, higher than it was at the lower temperature.

6.2   Measuring Temperature and Heat

Now that we know what heat and temperature mean, we can describe how to measure them.

Temperature Measurements

We call a device for measuring temperature a *thermometer.* Thermometers work by using some property of matter that changes with temperature. Properties often used are: the expansion of a material with increasing temperature and contraction with decreasing temperature; the pressure of a gas that changes with temperature; the electrical resistance (a property discussed in Chapter 10); and the color of the light given off by a body that is hot enough to glow.

The liquid-in-glass thermometer (Fig. 6.5) is probably the most familiar type. As the temperature rises, the liquid in the bulb at the bottom—usually mercury or alcohol—expands more than the glass. The only place for expansion is up the tiny tube. As the

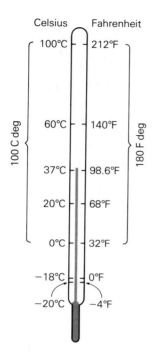

Celsius   Fahrenheit

100°C   212°F

60°C   140°F

100 C deg   180 F deg

37°C   98.6°F

20°C   68°F

0°C   32°F

−18°C   0°F

−20°C   −4°F

Figure 6.5 A liquid-in-glass thermometer, comparing the Celsius and Fahrenheit temperature scales.

temperature drops, the liquid contracts, moving down the tube. The top level of the liquid tells the temperature.

Temperature Scales

To assign a specific value to a temperature, we need some **temperature scale.** We can establish such a scale by assigning a number to any two points on the scale whose temperature can be reliably reproduced. Thus, we usually assign one number to the temperature at which water freezes, and another number to the temperature at which it boils. (The temperatures of these changes of state depend on the atmospheric pressure. In *this* chapter whenever we refer to the freezing or boiling points, we will be assuming sea level atmospheric pressure.)

Most of the world uses the Celsius scale, named after the Swedish astronomer Anders Celsius who introduced it in 1742. On this scale, the freezing point of water is assigned the temperature 0 degrees (0°), and the boiling point 100 degrees (100°).* For many years this was called the centigrade scale—"centi" standing for the 100 divisions between the two fixed points.

Until recently, we in the United States have used the Fahrenheit scale for all except scientific purposes. On this scale, the freezing point of water is 32°, and its boiling point is 212°. In setting up his temperature scale in about 1717, the German instrument maker Gabriel Fahrenheit chose one fixed point as the lowest temperature he could produce in his laboratory with a freezing mixture of ice and salt, and called it 0°. The other point was the normal temperature of the human body, to which he assigned the number 96°. The freezing and boiling points of water then came out to be 32°F and 212°F. When these later two points were accepted as standards and thermometers made more accurately, the normal body temperature turned out to be 98.6°F. Fahrenheit also invented the mercury-in-glass thermometer, and used it in defining his temperature scales.

Figure 6.5 shows a comparison between the Fahrenheit and Celsius temperature scales for some familiar temperatures.† Many

---

*Celsius first proposed a scale in which these two numbers were reversed—0° for boiling and 100° for freezing.

†The relationship between these two scales is given by the equation $C = 5/9(F - 32)$, where C and F are the temperatures in degrees Celsius and degrees Fahrenheit, respectively.

of us in the United States are just learning to "think Celsius." One way to make approximate conversions near familiar temperatures is to notice that 100 C deg spans the same interval as 180 F deg. This span means that a change of 1 C deg is about 2 F deg (actually 1.8 F deg). If you then remember a few Celsius-Fahrenheit equivalents, you can make conversions near these temperatures in your head. For example, you might remember that 68°F is the same as 20°C. Then each 2 F deg change near there gives a change of about 1 C deg. For instance, to the nearest degree, 70°F is 21°C. After a 10 F deg variation from the exact equivalent, your estimate is wrong by about ½°C.

[Note: To avoid confusion, we are following the common practice of writing Celsius degrees (C deg) when we are talking about a temperature *interval,* and degrees Celsius (°C) when talking about an actual temperature.]

Absolute Temperature

In laboratories that have facilities for taking substances to extremely low temperatures, experimeters have found that the closer they get to the temperature of −273.16°C, the more difficult it becomes to keep lowering the temperature. In fact, the temperature of −273.16°C has never been reached. Current theory predicts this value to be the lower limit of temperature, and thus not reachable at all.

It is sometimes useful to have a temperature scale for which the lowest possible temperature is defined as 0°. We call such a scale an *absolute* temperature scale, with the lower limit being called *absolute zero.* The *Kelvin* (K) temperature scale is an absolute scale that uses the same size degree as the Celsius scale. The melting and boiling points of water are then 273 K and 373 K, respectively.

Heat Measurements

Since heat is a form of energy, quantities of heat can be measured in any energy units—joules, foot-pounds, kilowatt-hours, for example. However, separate units of heat were defined before anyone knew that heat was a form of energy. These special heat units are still used, both because of tradition and because they are about the right size for many heat measurements.

Heat units are defined in terms of the heat needed to raise the temperature of water. Our basic heat unit is the kilocalorie (kcal), defined as follows: *1 kcal is the heat needed to raise the temperature of 1 kg of water 1 C deg.* (One liter—just over a quart—of water has a mass of 1 kg.) Similarly, 1 calorie (cal) is 1/1000 kcal and is the heat needed to raise the temperature of 1 g of water 1 C deg.

Confusion sometimes occurs because the Calorie (with capital C), used to give the energy value of foods, is the equivalent of the kcal, or 1000 cal. To avoid such confusion, many nutritionists now use the term "kilocalorie" rather than "Calorie."

In the British system of units, the heat unit is the British thermal unit (Btu), defined as: *1 Btu is the heat needed to raise the temperature of 1 lb of water 1 F deg.* Even after the American switch to metric units is complete, older heating and air conditioning systems rated in Btu will be in use for some time.

Table C-1, Appendix II gives a comparison between these and other energy units. From this comparison, we see that 1 kcal = 4190 J = 1.16 Wh = 3.97 Btu.

Example 6.1   How much energy is needed to heat the water to take a bath? Assume you use 100 liters (about 26 gal) of water that enters your house at 13°C (55°F), and you like your bath water at 43°C (110°F). If your water is heated with electricity costing 5 ¢ per kWh, how much does it cost to heat this water?

Since 1 kcal raises the temperature of 1 kg of water (which has a volume of 1 liter) 1 C deg, the total heat needed is:

heat needed = mass of water $\times$ 1 kcal/kg$\cdot$C deg $\times$ temperature change

$$= 100 \text{ kg} \times 1 \text{ kcal/kg} \cdot \text{C deg} \times 30 \text{ C deg} = 3000 \text{ kcal.}$$

We can convert this to kilowatt-hours:

heat needed = 3000 kcal $\times$ 0.00116 kWh/kcal = 3.5 kWh.

At 5 ¢ per kWh,

$$\text{cost} = 3.5 \text{ kWh} \times 5\text{¢/kWh}$$

$$= 17\text{¢.}$$

6.3   Change of Temperature; Specific Heat

Let's step back briefly and look at some of what we have said so far: (1) temperature, the hotness or coldness as determined by the average molecular energy, is changed by adding or removing energy in the form of heat; (2) temperature is measured on an arbitrary scale defined by assigning numbers for the temperatures of certain fixed points; and (3) heat can be measured in any energy units, but is usually measured in units such as kilocalories that are defined in terms of the heat needed to raise the temperature of water.

Because of the way we defined the heat units, we know how much heat it takes to warm a certain amount of water (or how much heat has to be removed to cool a certain amount of water) a fixed temperature difference. But what about heating and cooling other materials? For example, when you go to the beach in summer, the dry sand back from the water is very hot—you hurry across it on tiptoe, or wear something on your feet. Yet when you get near the water, the moist sand and the water itself are cool. Why is this true when both have been exposed to the same amount of the sun's radiation? Or, why does your freezer run longer (indicating more heat removal) after you plug it in if it's full of room-temperature food rather than full of room-temperature air? The answers to these questions are related to the property of matter called **specific heat.**

Specific Heat   *The specific heat of a substance is the amount of heat needed to raise the temperature of a unit mass* (for example, 1 kg) *of the substance 1 degree.* The values of specific heat vary widely from substance to substance. Table 6.1 presents some specific heat values for a few ordinary materials, given in kcal per kg per C deg:

Table 6.1   Specific Heat of Ordinary Substances

Material	Specific Heat (kcal/kg·C deg)	Material	Specific Heat (kcal/kg·C deg)
water	1.00	table salt	0.20
ice	0.50	hamburger	0.75
air	0.24	mercury	0.033
wood	0.42	steel	0.11
glass	0.20	aluminum	0.21
sand	0.2	gold	0.032

Knowing the specific heat and how much of a substance we have, we can calculate how much heat we must add to, or remove from, the substance to change its temperature a fixed amount. Since specific heat is the heat needed *per unit of mass* and *per degree temperature change,* we multiply it by both of these factors to get the total heat added or removed:

$$\text{heat to change temperature} = \text{mass} \times \text{specific heat} \times \text{temperature change.}$$

You'll notice that we in effect already used this relationship in working Example 6.1, but we'll illustrate it with another example.

**COUNT RUMFORD
BOILS WATER WITHOUT FIRE**

MUNICH, BAVARIA, (October 3, 1794)—Count Rumford, Inspector General of Artillery for the Army of Bavaria, took advantage of the beautiful autumn afternoon to hold another of his public demonstrations on boiling water without using fire. The Count had one of his machines for boring cannon immersed in a large tank of water. He astonished onlookers by showing that, just by turning the boring tool inside the cannon, he could heat the water to boiling and continue it boiling no matter how long he turned the tool.

In a news conference after the demonstration, Count Rumford explained that "Being engaged lately in superintending the boring of cannon in the workshops of the military arsenal at Munich, I was struck with the considerable heat which a brass cannon acquires in a short time in being bored. . . ." He went on to explain that his demonstration here today cast grave doubt on the views long held by philosophers that rubbing objects together heats them because it squeezes caloric fluid out of the objects. "How," he asks, "can you explain the inexhaustable supply of caloric in the cannon that is evidenced by the continued heat production, no matter how long I turn the boring tool?"

Count Rumford, born Benjamin Thompson in the British Colony of Massachusetts, entered the service of Elector Karl Theodor ten years ago. He was elevated to the rank of Count of the Holy Roman Empire seven years later. The Count chose for his title the name Rumford, the original name for Concord, New Hampshire where he married the wealthy widow of Colonel Benjamin Rolfe in 1772.

Rumford concluded his news conference by pointing out that this way to produce heat is not very efficient. Conceding that a great deal of heat is produced by the borer, he explained that if you burn the fodder fed to the horses turning the drill, the resulting fire could heat much more water.

Example 6.2   How much heat has to be removed from 50 kg of hamburger placed in a freezer at 20°C to lower the temperature to 0°C? Compare this with the heat needed to change the temperature of the same volume of air by the same amount. (Density of air = 1.2 kg/m³; that of hamburger = 830 kg/m³.)

$$\text{heat removed} \times \text{mass} \times (\text{specific heat}) \times (\text{temperature change})$$

$$= 50\text{kg} \times 0.75 \text{ kcal/kg} \cdot \text{C deg} \times 20 \text{ C deg} = 750 \text{ kcal.}$$

Since the mass of a given volume is proportional to its density, we can get the mass of the air from:

$$\frac{\text{mass of air}}{\text{mass of hamburger}} = \frac{\text{density of air}}{\text{density of hamburger}}$$

From this equation, the air's mass is 0.072 kg. Then, for air,

$$\text{heat removed} = 0.072 \text{ kg} \times 0.24 \text{ kg/kcal} \cdot \text{C deg} \times 20 \text{ C deg}$$

$$= 0.35 \text{ kcal.}$$

In Example 6.2 notice that the specific heat of hamburger is about three times that of air. At first glance you might expect the hamburger to take only three times as much heat. But since the heat needed depends on the mass (which is much higher for the hamburger), the heat needed is over 2000 times as much.

Question From the specific heats in Table 6.1, explain one reason why dry sand near the beach is so much hotter than the water, or even than the wet sand near the water. (The density of sand is roughly twice that of water.)

Water and its Specific Heat

The specific heat of water is exactly 1.00 kcal per kg per C deg. This handy figure is not just a fluke of nature, but comes from the way we defined the kilocalorie. From the definition of this unit, water must have this specific heat. This means that the specific heat of any substance is given by the ratio:

$$\frac{\text{heat needed to change temperature of substance 1 degree}}{\text{heat needed to change temperature of same mass of water 1 degree}}$$

(Compare this definition with the way we defined specific gravity. Since we find water almost everywhere, it is often easy to compare a certain property of other substances with the same property of water, and report the result as a ratio.)

The fact that the specific heat of this ever-present substance, water, is higher than for most ordinary materials is an important part of nature's design, and has a profound effect on your life. Some examples:

1. The surface of the earth is dominated by water. Water's high specific heat means that the earth can absorb or give off large quantities of heat without much temperature change, making the temperature difference between winter and summer acceptable for life.

2. Your body, which is largely water, can also absorb or give off a lot of heat without much temperature change.

3. Water, which can carry the most heat of any ordinary material, is readily available for carrying heat from furnace to user in applications such as home heating.

4. Areas near large bodies of water tend to have a more nearly constant year-round temperature. This is evident in the Great Lakes region and on the coasts, particularly the west coast of the United States where the prevailing wind is from sea to land.

5. This high specific heat has the disadvantage that heating water for your shower takes lots of fuel or electrical energy, most of which goes down the drain.

Energy Conservation   The energy conservation principle tells us that, when a hot and a cold substance are combined, the heat lost by the hot substance must be gained by the cold one. The use of this idea in studying specific heats (or the amount of hot or cold substance needed) is often called *calorimetry*. An everyday example will illustrate the principle.

Example 6.3   How much hot water do you have to mix with 1 liter of cold water (mass = 1 kg) in your wash basin to make it comfortable for washing your face on a cold day? Take the cold water to be fresh from the supply lines at 8°C (46°F) and the hot water to be at 60°C (140°F).

Since both hot and cold water come to the same final temperature, which we'll take as 43°C, we have:

heat lost by hot water = heat gained by cold water;

(mass of hot water) × (specific heat) × (temperature change of hot water) =

(mass of cold water) × (specific heat) × (temperature change of cold water);

(mass of hot water) × (1 kcal/kg·C deg) × (60°C − 43°C) =

(1 kg) × (1 kcal/kg·C deg) × (43°C − 8°C).

We can solve for the mass of hot water which comes out to 2 kg, or 2 liters of hot water.

Exercise   How much energy is needed to heat these 2 liters of water to 60°C if they enter the water heater at 8°C?   **Ans:** ½ piece boysenberry pie à la page 12

6.4 **Sources of Thermal Energy**

Most of the thermal energy we use comes from burning some kind of fuel. In heating our homes, many of us burn natural gas, fuel, oil, coal, or wood. These fuels release chemical energy as they burn. During the burning (oxidation) process, the original chemical form of the fuel is changed to one with lower chemical energy. The released chemical energy shows up as kinetic energy of the molecules—thermal energy. This energy heats the air and other surrounding materials.

Automobile engines burn gasoline to release heat that operates the engine. Most electrical power plants burn coal, gas, or oil to heat water and make steam for turning the turbines. Therefore, even homes or industries that use electrical energy for heating, obtain their energy, in reality, from burning fuel.

The food we eat basically serves as "fuel" for our bodies. The nutrients in the food are distributed to all parts of the body where they are oxidized—burned—to release energy. This energy is used to keep the body itself functioning, to do mechanical work, and to produce heat to keep the body warm. A healthy human body uses about 70 kcal/h even when sleeping. The same body when running fast uses about 600 kcal/h.

Example 6.4   If you eat a piece of a cheesecake containing 400 kcal of chemical energy, how far do you have to run if you want to exercise this energy off?

If you run fast enough to use 500 kcal/h, you need to run for 4/5 h. At a speed of 10 km/h (about half the speed a long-distance runner could run for that time), you need to run about 10 km/h × 4/5 h = 8 km or 5 miles.

The energy released by completely burning a unit of mass of a substance is called the **heat of combustion.** Table 6.2 gives the heat of combustion of some ordinary fuels. Included in Table 6.2 are some foods that are usually "burned" in a living body rather than a furnace.

Table 6.2   Energy Content of Certain Fuels

Fuel	Heat of Combustion (kcal/kg)	Fuel	Heat of Combustion (kcal/kg)
gasoline	11 500	sirloin steak	3 100
fuel oil	10 600	ice cream	2 100
natural gas	9 500*	apple	580
coal	7 100	peanut butter	5 800
charcoal	7 300	lemon meringue pie	2 600
oak wood	4 000	chocolate candy	5 000

*In kcal/m³ at standard atmospheric pressure.

Example 6.5   If coal sells for $140 per ton (mass = 910 kg) and oak firewood for $50 a pick-up truckload, which is cheaper to burn in your fireplace for the same amount of heat produced?

First calculate the total heat from a ton of coal and a load of wood.

Coal: heat (from coal) = heat of combustion $\times$ mass

$$= 7100 \text{ kcal/kg} \times 910 \text{ kg}$$

$$= 6.5 \times 10^6 \text{ kcal.}$$

Wood: A typical truckload, after stacking, has a volume of 1.75 m³. *The Handbook of Chemistry and Physics* gives the density of red oak wood as 660 kg/m³. If we allow about 25% for open space between the pieces of wood, the density is 500 kg/m³. Therefore,

$$\text{mass} = \text{density} \times \text{volume} = 500 \text{ kg/m}^3 \times 1.75 \text{ m}^3 = 875 \text{ kg};$$

then,

$$\text{heat (from wood)} = 4000 \text{ kcal/kg} \times 875 \text{ kg} = 3.5 \times 10^6 \text{ kcal.}$$

From the given costs,

$$\text{for coal: heat per \$}\ = \frac{6.5 \times 10^6 \text{ kcal/ton}}{\$140/\text{ton}} = 46\ 000 \text{ kcal/\$};$$

$$\text{for wood: heat per \$} = \frac{3.5 \times 10^6 \text{ kcal/load}}{\$50/\text{load}} = 70\ 000 \text{ kcal/\$.}$$

As we'll see in Chapter 8, for a fireplace most of the heat goes up the chimney, whatever you burn.

6.5   Expansion and Contraction with Temperature Change

Why does your toast pop out of the toaster when it's done? Why does your circuit breaker automatically turn off when you plug too many electrical appliances into the receptacle? Why do concrete highways have those annoying cracks every few car lengths that go thump, thump as you ride down the road? The answer to all these questions is related to the fact that most substances expand when their temperature rises, and contract when their temperature falls. We have already used this property in talking about liquid-in-glass thermometers—both the liquid and glass expand with increasing temperature, but the liquid expands more than the glass.

One prominent exception to this behavior occurs in water when its temperature is below 4°C. Water then *contracts* with temperature rise, and expands with temperature drop. This effect has some important impacts upon life on earth. However, since the

Figure 6.6 Does the hole in the washer get bigger or smaller as the temperature rises? Hint: Is the longer dimension parallel to, or perpendicular to, the dashed line?

reason for this behavior is closely related to the melting and freezing of water, we will discuss it later. In this chapter we will consider only those substances that expand with increase, and contract with decrease, in temperature.

When the temperature of a substance rises, the average kinetic energy of the molecules increases. Since the molecules then collide harder with each other, they spread out, causing the material as a whole to expand. (You can see a similar effect on a crowded dance floor when the tempo changes from a slow to a fast beat.) This effect holds in solids, liquids, and gases. A uniform cube of solid expands equally in all directions, but if a solid object is longer in one particular dimension, the overall expansion in that direction is greater than in the others. The amount of expansion or contraction for each degree of temperature change varies from substance to substance.

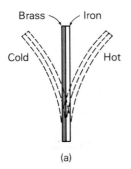

(a)

Question   For a flat washer (a flat piece of metal with a hole in the center) such as is shown in Fig. 6.6, does the hole get bigger or smaller as the temperature rises? Your instructor may need to demonstrate this effect before you are convinced of the right answer.

In the case of a liquid, only volume changes have meaning, since a liquid takes the shape of its container. In the liquid-in-glass thermometer, the liquid is confined to a tiny tube, so that the change is mostly in one direction.

Thermal expansion is both very useful and very much a nuisance. We will mention a few examples on each side—first, the useful.

**Bimetallic Thermometers and Thermostats*   Suppose we have thin strips of two different metals welded together as in Fig. 6.7(a). If one of these metals expands more for a given temperature change than the other, bending occurs as illustrated in the figure. For example, brass expands about 50% more than iron for a given temperature increase. When the temperature is low, the bimetallic (two-metal) strip bends to the left; when the temperature rises, the brass expands more, causing the strip to bend to the right. This device could be a crude thermometer.

To get a lot of bending from this simple strip, you need a large temperature change. The effect can be amplified by shaping the strip into a coil as in Fig. 6.7(b). Then, a fairly small temperature

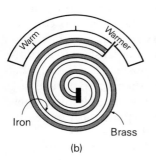

(b)

Figure 6.7 A bimetallic strip bends away from the metal that expands most as the temperature rises.

*Sections with the ∗∗ symbol are for illustration only and contain no *new* principles of physics.

change can produce a measurable movement of the pointer. Most thermometers now used on control units for heating and air conditioning systems have a coiled bimetallic strip.

Most *thermostats*, whether on home heating and air conditioning systems, electric blankets, or pop-up toasters, use some variation of the bimetallic strip. A simplified version of a thermostat used to control an electrical heater is shown in Fig. 6.8. As the temperature around the thermostat drops, the bimetallic strip bends downward, making contact with the terminal below. Electrical current can then flow from the power source, through the strip, through the heater, and back to the source. When the temperature rises, the strip bends upward, breaking contact so that no current can flow to the heater. For a large heating system, the current flowing through the thermostat when contact is made usually operates another switch (called a relay) that turns the system on. For controlling an air conditioner, the contact is on the other side of the bimetallic strip, so that contact is made and current flows when the temperature *rises* above a certain point.

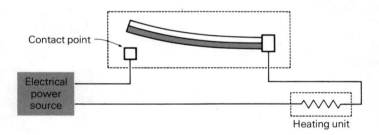

Figure 6.8   A bimetallic strip is used as a thermostat to turn heating and cooling systems on and off.

Real thermostats are usually more complicated than we've shown. There is usually an adjustment so that contact is made at whatever temperature you want. Often the strip is in the form of a coil, such as in the thermometer of Fig. 6.7(b). In most cases, the thermostat merely turns the system on or off at some preset temperature rather than regulating the rate at which heat is produced or removed.

Question   Suppose you come into a cold house and are in a hurry to get the temperature up to 20°C. Will the temperature get there faster if (a) you set the thermostat on 30°C and then set it back to 20°C when the temperature rises, than if (b) you set it on 20°C at the start?

Many household appliances use bimetallic strip thermostats to turn the heating element on and off in order to control their temperatures within a given range. These include electric frying pans, percolators, hair driers (as a safety device to prevent too much heat), pop-up toasters (to activate an electromagnet that releases the racks), tabletop ovens, irons, electric blankets, and so forth.

**Unwanted Thermal Expansion   We mentioned earlier that thermal expansion can be a nuisance. As Fig. 6.9 shows, it also can be hazardous if not properly allowed for. In concrete highways or sidewalks, gaps must be left every few meters to allow for expansion and contraction as the temperature changes. Bridges also need room to expand and contract with temperature change (Fig. 6.10). Power lines, if stretched tightly in summer, would break in winter. A cold glass bowl may break when filled with hot water because the layer of glass that gets hot first expands faster than the cold layer below, resulting in fracture.

Figure 6.9   The summer heat of 1966 in Stewartville, Minnesota, caused the buckling due to thermal expansion that you see in these tracks, posing danger to the oncoming train.

Figure 6.10   Expansion joints (such as this one from the George Washington Bridge) are provided at each section of a bridge.

Expansion of Gases   Gas molecules bounce around freely and fill up the complete volume of their container. When a gas is heated, it expands only when its container increases in volume. If the volume of the container doesn't change, the gas cannot expand. Instead, the molecules of the gas merely pound harder against the walls of the container and against each other, causing an increase in the pressure of the gas.

Suppose a gas is confined in a cylinder with a moveable piston (Fig. 6.11). The weight on the piston keeps constant pressure on

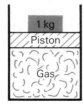

Figure 6.11   A gas expands when heated only if it is in a container whose volume increases.

the gas while letting it expand as heat is added. Work is done in increasing the potential energy of the piston and 1-kg mass. If the piston were held fixed, the gas could not expand upon heating; the pressure would rise instead.

As an illustration, consider what happens to the air pressure inside an automobile tire when it turns on a hot highway. The air inside the tire absorbs heat from the road. Even if the road were not hot, the air would be heated from the friction of the continuous flexing of the tire, especially if the pressure is low. Since the tire stretches very little, the air pressure must increase. That's why tire manufacturers recommend checking the pressure when the tires are cold, and *not* letting air out if the pressure goes up while on the road.

As a second example, consider this question: How do you keep the bounce in your tennis balls when they lie around unused for a long time? Most tennis balls come pressurized in a can to maintain high pressure inside the balls and give them a lively bounce. After the can is opened, the gas in the balls leaks out gradually, reducing the pressure inside so that the balls become "dead."

One way to prevent this leakage is to buy a gadget from the sporting goods supplier for storing the balls under pressure when not in use. There's a simpler way: Reduce the pressure *inside the balls* when they're not in use so that the air doesn't try to get out! That's really not so hard: Store the tennis balls in the freezer compartment of your refrigerator. As the temperature drops, the pressure drops accordingly. (The balls won't bounce well until they warm up after you take them out.)

Key Concepts

Temperature, a measure of the hotness or coldness of a substance, is determined by the *average* energy of the molecules of the substance. **Thermal energy** is energy associated with the random molecular motion. **Internal energy** is the total energy associated with all the molecules of the substance. **Heat** is internal energy being added to, or removed from, a substance.

If two bodies at different temperatures are placed in contact, heat always flows from the warmer to the cooler body and never the reverse. Two bodies are at the same temperature if no heat flows from one to the other when they are placed in contact. Adding heat to a substance raises its temperature; removing heat lowers its temperature (if the substance does not change its state).

A **temperature scale** is defined by assigning a number (temperature) to two fixed points and dividing the interval into equal parts. Heat, a form of energy, can be measured in any energy units, but is often measured in kilocalories—the amount of heat needed to raise the temperature of 1 kg of water 1 C deg.

The **specific heat** of a substance is the heat needed to raise the temperature of 1 kg of the substance 1 C deg. It is numerically equal to the ratio of the specific heat of a substance to that of water. The **heat of combustion** of a substance is the energy released in the burning of unit mass of the substance.

Most substances expand when heated and contract when cooled.

Questions

Temperature and Heat

1. When it's cold outside and your little brother leaves the door of your house open, does cold come in or does heat go out?

2. Which does an air conditioner remove from a room, temperature or heat?

3. If the nurse removes a thermometer from your mouth and it reads 37°C (98.6°F), do you have a temperature?

4. True or false: Any two pans of water at 50°C both contain the same amount of heat. Explain the reasons for your answer.

5. Lucy tells Charlie Brown that the water in her pool is warmer than his because *all* the molecules in her water have higher kinetic energy than any of his. Is she right, even if her water is warmer?

6. When two bodies at different temperatures are put in contact, why does the warmer one get cooler, and the cooler one get warmer? Explain the heat flow in terms of molecular activity.

7. Father Eskimo puts a hot 25-kg stone in bed with him to keep his feet warm; Baby Eskimo does the same with a 10-kg stone. As they heat the stones from −45°C to +45°C near the fire, which stone gains more heat? Which gains more internal energy? For which does the average molecular kinetic energy change the most?

8. After the stones of Question 7 have reached 45°C, which stone has more internal energy? For which is the average molecular kinetic energy higher?

Temperature and Heat Measurements

9. From the facts that 68°F is 20°C and a temperature difference of 2 F deg is roughly 1 C deg, what are the approximate temperatures in °C of 66°F, 70°F, and 72°F?

10. Dr. Frankenstein sets up a temperature scale on which he assigns the value of 0° to the temperature of a freezing human body and 100° to the temperature of boiling blood at sea level atmospheric pressure. Is his scale as legitimate as the scale of Celsius?

11. Why does the diameter of the bulb that holds most of the mercury at the bottom of a mercury-in-glass thermometer have to be larger than the tube running up the length of the thermometer?

12. Why do you have to shake a clinical thermometer (Fig. 6.12) to make the reading go down after it's used?

13. What are some advantages and disadvantages of defining units of heat on the basis of the temperature rise of water?

Specific Heat

14. If you keep your feet warm at night by taking a hot object to bed with you, would you be better off taking a steel cannon ball or a piece of wood the same size and at the same temperature? Why?

15. Why does your freezer run longer after plugging it in if it is full of room-temperature hamburger than if it is full of room-temperature air?

16. Antifreeze has a specific heat less than 1. What disadvantage would there be to using pure antifreeze rather than a mixture of anti-freeze and water in your car's cooling system?

17. Why is water a good medium for carrying heat to baseboards or radiators in a home heating system?

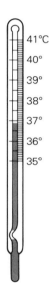

Figure 6.12 Notice the wiggle in the tube of the clinical thermometer.

18. If Snoopy is to be warmest in his outdoor doghouse, should the house be large and roomy, or just big enough for him to fit in comfortably? (The house is being heated by Snoopy's body heat only.)

19. Why does the temperature in Omaha vary so much more between winter and summer than the temperature in San Francisco?

20. Does a bridge built in January in San Francisco need as large an expansion joint as one built at the same time in Chicago?

Thermal Energy Sources

21. Estimate how far you would need to run to work off the energy in a 30-gram Hershey bar.

Thermal Expansion

22. Why can you loosen a "stuck" peanut butter jar lid by pouring hot water over the lid?

23. Which is more likely to crack if you pour hot water into it—a thick glass or a thin one?

24. Which, do you think, would expand more for each degree change in temperature—Pyrex or ordinary glass?

25. When you turn on a hot water faucet part way, the flow will slow down and sometimes almost stop after a brief time. This doesn't happen with the cold water. Why?

26. When a hot water heating system first comes on, why do you hear creaking in the pipes, or in the radiators or baseboards?

27. A machinist wants to fit a round shaft into a hole the same size, and have a tight fit so that the shaft will not slip out. How can he use thermal expansion to aid his assembly?

28. Since dentists successfully use gold for crowns and bridges on teeth, how would you expect the thermal expansion properties of gold to compare with those of tooth enamel?

29. Overheating of automobile brakes causes the brake drums to expand (see Fig. 5.8). Why would such expansion decrease braking ability?

30. When you change the temperature setting of a surface heating unit on an electric range, you shift a moveable electrical contact closer to, or farther from, the end of a bimetallic strip (Fig. 6.13). This strip turns the current on and off periodically, with the percent of ON time determining the temperature of the heater. For high heat, should the contact be closer to, or farther from, the end of the strip? Assume high temperature causes the strip to bend away from the contact. (Note: Some ranges use a different type of mechanism.)

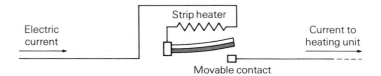

Figure 6.13 Electric current flows to the surface heating unit of an electric range only when the bimetallic strip touches the moveable contact. The percentage of ON time is set by the distance of the contact from the strip.

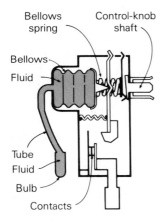

Device on

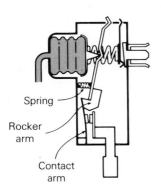

Device off

Figure 6.14 A bellows thermostat.

31. Draw a design that you make up for the mechanism of a simple pop-up toaster, using a bimetallic strip to release the toast elevator. Include a darkness control.

32. When you fill your car tank with gasoline on a warm day, why does the gasoline overflow after sitting for a while, even though you didn't fill it to the top?

33. Why is the door of a freezer harder to open after it has been closed a few minutes than right after it's closed? Hint: The effect is more noticeable if the freezer is nearly empty.

34. Refrigerators, ovens, oil-fired water heaters, and various other devices use a bellows type thermostat (Fig. 6.14). The bulb is in the region where the temperature is to be sensed, and connected by a long tube to the bellows. As the fluid in the bulb expands and contracts, the end of the bellows moves back and forth, opening and closing the contacts of an electrical switch. Is the thermostat shown in the figure intended for a refrigerator or for an oven? How would you modify the thermostat to make it work for the other device?

Problems

1. On January 28, 1977 the temperature in Cuckoo, Virginia dropped 40 deg from 48°F to 8°F in just a few hours' time. How much was this temperature *change* in C deg?   **Ans:** 22 C deg

2. I like to shower in water at a temperature of about 110°F. What is this temperature in °C?   **Ans:** 43°C

3. Refer to Example 6.1. How does the energy needed to heat the water for a bath compare with the energy needed to pump this much water from a 60-m-deep well?

   Ans: 210 times as much to heat

4. The cook at the Denver Hilton boils potatoes at 95°C. How much heat is needed to get 40 liters of water from 10°C to this temperature?   **Ans:** 3400 kcal

5. Assume that the range used by the cook in Problem 4 is 50% efficient in getting heat into water. How much does it cost to heat this water, at 5¢ per kWh?   **Ans:** 39¢

6. If 3 liters of water are heated in a 1-kg aluminum pot from 10°C to 100°C, how much heat goes to the pot? How much goes to the water?   **Ans:** 19 kcal; 270 kcal

7. The River City Boys Band has collected 2000 metric tons (1 metric ton = 1000 kg, and weighs 2200 lb) of old steel automobile engines for melting down and recycling. How many kWh of energy are needed to get this much steel from 20°C to the temperature at which it starts to melt (1430°C)?

   Ans: 310 000 000 kcal or 360 000 kWh

8. An auditorium 60 m long by 20 m wide by 10 m high holds 3000 people. The average healthy person burns up about 100 kcal of food energy each hour while sitting still. If each person continued to produce heat at this rate, and no heat were removed or leaked out, how much would the temperature of the air in the auditorium rise in 2 hours? The density of air is 1.2 kg/m³.   **Ans:** 170 C deg

9. For a properly operating fireplace, about 15 m³ per minute of air is pushed up the chimney. If this air filters in from all over the house, and is heated by the central heating system from −10°C to 20°C before it gets to the fireplace, how much heat is wasted in 8 hours of heating this air?

   Ans: 62 000 kcal (72 kWh)

10. Superman found that 26 000 kcal were needed to change the temperature of 20 kg of kryptonite from 0°C to 60°C. What is the specific heat of kryptonite?   **Ans:** 22 kcal/kg·C deg

11. If you leave the room briefly, should you turn the light off or leave it on? An ordinary incandescent light bulb uses about 10 times as much power at the instant it is turned on as it does after the filament is hot. Assume that during the heating process, a 60-W bulb generates heat at an *average* rate of five times its rated power, or 300 W. If the mass of the tungsten filament is 15 mg (1 mg = 10^{-6} kg), how much heat is needed to raise the temperature of the filament from 20°C to 2520°C? If we neglect the energy radiated from the filament during this time, how long would it take to heat this filament? For what maximum length of time would you save electrical energy by leaving the light on rather than turning it off? The specific heat of tungsten is 0.037 kcal/kg·C deg.

    Ans: 0.00139 kcal; 0.0194 s; 0.078 s

12. If 1% of the chemical energy stored in a ½-kg sirloin steak could be converted to doing mechanical work, how many 40-kg sacks of mail could your mail carrier lift onto the 1-m-high mail truck after eating such a steak?   **Ans:** 166

13. A power company can buy coal at $140 per metric ton (1000 kg) or fuel oil at 19¢ per liter (1 liter of fuel oil has a mass of 0.94 kg). Assume the company can burn both at the same efficiency (same fraction of chemical energy converted to electrical energy). Which would be cheaper?

    Ans: coal (about 3% cheaper)

14. A Labrador retriever eats four 440-g cans of Ken-L-Ration a day. She lies in the sun all day and in her doghouse all night. Assume an energy content of 4000 kcal/kg and that her body converts 75% of what she eats to heat. How much heat is available to heat her doghouse during a 10-h night?

    Ans: 2200 kcal

Home Experiments

1. Take two identical pans and put twice as much cold water in one as the other. Compare the time required for them to reach the boiling temperature when they are placed on identical heating units.

2. The next time you bake potatoes put a rock about the size of one of the potatoes in the oven at the same time. When the potatoes are done, compare the time needed for the rock and the potato to cool to about the same temperature. What can you say about the relative specific heats of the two? (The rock's mass will be about 2.5 times that of the same-sized potato.)

3. Find out how much heat you lose from your water heater, even if you use no hot water. At the nearest faucet to the heater, run the water until it is as hot as it will get. Using a thermometer that registers up to 100°C, measure the temperature of some of this water. When no hot water has been used for a while (so that the heater is completely full of hot water), turn off the heater for eight hours. Remeasure the temperature as before. From the temperature drop and the size of the heater tank, you can now calculate the heat loss. From the cost of your energy supply, you can then estimate how much money you can save by better insulating your water heater.

4. Contraction with *increasing* temperature can be observed by tying an ordinary rubber band to a weight that is heavy enough to stretch the band slightly. Move a burning match or candle up and down near the band. The band will contract and raise the weight a noticeable amount. Put a stationary object beside or just below the hanging object to make the movement easier to see.

References for Further Reading

Brown, S. C., *Count Rumford Physicist Extraordinary* (Anchor Books, Doubleday & Co., Garden City, N.Y., 1962). An account of a man who was a gifted experimenter, prolific inventor, and social reformer, but who at the same time was a scheming adventurer, cynical soldier of fortune, and spy.

Dyson, F. J., "What Is Heat?" *Scientific American,* Sept. 1954, p. 58.

7

Change of State

Snow melts; water in the ice tray freezes; puddles of water evaporate; frost appears on the grass on a cool autumn morning; and the water in the pot of beans boils away if we leave the pot on the stove too long—we're all familiar with water's changes of state.

But many other everyday substances also change state. Moth balls gradually vaporize and disappear, the vapors keeping the moths away. When we solder metal pieces together, we melt solder that later solidifies to hold the parts together. Industries melt iron and aluminum to form them into many of the mechanical gadgets we use. Refrigerators and air conditioners work by continuously evaporating and condensing refrigerant inside the coils. Our bodies are cooled on a hot day by the perspiration evaporating from our skin.

In this chapter we will study how substances change state, and how these changes affect our lives. After summarizing the properties of the states of matter, we'll discuss the specific changes that can occur between states, and the energy transfers accompanying these changes. We'll find out why water expands rather than contracts when it freezes, and we'll consider some of the external influences on the temperature at which the state changes. Finally, we'll discuss a topic very relevant to our comfort: humidity.

7.1 The States of Matter

You learned in grade school that matter at ordinary temperatures exists in one of three states—solid, liquid, or gas. Before we discuss what's needed for changing the states of matter, let's briefly explore the properties of each of the states, and the ways in which they differ.

Solids   In **solids,** the molecules are held close together by forces that keep any given molecule in the same place, on the average, relative to its neighbors. Most solids have a crystalline structure, in which the molecules are arranged in some regular pattern (Fig. 7.1). The molecules vibrate back and forth around their average positions, as if they were held together by tiny springs. As the temperature is raised, the molecules jiggle around faster and move farther from home in their back-and-forth motion, but nonetheless they keep their same average positions.

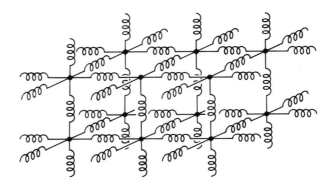

Figure 7.1   The molecules in a crystalline solid are arranged in a regular pattern. The forces holding them together behave like those from tiny springs. (The springs, of course, do not actually exist.)

If you try to compress a solid, the springlike forces "give" some, allowing the body to compress slightly. Similarly, these forces let you stretch a solid a small amount without a permanent deformation or break. We talked about this elasticity of solids—their ability to return to their original shape when the deforming force is removed—in connection with elastic potential energy in Chapter 1, and with elastic collisions in Chapter 3.

Liquids   In **liquids,** the molecules still exert fairly strong forces on each other, but not strong enough to hold them in fixed positions. The molecules bounce around freely in all directions as they collide with each other and with the walls of the container (Fig. 7.2). Even though the analogy is not perfect, the relationship between solid and liquid is similar to the relationship between zone and man-to-man defense in basketball. In zone defense (solid) each player (molecule) "bounces" around within a certain region as he interacts with other players. In man-to-man (liquid), the players move around freely throughout the court, interacting with other players, but staying within the court boundary. High pressure can crowd the molecules of a liquid only slightly closer together. That's why we said in Chapter 5 that liquids are "almost" incompressible.

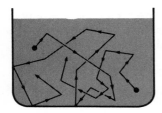

Figure 7.2   In a liquid, the molecules bounce around at random within the bounds of the surface of the liquid and the container.

Gases   In **gases,** the molecules are far apart so that the force between them is quite small except during collision. Gas molecules spread out to fill the entire volume of their container. Our basketball analogy is not quite as good here, but the gaseous state is somewhat like the end of the game, when the players spread out over the entire town, only occasionally interacting with each other.

If this were a detailed study of the gaseous state, we would distinguish a gas from a vapor. For our purposes, however, we will consider a gas and a vapor to be the same. For example, we'll consider steam and water vapor both to be the gaseous state of water.

7.2   The Processes of Changing State

Materials such as glass are mixtures of different elements, and do not have a crystalline structure with a regular ordering of molecules. As you heat glass, it gradually becomes softer and softer until it is a free-flowing liquid. There is no clear-cut condition for which we can say the glass has changed from the solid to the liquid state. Many plastics behave the same way. Such materials are said to be *amorphous.*

Most substances are crystalline in the solid state, and for these, a definite temperature exists at which the transition from one state to another occurs. In discussing change of state, let's focus our attention on water, since water is familiar to most of us in all three of its states. The change of state of other crystalline substances is similar to that of water in most respects. We're using the word "water" here in the general sense to include solid (ice), liquid, and vapor (steam).

Melting and Freezing   When ice or some other solid melts, the molecules must be pulled apart, freed from their characteristically regular pattern so that they can move around as a liquid. Work must be done in breaking down this regular structure, and therefore energy must be supplied. We usually provide this energy in the form of heat. Here, the heat does the work of overcoming the bonds that hold the molecules in the solid state, rather than changing the temperature of the material. For crystalline solids, this process happens at a particular temperature called the *melting point.* For ice at sea level atmospheric pressure, the melting point is 0°C.

Suppose you have a well-stirred glass of iced water with lots of ice. As this water sits in a warm room, some heat will enter the glass from the room and melt a little of the ice. This heat goes to the work of changing some of the ice to water, *not* to changing its temperature. Thus, the temperature of the *water* after melting is the same as the temperature of the *ice* before melting.

Freezing is just the reverse of melting. Since ice absorbs energy when it melts, water gives up energy when it freezes. If we put our glass of iced water in a freezer, heat will leave the glass to the colder surroundings, causing some of the water to freeze as it gives up this energy.

When ice melts by using heat from its surroundings, the loss of heat cools the surroundings. On the other hand, freezing adds heat to the surroundings.

Question   Why do people sometimes put a tub of water in their fruit cellar on bitterly cold nights?

The heat required to change 1 kg of ice into water is called the **latent heat of fusion,** or just the *heat of fusion,* of water. It has this name because exactly the same amount of heat is released as 1 kg of water fuses to ice. The heat of fusion of water is 80 kcal/kg. That is, it takes 80 kcal of heat to melt 1 kg of ice at a temperature of 0°C. If you want to know the total amount of heat needed to melt a certain mass of ice, multiply the total mass by the heat of fusion. Thus,

$$\text{heat to melt} = \text{mass} \times \text{heat of fusion}.$$

As the same quantity of water freezes, exactly the same amount of heat is given off.

Example 7.1   How much heat must be removed from the water in an ice tray to freeze it, once the temperature is at 0°C. A typical ice tray holds about ½ liter (about a pint), or a mass of ½ kg.

$$\text{heat removed} = \text{mass} \times (\text{heat of fusion})$$

$$= \tfrac{1}{2} \text{ kg} \times 80 \text{ kcal/kg}$$

$$= 40 \text{ kcal}.$$

Exercise   How much heat was removed from the water in the ice tray of Example 7.1 to get its temperature from 18°C to 0°C? Compare your answer with the result of Example 7.1.   **Ans:** 9 kcal

---

Evaporation   You know that a puddle of water slowly evaporates after a rainfall. So does the water in clothes hung on a clothesline,

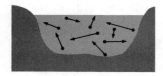

Figure 7.3 The molecules of water do not all have the same kinetic energy. The high-energy ones can escape through the surface to the air above and become vapor.

perspiration on your skin, and alcohol rubbed on your tired feet. The arrows in the water puddle of Fig. 7.3 indicate the directions of motion of a group of molecules at any instant, the length of the arrows being proportional to their speeds. Not all molecules have the same speed—some move very slowly, some very fast, with a continuous distribution in between. Molecules near the surface feel a downward force from the other molecules in the water. But the faster-moving ones have enough kinetic energy to break through this surface layer and escape into the air above. Being widely separated from the others, these molecules are then in the vapor state.

Since only high-energy molecules escape, the *average* energy of those remaining is lower. This condition means that the temperature of the remaining water is lower. *Evaporation is,* therefore, *a cooling process.*

If you've ever spilled gasoline or rubbed alcohol on your skin, it feels very cool. Both of these liquids evaporate very rapidly, leaving the remaining liquid, and your skin, cooler. Your body uses this mechanism to eliminate heat. You are cooled by perspiring only when the perspiration evaporates.

Boiling   As the temperature of a liquid is raised, the average kinetic energy of the molecules increases. Therefore, the fraction of the molecules having enough kinetic energy to get through the surface increases, which in turn increases the rate of evaporation. When the temperature is raised until the average molecular energy is equal to the work needed to separate the molecules of the liquid from each other, *boiling* occurs. The temperature at which this happens is called the *boiling point.* For water at sea level atmospheric pressure, the boiling point is 100°C.

Any heat added to the liquid at the boiling point produces vapor that forms bubbles throughout the liquid. The vapor molecules must be moving fast enough so that the pressure in a bubble exceeds the pressure from the atmosphere and the liquid above it. Otherwise, the bubble would collapse. Boiling starts when this condition is first met.

The heat needed to convert 1 kg of water to 1 kg of steam at the same temperature is called the **heat of vaporization.** At sea level atmospheric pressure, the heat of vaporization of water is 540 kcal/kg. This value is almost 7 times the heat of fusion and over 5 times as much heat as is needed to raise the temperature of the water from 0°C to 100°C. We can calculate total heat, using the heat of vaporization, in the same way we did using the heat of fusion.

Condensation   Condensation is the reverse of boiling. When the temperature of water vapor is at the boiling point, and heat is removed, the vapor goes back to the liquid water state. In doing so, it gives up the heat of vaporization in exactly the same amount as was needed to boil it.

Condensation can also take place at a slower rate, the reverse of evaporation. When water vapor molecules above the liquid hit the liquid surface, they can be captured by this surface and returned into the liquid state. The more vapor there is above the liquid, the faster the rate of condensation.

Example 7.2   A typical steam radiator can hold about 0.1 cubic meter of steam, which would have a mass of about 80 g. How much heat is given up to the room as this much steam condenses to water?

$$\text{Heat lost} = \text{mass} \times (\text{heat of vaporization})$$

$$= 0.080 \text{ kg} \times 540 \text{ kcal/kg}$$

$$= 43 \text{ kcal.}$$

We mentioned earlier that the molecules in a vapor (or gas) are quite far apart compared with their separation in the liquid state. This fact means that when a liquid is boiled, it expands tremendously. For example, for every liter of water boiled at 100°C, you have about 1700 liters of steam generated if the steam is confined and kept at atmospheric pressure. (About ⅔ cup of water can produce the steam in the radiator of Example 7.2.)

Question   Why does an old-fashioned steam heating system need only one pipe running to the radiators (see Fig. 7.4)?

Sublimation   Even in the solid state, a small fraction of the molecules of a substance have enough kinetic energy to break the bonds of the crystal and escape. Some of those near the surface may do so, thereby going directly from the solid to the vapor state. This process is called *sublimation*.

Sublimation is what happens to moth balls as they gradually disappear. At ordinary temperatures and atmospheric pressure, dry ice (solid carbon dioxide) sublimates directly to the vapor state. That's the reason it's called *dry* ice. If you live in a climate where it snows, you've probably noticed that snow gradually disappears even if the temperature never gets to the melting point. This is another example of sublimation.

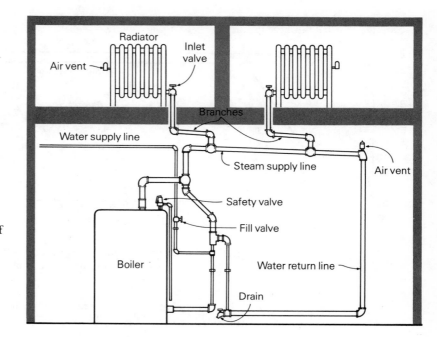

Figure 7.4 A diagram of a steam heating system such as those found in older houses. Notice that in this one-pipe system, the pipes have to slope downward from the radiators.

Question How would you expect the temperature of a moth ball to compare with the temperature of the surrounding air?

7.3 The Energy to Change the Temperature and State of a Substance

We can now summarize the result of adding heat to a substance in any particular state. For definiteness we will use water, but the basic ideas apply to most pure substances.

Suppose you have 1 kg of ice at a temperature of −40°C and you start adding heat to it slowly. The graph of Fig. 7.5 shows the temperature after various amount of heat have been added. From the heats of fusion and vaporization (80 and 540 kcal/kg, respectively), and from the specific heats of ice, water, and steam (0.5, 1.0, and 0.5 kcal/kg·C deg, respectively) you can verify the numbers on the graph. Notice some of the important features:

1. Once the melting or boiling point is reached, the temperature does not rise until all of the substance has changed state. The added energy goes to change the state.

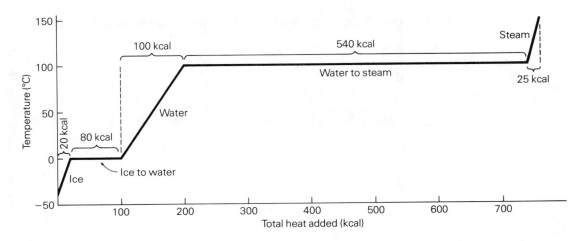

Figure 7.5   A graph showing the variation in temperature with total heat added to 1 kg of ice at −40°C until it is changed to steam at 150°C.

2. The temperature of the ice just as it starts to melt, and the temperature of the water just after it has all melted, are the same—0°C. The same is true of the water and steam just before and just after boiling.

3. By far the greatest part—70%—of the energy is needed to change the state from liquid to vapor. Once the water has vaporized, the steam must be confined in a container in order to add more heat to it.

4. If heat is removed from 150°C steam, the graph looks the same. The substance goes from vapor to liquid to solid with exactly the same quantities of heat being removed as were added in going the other way.

Question   When the outside temperature is dropping and gets to 0°C, why does it often stay at about this temperature for some time before getting any colder?

If a hot substance and a cold substance are mixed, we can use the principle of conservation of energy to determine the overall effect on the system. If no heat is lost to the outside, the heat given up by the hot substance equals the heat gained by the cold substance. Let's illustrate this concept with an example.

Example 7.3   A large scoop of crushed ice at 0°C is added to ½ glass (about 250 g) of tea at a temperature of 80°C. How much of the ice melts?

Since the tea is mostly water, its specific heat will be about the same as that of water. There is plenty of ice, so the tea will come to the temperature of the ice as the ice melts; that is, its temperature changes by 80 C deg.

Heat lost by tea = (mass of tea) × (specific heat) × (temperature change)

$$= 0.25 \text{ kg} \times 1 \text{ kcal/kg} \cdot \text{C deg} \times 80 \text{ C deg}$$

$$= 20 \text{ kcal.}$$

Assuming no heat lost to the outisde, this amount of heat is absorbed by the ice as it melts. Thus,

$$\text{heat gained by ice} = (\text{mass of ice}) \times (\text{heat of fusion of ice});$$

$$20 \text{ kcal} = (\text{mass of ice}) \times 80 \text{ kcal/kg.}$$

Thus,

$$\text{mass of ice} \quad = \frac{20 \text{ kcal}}{80 \text{ kcal/kg}} = \frac{1}{4} \text{ kg} = 250 \text{ g.}$$

7.4   The Unusual Expansion of Freezing Water

When most substances freeze from liquid to solid, the molecules arrange themselves in a crystal lattice so that they are, on the average, closer together than in the liquid state. Such substances, therefore, contract when they freeze and expand when they melt. But for water, just the opposite happens: Water expands when it freezes and contracts when it melts.

If you live in a climate where the temperature goes much below 0°C, you don't leave water pipes exposed to outside temperatures—the water will freeze and burst the pipes. In whatever climate you live, you know that you do not put water or food (mostly water) in a freezer in a glass container—it will expand and break the glass. Or perhaps you have frozen some water in a plastic milk bottle to use on a camping trip, and found that, even though you didn't completely fill the bottle with water, it still overflowed when it froze.

Water expands when it freezes because of the way the molecules arrange themselves to form the ice crystal. The molecules of ice are arranged in a hexagonal lattice (Fig. 7.6). In this arrangement they are farther apart than they are in the liquid state, where they bounce around freely. When ice melts, the

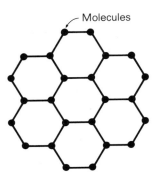

Molecules

Figure 7.6 The molecules of ice are arranged in a hexagonal lattice that takes up more space per molecule than when the molecules bounce around at random in the liquid state.

molecules get closer together and the substance as a whole contracts. On freezing, the water expands as the molecules spread out and form the hexagonal lattice. (If you ever made a Ferris wheel out of Tinker Toys, you certainly know that the wheel took up more space than when the pieces were stuffed into the can they came in.)

The questions at the end of the chapter point out some of the other problems caused by the expansion of freezing water. This behavior leads to an important benefit that we'll consider in Chapter 8: Deep lakes freeze only near the top, thus allowing marine life to continue to live in the water below.

7.5 Changing the Melting and Boiling Points

Several times we have mentioned that the melting and boiling points of water are 0°C and 100°C *at sea level atmospheric pressure.* As this remark implies, these melting and boiling points are different at other pressures. They also change if you add some impurity such as salt to the water.

The Effect of Atmospheric Pressure

Boiling occurs when the molecules of vapor in a bubble formed below the liquid surface have enough kinetic energy to make the pressure in the bubble at least as high as that in the surrounding liquid. As the outside pressure on the liquid increases, the temperature of the liquid must be higher in order to boil. That makes the boiling point increase with increasing pressure, and decrease with decreasing pressure on the liquid.

At high altitudes, where the atmospheric pressure is low, water boils at a lower temperature. Since the rate at which foods cook depends upon their temperature, cooking food in boiling water takes longer as you go higher in altitude. In Sopchoppy, Florida you can hard-boil an egg in 12 minutes. On Mt. Mitchell, North Carolina it takes about 20 minutes, and on Mt. Whitney, California about 35 minutes. On Mt. McKinley, Alaska you might as well give up and have corn flakes.

On the other hand, higher than normal atmospheric pressure increases the boiling point. The steam pressure in a steam heating system is usually at about 1½ times atmospheric pressure, making its temperature about 110°C. A pressure cooker confines the steam until the internal pressure builds up to about twice atmospheric pressure. The water reaches approximately 120°C, thereby bringing the food to that temperature for faster cooking.

Since the energy needed to separate the molecules of a liquid increases with increasing pressure, the heat of vaporization also changes with change in pressure. But for pressures near standard atmospheric, the heat of vaporization varies only slightly.

The pressure on a substance also influences its melting point. As we saw in Section 7.4, water expands when it freezes. If high pressure is placed on ice, it can force the crystals to collapse and melt at a temperature below 0°C.

The reason your ice skates glide so easily over ice is that the narrow blade puts enough pressure on the ice to melt it, letting the skate slide on a layer of water. Once the skate moves off, decreasing the pressure, the water refreezes.

Many of us who live in climates where snow comes periodically, but usually melts within a week or two, often complain that, "People around here just don't know how to drive in snow. Up North they drive on a layer of snow all winter without a bit of trouble." The fault may not always be with the drivers. Driving over snow that is only slightly below the freezing point can cause the snow to melt from the pressure under the tires, forming a slick layer of water. This water then refreezes into solid ice as the tire moves off. The difficulty of removing this layer of packed snow and ice may have some bearing on another of our frequent complaints: "They just don't have the equipment to clear the roads around here like they do up North where it snows all the time." In colder temperatures, the snow remains powdery, which makes it both easier to remove and easier to drive over without slipping.

The Effect of Impurities

Salt or other chemicals are scattered on snow-covered highways to speed the melting of snow and ice. We put antifreeze in our car radiator and windshield washer tank to lower the freezing point and prevent freezing of the water. In general, adding impurities to water changes its melting point.

These impurities affect the freezing by mixing molecules of a different type with the water molecules, thereby impeding the formation of the hexagonal lattice. With additives, the water must be brought to a lower temperature before the crystal lattice forms.

Impurities such as salt or antifreeze also raise the boiling point of water. Water molecules are attracted to those of the salt, and therefore need a higher kinetic energy to escape the liquid.

**The Ice Cream Freezer   The freezing of "homemade" ice cream excellently illustrates several of the processes we have talked about. The ice-cream freezer, (Fig. 7.7) consists of an inner metal container with a fairly large gap between it and the outer bucket. The ice-cream mixture is placed in the inner container and a mixture of ice and salt is placed in the outer.

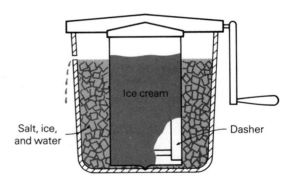

Figure 7.7   An ice-cream freezer illustrates many of the characteristics involved in changes of state between liquid and solid.

For the ice cream to freeze, its temperature must be lowered to the freezing point (around 0°C), and then more heat must be removed to change its state. Pure ice in the outer container would lower the temperature to 0°C, but no more heat would leave the ice cream because it would have the same temperature as the ice. When salt is added to the ice, however, it mixes with the outer layers of the ice chunks, lowering the ice's melting point and causing it to melt. But melting requires heat. This heat is taken from the remaining ice and surrounding water, which reduces their temperature. Heat then passes from the ice cream to this lower-temperature mixture of ice, water, and salt, thus causing the ice cream to freeze. Further melting of the ice absorbs this heat.

From the standpoint of overall energy transfer, the heat given up as the ice cream freezes is absorbed by the ice as it melts. Turning the crank merely turns the inner container while the dasher and outer bucket stay fixed, thereby keeping everything well mixed to speed up the heat transfer and give a smooth texture to the ice cream.

7.6   **Humidity**

The earth's surface contains vast amounts of water, with some of it continually evaporating into the air. But the amount of water vapor in the air, the *humidity,* varies from time to time.

In a pan of water such as the one in Fig. 7.8(a), high-energy molecules near the surface continually escape from the liquid into the air above it. At the same time, many vapor molecules hit the water from above, get captured, and become part of the water again.

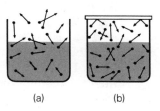

(a)   (b)

Figure 7.8   When the air above a liquid gets saturated, the same number of molecules return as leave the liquid per minute.

If we put a lid on the pan, as in Fig. 7.8(b), the vapor molecules will be trapped, and will increase in density until the number leaving the water equals the number returning. We say that the air is then **saturated**—it holds all the water vapor it *can* hold. If we increase the water temperature, more molecules have enough kinetic energy to leave the liquid, and more water vapor will be in the air when it gets saturated.

The air temperature also limits the amount of water vapor the air can hold. When vapor molecules collide at low speeds, they may stick together forming tiny water droplets. At higher speeds they are more likely to glance off each other and go their separate ways. The higher the temperature, the higher the fraction of molecules *not* sticking together when they collide, and therefore the more water vapor the air can hold. Table 7.1 gives the amount of water vapor in saturated air at various temperatures.

Table 7.1   Water Vapor in Saturated Air

Temperature (°C)	Water Vapor (grams per cubic meter)
−8	2.74
0	4.84
8	8.21
16	13.50
18	15.70
20	17.12
24	21.54
32	33.45

The *absolute humidity* is the actual amount of water vapor in the air at any instant. The term used more often is the **relative humidity,** the ratio of the amount of water vapor in the air to the amount it could hold if saturated at that temperature. For example, if the temperature is 16°C and the air contains 8.0 g/m³ of water vapor, the relative humidity is (8.0 g/m³)/(13.50 g/m³), which is 0.59 or 59%. (The 13.50 comes from Table 7.1.)

Relative humidity tells how fast you can expect evaporation to occur. For example, if the relative humidity is 100%, you know that the rate of condensation is exactly equal to the rate of evaporation, so that there is no net evaporation. On the other hand, if the relative humidity is quite low, say 10%, you can expect rapid evaporation. This evaporation rate has a pronounced effect on your comfort. If it's a hot day and the humidity is low, perspiration evaporates quickly from your body, using up body heat to change the state of the perspiration. You can then be quite comfortable at fairly high temperatures. But if it's a humid (muggy) day, even moderately high temperatures make you uncomfortable because little net evaporation of perspiration occurs. Perspiration may pour off your body, but unless it evaporates, it doesn't cool you. That's why an air conditioner needs to remove water vapor as well as heat from the room air.

Question   Why does water drip from the outside end of a working room air conditioner, especially on a humid day?

In winter, the problem is usually too *little* humidity inside the house. If the humidity is low, any perspiration evaporates immediately, taking heat from your body. Therefore, a higher temperature is needed for comfort. But your comfort is affected in another important way by low humidity. Rapid evaporation can cause excessive drying of mucous membranes in the nose and throat. The resulting swelling and irritation can cause sore throats for anyone, and even worse difficulties for people with respiratory problems. That's why doctors recommend the use of humidifiers during certain illnesses and for people with allergy problems. Around 35 to 45% relative humidity is considered best.

But why *is* the inside humidity so low in winter? Except for moisture added inside the house, the actual amount of water vapor in a cubic meter of air inside is the same as that in a cubic meter of air outside. Even if the relative humidity is high outside, the higher inside temperature means a much lower relative humidity inside. For each 10 C deg we increase the air temperature by heating, the relative humidity is approximately cut in half. Let's illustrate with a specific example.

Example 7.4   Suppose the outside temperature and relative humidity are −8°C and 70%, respectively. For the same absolute humidity, what is the relative humidity inside a house where the temperature is 20°C (68°F)?

Table 7.1 gives the water vapor in saturated air at −8°C as 2.74 g/m³. At 70% relative humidity,

$$\text{absolute humidity} = 0.70 \times 2.74 \text{ g/m}^3 = 1.92 \text{ g/m}^3.$$

The absolute humidity inside is the same. But at 20°C, the inside air can hold 17.12 g/m³ of vapor. Therefore, inside

$$\text{relative humidity} = \frac{\text{absolute humidity}}{\text{absolute humidity at saturation}}$$

$$= \frac{1.92 \text{ g/m}^3}{17.12 \text{ g/m}^3}$$

$$= 0.11, \text{ or } 11\%.$$

Question   Why does the use of a "cool-mist" humidifier in a room cause the room to cool somewhat? (A cool-mist humidifier, the kind most doctors recommend, sprays tiny droplets of *room temperature water* into the air.)

Dew and Fog   You are no doubt familiar with the dew often found on grass and elsewhere outdoors in the morning. Where does dew come from?

Suppose that during the day the outside temperature is 32°C and the relative humidity is 40%. The air then contains 13.4 g/m³ of water vapor. If this amount of water vapor does not change much as the temperature drops at night, the air will be completely saturated (100% relative humidity) when the temperature falls to 16°C. If the temperature drops more, the air must give up some of its humidity. Some of the water vapor condenses onto grass and other cool objects. The temperature at which dew starts to form is called the *dew point*. The dew point depends upon the actual amount of water vapor in the air.

If the temperature of the cool surfaces is below the freezing point before the air gets saturated, the vapor sublimates out as frost rather than dew.

Question   Why does frost build up on the inside of a freezer or the freezing compartment of a refrigerator? Does the rate of frost buildup have anything to do with how often the door is opened? (You don't see frost in a "frost-free" model because the coils are periodically heated enough to melt the frost, but not enough to thaw the food.)

When warm moist air moves into a cool region, the air may be quickly cooled below its dew point. The vapor then condenses into little water droplets that we call *fog*. This process often happens as a warm breeze blows from a large body of water over the cooler land. If the cooling and condensation occurs high in the earth's atmosphere, we call the result a *cloud*.

Key Concepts

In a **solid,** the molecules are held tightly in a fixed average position relative to each other. They vibrate back and forth around this average position. In a **liquid,** the force of attraction is weaker so that the molecules bounce around at random throughout the liquid. In a **gas** (vapor), the molecules are far apart and experience little force except during collisions with other molecules.

Energy must be added in the form of heat to change a solid to a liquid, and a liquid to a vapor. The same amount of energy is removed in the reverse processes. The **(latent) heat of fusion** is the heat needed to melt 1 kg of a substance, and is the heat given up as 1 kg of the substance freezes. The **heat of vaporization** is the heat needed to change 1 kg of a liquid substance to vapor, and is the amount of heat given up as 1 kg of the substance condenses.

Evaporation is a cooling process because high-energy molecules escape the liquid and become vapor molecules. **Boiling** occurs when the average molecular energy equals the work needed to separate the liquid molecules from each other. **Condensation** is a change from the vapor to the liquid state. **Sublimation** is a change directly from the solid to vapor state, or vice versa.

Water expands when it freezes, and contracts when it melts.

Melting and boiling points depend on the pressure on the substance, and on the amount of impurities in the substance. The higher the pressure, the higher the boiling point. For water, which expands as it freezes, increased pressure decreases the melting point.

Relative humidity is the ratio of the amount of water vapor in the air to the amount it can hold at saturation. The air is **saturated** when the rate at which molecules go from the liquid to the vapor state equals the rate at which they return to the liquid state. The amount of water vapor in the air at saturation increases with increasing air temperature.

Questions

Change of State

1. What happens to the energy that is added to ice to melt it?

2. Is energy absorbed or released as water freezes?

3. Explain why evaporation is a cooling process.

4. True or false: When we perspire, our bodies are cooled because the perspiration gives up heat as it evaporates.

5. Grandpa dips his shirt in the creek and wears it wet while working in the hot corn field. His "city slicker" son tells him he's fooling himself because the wet shirt just gets hot and doesn't keep him any cooler. Is the son right?

6. Wet tennis courts will dry much quicker if you sweep the water out of the puddles and spread it out. Why?

7. Is boiling merely fast evaporation?

8. When water is boiling, does boiling it faster make it any hotter? Does this suggest a way to conserve useable energy while cooking?

9. What is the stuff you see rising from a pan of boiling water? (It is not steam; steam is invisible.) Why does it disappear after rising a small distance?

10. Why does coffee rise in a percolator stem?

11. Why can you see your breath in cold weather? Why can you not see it in warm weather?

12. Does snow have to melt to go away after it falls on the ground?

13. Why do moth balls get smaller with age?

Changes of Temperature and State

14. Why does it take so much longer to boil away a pan of water than to raise its temperature to the boiling point?

15. A CIA agent is being tortured by a foreign government in an attempt to extract national secrets from him. He is given a choice of having his hand placed for 2 seconds in boiling water or in steam at the same temperature. Which would you advise him to choose, and why? (One choice will be much worse.)

16. Some people claim that you get ice quicker if you start with hot water rather than cold. Do you believe this claim? (See the Kell reference.)

17. A heavy snowfall occurred in a certain southern town unaccustomed to snow. The town is unequipped to remove it, so the town officials decided to hose it off the streets with water. They ended up with a solid sheet of ice on the streets. Are there weather conditions under which you would expect such a scheme to work?

Freezing of Water

18. Why is it important, when pouring the concrete footings making up the foundation of a house, that the footing bottoms be underground, below the frost line? (The frost line is the lowest depth at which the water in the ground freezes.)

19. If a door swings with very little clearance above an outdoor concrete walkway, why would such a door hit the concrete and not open in extremely cold weather?

20. Why does ice float at the top of a glass of water?

Changing the Melting and Boiling Points

21. True or false: Food cooks faster in a pressure cooker because the increased pressure increases the speed of the cooking process.

22. Is it more important that you take your pressure cooker with you on a camping trip to the beach or to the high mountains?

23. Why is it possible to make a snowball? Why is it *not* possible to make a snowball if the snow is too cold?

24. Why is ice skating not possible in extremely cold temperatures?

25. Why can't you skate on glass, since it is much smoother than ice?

26. Special waxes are put on snow skis to make them slide easier. Why is such waxing more important at low temperatures than when the temperature is near the melting point?

27. Why does snow squeak when you walk on it if the temperature is extremely cold? It happens at about −25°C or colder.

28. Why does antifreeze keep the water in a car radiator from freezing?

29. Could you use antifreeze in place of salt in making homemade ice cream?

Humidity

30. Why is a humidifier needed so much more in the winter than in the summer?

31. If you take a shower in a chilly house, why are you warmer if you stay in the steamy bathroom to dry off than if you step into the next room, which is at the same temperature? Why do you *not* feel so much difference once you're dry?

32. Why does a cold glass of lemonade sweat?

33. Why do the insides of windows sometimes get wet in winter, especially if they are single thicknesses of glass without storm windows?

34. House insulation usually has a vapor barrier on one side that water vapor cannot penetrate. Why should this barrier be placed on the interior side of the insulation?

35. The electric fan was still running in the first floor apartment when the landlady found the body. The police know the murder occurred either July 25 or July 26. Sherlock Holmes said that, because the humidity was 18% on July 25 and 98% on July 26, he was sure he knew the correct day. Does he have grounds for being so sure?

Problems

1. How much heat has to be removed from 50 kg of hamburger to freeze it once its temperature has reached the freezing point? The heat of fusion of hamburger is about 50 kcal/kg. Compare your result with the result of Example 6.2 for the heat removed in lowering the temperature to the freezing point.   **Ans:** 2500 kcal

2. As much as 40 million metric tons of snow can fall in a snowstorm. How much energy is needed to melt this snow? If it takes 30 000 kWh of heat to heat a house in winter, how many such houses could be heated with this amount of energy? (1 metric ton = 1000 kg; 1 kcal = 0.00116 kWh.)

    Ans: 3.2×10^{12} kcal; 124 000

3. How much heat is absorbed as a 30-g ice cube melts?

    Ans: 2.4 kcal

4. How much heat is needed to melt the iron that goes into one 200-kg automobile engine, once the iron has reached the melting point? The heat of fusion of iron is 12 kcal/kg.

    Ans: 2400 kcal

5. How much heat is needed to boil away 1 pint of water (350 g), once the water has reached the boiling point?   **Ans:** 290 kcal

6. When the baby fell off the bed, the harried father forgot the potatoes he had left boiling in a pan with 2 liters of water. If the stove supplies 30 kcal/min of heat, how long will it take the pan to boil dry?   **Ans:** 36 min

7. Suppose 5 cm of rain falls in one week on the state of Rhode Island, an area of 3.1 billion square meters. How much energy is released in the clouds as water vapor condenses to form this rain? The people of Rhode Island consume about 100 billion kWh of useable energy for all purposes in one year. How does this compare with the energy released during the 5-cm rainfall?   **Ans:** 8.4×10^{13} kcal or 97 billion kWh

8. During the night a "cool-mist" humidifier sprays 6 liters of water into the air in the form of tiny water droplets. How much heat is taken from the air as this water vaporizes? As a result, how much would the temperature of the air drop in a room 4 m by 4 m by 2.5 m, *if no heat were added or removed* from the air? (The density and specific heat of air are 1.2 kg/m^3 and 0.24 kcal/kg·C deg.)   **Ans:** 3200 kcal; 280 C deg

9. How much heat is needed to change 40 kg of ice at −20°C to steam at 130°C? (The specific heat of steam is 0.5 kcal/kg·C deg.) How much heat is removed in changing it back to ice at −20°C?   **Ans:** 30 000 kcal

10. Suppose that for your Fourth-of-July picnic with your friends, you put twenty-four 280-g bottles of Coke in an insulated tub and dump in lots of ice at 0°C. If the Coke temperature is 30°C, how much ice will be melted in bringing the Coke down to a temperature of 0°C? (Neglect the heat absorbed by the bottles and the tub, and take the specific heat of Coke to be about the same as that of water).   **Ans:** 2.5 kg

11. Assume a wooden tub in Problem 10 and make some reasonable assumptions about the mass of the tub and bottles. Then estimate their effect on the amount of ice melted. (See Table 6.1, Chapter 6.)

12. If, at a temperature of 24°C, the air contains 13.5 g/m³ of water vapor, what is the relative humidity? What will the relative humidity be if the temperature drops to 20°C? At what temperature will dew start to form?   **Ans:** 63%; 79%; 16°C

13. If a glass of iced lemonade is not to sweat, what is the highest possible value of the relative humidity when the temperature is 24°C?   **Ans:** 22%

14. If the outside relative humidity and temperature are 60% and −8°C, what is the relative humidity inside a house where the temperature is 24°C and no extra water vapor is added inside? What is the relative humidity inside if the temperature is 18°C?

    Ans: 7.6%; 10.5%

15. The dew point is 18°C when the temperature is 32°C. What is the relative humidity?   **Ans:** 47%

Home Experiments

1. Pour some water into an ordinary paper cup and hold it over the flame of a candle. You can hold it there until the water boils and the cup will not burn. Why?

2. Tie a very thin wire around a block of ice and suspend the block by the wire. The wire will work its way through the ice, melting from the pressure above the wire, refreezing below the wire as the pressure is relieved. The wire will pass completely

through the ice, leaving the block intact. (You can get a block of ice by freezing water in an old plastic or paper milk carton. Before starting the experiment leave the ice at room temperature until the outside starts to melt. This process will insure that the ice is near the melting point and the experiment will go faster.)

3. Place an ice cube in a glass of water. Wet one end of a thread or string and lay it across the ice cube. Sprinkle some salt on top of the cube. After a few seconds, lift the string by the end you did *not* wet and you will lift the ice cube out of the water. Why does this work?

References for Further Reading

Kell, G. S., "The Freezing of Hot and Cold Water," *American Journal of Physics,* May 1969, p. 564. A study of which freezes faster, hot or cold water.

Plumb, R. C., "Squeak, Skid and Glide—The Unusual Properties of Snow and Ice," *Journal of Chemical Education,* March 1972, p. 179. Why snow and ice behave as they do when very cold.

Pounder, E. R., *The Physics of Ice* (Pergamon Press, New York, 1965).

Thompson, D. T. and R. O'Brien, *Weather* (Time-Life Books, New York, 1968). Beautifully illustrated with color photographs and overlays that supplement the description of all aspects of weather.

Tufty, B., *1001 Questions Answered about Storms and Other Natural Air Disasters* (Dodd, Mead, & Co., New York, 1970). Many of these questions involve everyday changes in the states of matter.

8

Transfer of Heat

When you stand in front of an open fireplace, you expect heat from the fireplace to warm your body. When you put a pot of beans on the stove to cook, you expect heat to go from the burner into the beans. When you sit in front of a fan, you expect to be cooled as heat leaves your body.

In many of our daily activities we expect heat to be transferred from one place to another. But how does it get transferred? What happens within objects, and between objects, that causes heat to move from one place to another?

Heat may be transferred by three different mechanisms that we call *conduction, convection,* and *radiation.* We will discuss each of these mechanisms separately, with examples from our everyday lives, and then show how, in most situations, heat travels by a combination of the three means.

8.1 Conduction

If you've ever scrambled eggs in a frying pan with a metal handle, you have probably found that, by the time the eggs are done, the handle is too hot to hold in your bare hand. Or you may have held the end of a poker in a fire for a while, and found that the end you were holding became hot. In both of these examples, heat traveled through the metal by a method called **conduction.**

As Fig. 8.1 shows, when one part of an object is held in a hot region, such as in a flame, the molecules in that part of the object gain kinetic energy; that is, the part held in the heat gets warmer. These molecules then collide harder with their neighbors, passing some of their extra kinetic energy to them. These neighboring

Figure 8.1 The molecules in the end of the rod in the flame gain extra energy, which is passed on to their neighbors as they collide with them.

molecules in turn pass kinetic energy on to *their* neighbors on the other side in a "bucket brigade" fashion. In this way, heat is transferred—conducted—from the warmer to the cooler part of the object. *Conduction is the transfer of heat by means of collisions from molecule to molecule in the material.*

In Chapter 6 when we described what happens in a situation such as the one in Fig. 6.4, where two bodies at different temperatures are in contact, we were describing the process of conduction. Conduction can take place not only within a body whose parts are at different temperatures, but between two contacting bodies at different temperatures. The better the contact (that is, the greater the area actually touching), the more heat will be conducted.

The ability of materials to conduct heat varies widely from substance to substance. This ability is described by a property called **thermal conductivity.** The higher its thermal conductivity, the better a substance can conduct heat. Table 8.1 gives the thermal conductivities of a few ordinary materials. We'll talk about what the units mean in a moment; for now, just compare the relative size of the thermal conductivities.

Generally, solids are better conductors than liquids or gases. The reason for this situation is that the molecules in solids are closer together and interact strongly with each other. Gases are poor conductors because the large distances between molecules permit fewer collisions.

Table 8.1   Thermal Conductivities of Materials

Material	Thermal conductivity, k (kcal/m·s·C deg)	Material	Thermal conductivity, k (kcal/m·s·C deg)
silver	0.10	brick	0.00015
copper	0.092	pine wood	0.000028
aluminum	0.051	plaster	0.0001
stainless steel	0.0033	ice	0.00052
ordinary steel and cast iron	0.011	snow (compact)	0.000051
glass (window)	0.00025	water	0.00014
porcelain	0.00025	air	0.0000055

As Table 8.1 shows, thermal conductivities vary widely, even among solids. In general, metals are better conductors than nonmetals because the atoms of metals have free (valence) electrons that can move easily throughout the body. Experiments show that the electrons rather than the molecules as a whole are responsible for most of the conductivity of metals.

Question   If you were concerned only with getting heat most efficiently from the stove burner into the food, which of the materials in Table 8.1 would you choose for constructing a frying pan?

    The amount of heat conducted through a body depends upon more than just its thermal conductivity. Figure 8.2 illustrates the factors that influence the rate of heat flow. By rate of heat flow, we mean the amount of heat flowing through the object per unit of time (for example, kcal/s). We are not talking about how fast the heat actually moves through the material, but *how much* heat gets through in a given amount of time. (For comparison: When you're running water in your bathtub, you don't worry about how fast the water molecules are moving through the pipe on their way to the faucet; you are only concerned with how much water comes out of the faucet per unit time.)

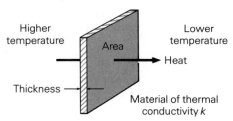

Figure 8.2   The rate of conduction of heat through a material depends upon all the factors shown here.

    In Fig. 8.2 the slab of material of thermal conductivity k separates a high-temperature region from a low-temperature region. If you think of the conductor as a sort of pipe that carries heat, you can probably predict correctly how the amount of heat conducted depends on the factors shown in the figure. You probably would expect the rate to be higher for a bigger temperature difference across the slab, and for a larger area. Also, you probably would expect the rate to be lower the farther the heat must travel— meaning, the thicker the slab is. If so, your expectations are all correct, and can be put in equation form as

$$\text{heat conducted per second} = \frac{k \times \text{area} \times (\text{temperature difference})}{\text{thickness}}.$$

Think about this equation and convince yourself that increasing any of the factors on the right has the expected effect on the heat conduction.

**An Illustration: Design of Cooking Utensils   Cooking utensils hold food during cooking and conduct heat from the stove burner into the food. The chemical changes involved in cooking take place only at high temperatures. We will pose a few questions for you to answer on the basis of what we have said about heat conduction.

1. How would doubling the thickness of the bottom of an aluminum pan influence the rate of heat transfer through the bottom?

2. Since stainless steel is much stronger than aluminum, stainless steel pans can be made thinner than aluminum ones. How much thicker could you make an aluminum pan and still have the same rate of heat conduction through the bottom?

3. Most stove burners are hotter in some spots than others. To achieve uniform heat throughout the inside of the pot, should the bottom be thick or thin? Should the conductivity of the material be high or low?

4. Why do some manufacturers put a layer of copper over the bottoms of their stainless steel pans?

5. The specific heat (see Chapter 6) of aluminum is about twice that of cast iron, but cast iron is about three times as dense as aluminum. Which would stay hot longer—a cast iron or an aluminum pot?

6. If a pan is to be used on the perfectly flat heating element of an electric range, how will any nonflatness of the pan bottom affect the rate of heat transfer?

Calculations of Heat Conduction   We will work an example that illustrates the use of our heat conduction equation for calculating the rate of heat transfer. First, let's take another look at the meaning of the thermal conductivity k. We said that the rate of heat transfer is proportional to the thermal conductivity. In the units in Table 8.1, the thermal conductivity is the amount of heat in kcal conducted in 1 s through an area of 1 m² of 1-m-thick material across which there is a temperature difference of 1 C deg. To get the total heat transferred per second, we multiply the total area and temperature difference, and divide by the thickness.

Example 8.1   A window 120 cm by 180 cm is made of glass 2.0 mm thick. If the inside surface of the window is at 20°C (68°F) and the outside surface is at −18°C (0°F), what is the rate of heat flow through the window?

$$\text{Heat conducted per second} = \frac{k \times \text{area} \times (\text{temperature difference})}{\text{thickness}}$$

$$= \frac{0.00025 \text{ kcal/m} \cdot \text{s} \cdot \text{C deg} \times 1.2 \text{ m} \times 1.8 \text{ m} \times 38 \text{ C deg}}{0.0020 \text{ m}}$$

$$= 10 \text{ kcal/s}.$$

Based on the answer in Example 8.1, this window would conduct about 290 000 kcal or 340 kWh of heat to the outside in 8 hours. At a price of 5¢/kWh, that would cost $17 per window, or $340 per 8 hours in a house with 20 windows! Obviously, something is amiss in what we did.

Even though your hand is an unreliable thermometer, it's a good enough instrument to give you a hint at where we went astray. If you touch the inside of a single-glass window on a cold day, it is definitely much colder than the room temperature. In the example, our implicit assumption that the glass surfaces were at the same temperature as the inside and outside air is not valid. The air next to the window on both sides must transfer heat to and from the window, and air is a poor conductor of heat. The inside and outside glass surfaces differ little in temperature, and are somewhere near the average of the inside and outside air temperatures. However, the glass temperatures and the overall heat transfer rate depend strongly on how much the air itself moves, bringing cold or warm air into the vicinity of the window. That brings us to our next method of heat transfer: *convection*.

8.2   Convection

You probably have heard ever since you were in elementary school that "hot air always rises." Let's examine this "old wives' tale" and see why it's a distortion of the facts.

Consider a room that has a heater on one wall and a window on the opposite wall (Fig. 8.3). If the outside air is cold, heat will be conducted through the window, cooling the air nearby. As this air cools, it will contract and become more dense—a given volume will weigh more. This cooler, heavier air will push aside the warmer air below it and move down to exchange places with the warmer air. Thus, the warmer air will be *pushed* up by the cooler air.

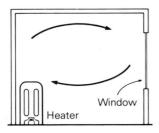

Figure 8.3 Heat is transferred by convection when warmer air is forced up by the cooler air that is more dense.

On the other side of the room, where the heater is, the air near the heater is being warmed. Here, therefore, the air expands so that a given volume is lighter. This lighter air is then *pushed* up by the colder air above it that comes down to take its place. In both cases, the warmer air does *not rise* of its own accord. It still has weight and is being pulled toward the earth. But the cooler air has more weight per unit volume and bullies its way in to be closer to the earth. Hot air rises *only if* there is cooler air above it, and then only because it is pushed up by the cooler air.

As a result, there is a natural circulation of air in the direction of the arrows in Fig. 8.3. The lower part of the room always contains cooler air, the upper part, warmer air. A room arranged this way can easily have a temperature difference of 10 C deg (18 F deg) between head and foot level.

Question   Where should the heater be placed for the most even distribution of temperature in the room? Why? Hint: Where are heaters usually located?

The process of heat transfer we have described is an example of **convection.** Convection is a transfer of heat by movement of the heated medium itself—in this case, air.

A similar process happens in liquids. Suppose a flame is held under one side of a pan of water as in Fig. 8.4. When the water directly above the flame is heated by conduction and expands, it is forced up by the cooler denser water. Thus, heat is transferred from the bottom to the top of the pan by convection.

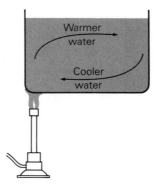

Figure 8.4 Any fluid— liquid or gas—can transfer heat by convection. Here the cooler, denser water pushes the warmer water to the top.

Many houses built 30 to 40 years ago used this natural convection effect to distribute heat to "radiators" placed throughout the house. In Fig. 8.5, water heated in the boiler is less dense and is

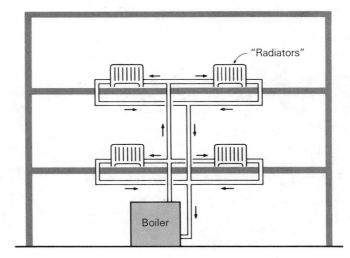

Figure 8.5 A gravity-fed hot water heating system. The arrows indicate the direction of water flow.

forced up into the "radiators" by the denser water that has lost its heat to the air in the rooms. Heat is distributed throughout the room by convection in the room air.

Rather than relying purely on the natural convection from the expansion of heated fluid, heating systems sometimes rely on forced convection. For example, hot water systems installed now usually include a circulator to pump the water to, and through, the room heaters and back to the boiler for reheating. Hot-air central heating systems usually have a fan that blows air from the furnace through ducts to the rooms. Return ducts bring the cooled air back to the furnace.

Movement of air in the earth's atmosphere—winds—are merely convection currents. The sun heats the earth more in some places than in others. As the air above these places gets warmed (by conduction), convection causes a redistribution of air. The cool daytime breeze coming in from large bodies of water is caused by convection. Since the land is warmer than the water, the air above the land gets warm. As Fig. 8.6 shows, the cooler air above the water moves in and replaces this warmer air.

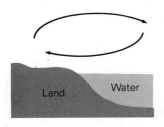

Figure 8.6 Convection currents between the warmer land and the cooler water cause the cool daytime breeze felt near large bodies of water.

**The Yankees Knew about Convection—a Civil War Example
In the Battle of Petersburg, Virginia—the last large battle of the Civil War before General Lee surrendered to General Grant at Appomattox—the Union and Confederate armies had been at a standoff for months. In an attempt to penetrate the Confederate lines, the Forty-eighth Pennsylvania Regiment, made up mainly of coal miners, dug a 511-foot tunnel that ended under the Confed-

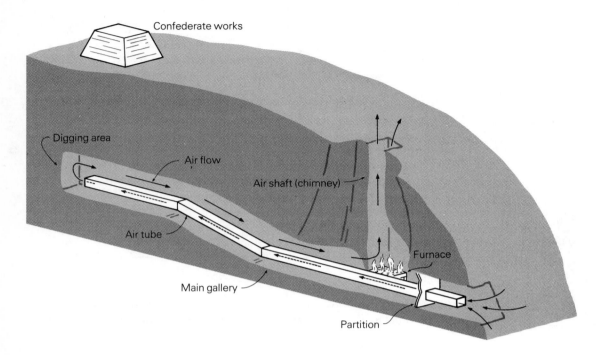

Figure 8.7   The tunnel dug during the Battle of Petersburg, Virginia, during the Civil War was ventilated by convection currents from a fire built near the entrance. (Courtesy of National Park Service, Petersburg National Battlefield.)

erate Pegram's Battery (Fig. 8.7). They also dug two 38-foot branches to the right and left at the end, and packed them with 8000 lb of powder. On July 30, 1864, 35 days after the project began, they exploded the powder, completely destroying Pegram's Battery. According to Col. Henry Pleasants, commander of the Forty-eighth, "the crater formed by the explosion was at least 200 feet long, 50 feet wide, and 25 feet deep."*

While digging the tunnel, the soldiers needed some means to get fresh air to the digging area. To get fresh air, they ran an 8-inch square wooden tube from the entrance to near the end of the tunnel, and extended it as the tunnel was lengthened. They sealed the entrance, except for the tube opening, with an airtight partition. Beyond this partition, they built a vertical chimney to the surface, and kept a fire burning below the chimney (see Fig. 8.7).

_____

* From the *Official Records of the War of Rebellion,* Vol. 40.

Convection currents from the heated air forced fresh air through the tube to the far end of the tunnel, with air from there traveling back to the fire and up the chimney.*

The convection currents set up in this tunnel "fireplace" are like those from a home fireplace. The fire gets fresh air needed for combustion from convection, which brings in new air to replace the heated air and combustion products that are pushed up the chimney (Fig. 8.8). A fireplace passes up to 17 cubic meters of air per minute—enough to empty a 5 m by 6 m (about 15 ft by 20 ft) room in 5 minutes. Unless there is a special vent to the outside, this air must come from the room which, in turn, must gets its air from the outside through (usually small) cracks around windows, doors, and other places in the house. A fireplace could not operate in a completely leakproof house.

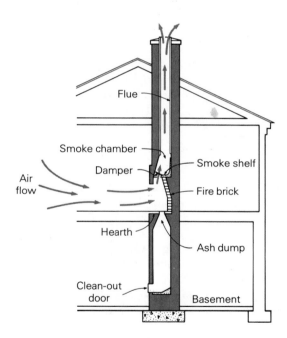

Figure 8.8 A fireplace needs convection currents for the fire to burn.

---

* The results of the effort were disastrous for both sides. Not only were hundreds of Confederates killed during the explosion but, during the melee that followed, masses of Union soldiers, horses, and equipment poured up the hill and tumbled into the crater left by the explosion, where they were either crushed by others falling in on top of them or later fell easy prey to Confederate guns.

Question   Suppose a fireplace gets its air by infiltration from the outside through small cracks and openings throughout the house, as most do. If the thermostat for the rest of the house is set to its usual value, so that the central heating system must heat this air before it gets to the room with the fireplace, is the homeowner likely to save on his fuel bill by using his fireplace? (Notice that the draft doesn't stop when the fire dies down. Much of it continues to flow as long as the fireplace damper is open.)

Exercise   If you haven't worked Problem 9, Chapter 6, now would be a good time to work it.

8.3   Combinations of Conduction and Convection

Many important heat transfer processes are combinations of conduction and convection. We've already mentioned some of these in our comments on convection. The air touching a cold window is cooled by conduction of heat to the glass. In Fig. 8.4, the heat passes through the container and into the water above by conduction. In the "radiators" of Fig. 8.5, heat passes from the water in the radiator to the air outside by conduction. Warm earth heats the cooler air above it by conduction.

Once heat travels by conduction from a solid to a fluid, convection usually dominates the transmission in the fluid. But, as we discussed earlier, the rate of conduction between the solid and the fluid depends on the temperature difference between the two. Look again at Example 8.1, and the comments concerning why the answer for the heat transmitted through a window was much too high. Once the air near the window gets warmer, less heat will be conducted to it because of the smaller temperature difference between window and air. As that air moves away by convection, cooler air takes its place to again increase the rate of conduction.

This effect is greatly increased if the wind is blowing, because large quantities of cool air are brought in contact with the window to absorb its heat by conduction. Therefore, aside from the increase in the leaking of cold air through the cracks around doors and windows when the wind blows, more heat also is lost by conduction. Adding a storm window typically cuts the heat transfer rate about in half by providing a layer of air between the two sheets of glass that confines convection currents to that space. Conduction in the air is extremely slow.

A more personal example demonstrates this same effect. Your

body is usually warmer than the surrounding air, so it loses heat by conduction to the air. If the air is still, a layer of warmed air surrounds your body and reduces the rate of conduction. In a breeze, this warm air is swept away and cool air brought in to pick up heat.

The harder the wind blows, the more heat can be transmitted. Wind chill factors, often given in the weather report in winter, are a measure of the effect we are describing. The *wind chill factor* is the temperature in still air that has the same cooling effect on the body as that particular combination of wind and temperature. Table 8.2 gives some wind chill factors for various temperatures and wind speeds. Wind speeds above 60 km/h have little additional effect.

Table 8.2 Wind Chill Factors (equivalent temperatures in °C)

Air Temperature (°C)	Wind Speed (km/h)				
	10	20	30	40	60
0	−3	−9	−13	−16	−19
−5	−8	−16	−20	−23	−27
−10	−13	−21	−26	−30	−34
−20	−25	−35	−42	−46	−59
−40	−47	−61	−69	−76	−81

**Insulation   Why does insulating a house result in less heat loss through the walls? Figure 8.9(a) shows a cross section of the wall of a house that is not insulated. For most houses, the walls have a space that is hollow except for studs placed every 16 or 24 inches. The air in this space next to the warmer side is heated; the air next to the cooler side is cooled. The result is a convection current in the direction of the arrows, carrying heat from the warmer to cooler side. The heat transfer from convection in this air space is many times more important than the transfer from conduction.

Figure 8.9 Cross section of uninsulated and insulated wall. When insulation is placed in a wall, heat transfer by convection is greatly reduced because air can circulate only in small pockets.

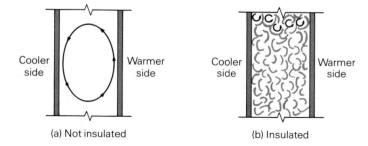

The purpose of insulation is to reduce convection. When the air space is insulated, it is broken up into small pockets, as shown in Fig. 8.9(b). Heat transfer by convection is drastically reduced because it can occur then only in small loops, as shown by the arrows near the top of the figure. The effect of conduction stays small because the insulation is fluffy, leaving the space still filled mostly with air, which has low conductivity.

Question   Why does wearing thick clothes keep you warm?

Commercially available insulation is rated with an "R" value that stands for resistance to the flow of heat. For example, 6-inch-thick fiberglass is rated R-19, as marked on the bundle. The higher its R value, the more effective the insulation.

The R rating basically combines the thickness and thermal conductivity factors in our conduction equation. The effect of any convection is also lumped into the R value. The total rate of heat transfer is

$$\text{heat transfer rate} \quad = \frac{\text{area} \times \text{temperature difference}}{\text{R rating}}.$$

Commercial R ratings like the R-19 are in units of $(\text{ft}^2 \cdot \text{F deg})/(\text{Btu/h})$. To use R ratings directly, you need to express area in square feet and temperature difference in Fahrenheit degrees. The heat transfer rate then comes out in Btu/h.

Conduction, Convection, and Freezing Lakes

The low thermal conductivity of ice and water, and the natural convection in water play an important role in the freezing, or lack of freezing, of deep lakes. To understand what happens, we first need to know the expansion properties of water. In Chapter 7 we discussed the fact that water expands when it freezes. But water *above* the freezing point has unusual expansion properties also.

Figure 8.10 shows how the volume of 1 kg of water changes with temperature between 0° and 100°C. The inset for the low-temperature region shows that the volume is least at 4°C. Above that temperature, water expands with temperature rise, as do most substances. But below 4°C, water expands with temperature *drop*. As water freezes, it expands even more, so that the volume of 1 kg of ice is larger than that of 1 kg of water at 0°C.

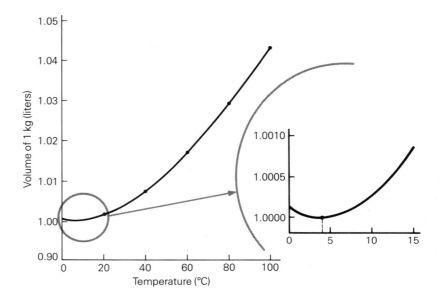

Figure 8.10 The variation in volume of 1 kg of water between 0° and 100°C. The low temperature region is shown in detail in the inset. Notice that the zero of the volume scale would be far below the bottom of the page, so that the overall fractional change in volume between 0° and 100° is only about 4%.

As we said in Chapter 7, the collapse of the hexagonal crystal structure of ice (Fig. 7.6) causes the contraction that water undergoes when it melts. But after melting, certain groups of molecules keep this hexagonal structure. As the temperature rises, more and more of the structures collapse. At 4°C the contracting effect of the collapsing crystals is overcome by the ordinary expansion due to higher molecular kinetic energy. At higher temperatures the expansion has the larger effect.

If you were a fish, you would be extremely glad that water behaves as it does near the freezing point. Suppose the air above the deep lake in Fig. 8.11(a) is warm—say 35°C—and that the temperature of the water near the top is only a few degrees lower. As the air temperature drops below the water temperature, the air cools the water by conduction. As the surface water cools, it contracts. Being more dense, it sinks, pushing the warmer water to the top where it gets cooled. As the air temperature falls to near 0°C, this convection process continues until the whole lake is at 4°C. Further cooling of the water at the top causes it to expand. The

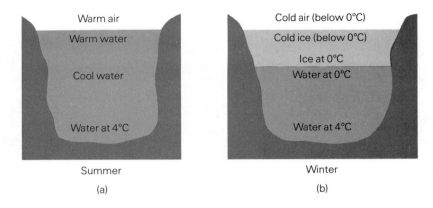

Figure 8.11   Even in extremely cold climates, deep lakes do not freeze all the way to the bottom. The water in the lower part stays at 4°C year-round.

water is then less dense—that is, a given volume is lighter—than the water below it, so it stays on the top.

As the water temperature drops below the freezing point of water, the water at the surface freezes. In freezing it expands more, becoming less dense still and therefore staying on top as shown in Fig. 8.11(b). The water just below the ice is colder than that further down. It, therefore, can freeze only by losing heat in conduction through the ice to the cold air above. But ice is a poor conductor of heat. Therefore, the thickness of the ice layer increases very slowly. If the lake is deep, the heat cannot transfer through the ice fast enough to freeze the entire lake before spring comes and the ice starts to melt. In fact, the bottom portion remains at 4°C, the temperature at which the water is most dense.

In summer, as the temperature of the water at the surface goes above 4°C, it expands and stays on top. To warm the lower water, heat must be conducted through this upper layer of water. For a deep lake the bottom then stays at 4°C year-round. Water either warmer or colder than this is less dense and therefore moves no heat downward by convection. Only conduction occurs, and water is an exceptionally poor conductor of heat.

Questions   1. If water continued to contract on down to the freezing point and also contracted upon freezing, why would all lakes freeze solid once the temperature above them got below the freezing point?

2. What would have been the effect on the evolution of animal life if water had not been designed to have this anomalous expansion property?

8.4 **Radiation**

The two methods of heat transfer we have talked about so far both involve a medium. In conduction, heat is transferred through a stationary medium by the collisions of the molecules. In convection, heat is transferred by movement of the heated medium itself.

But heat is transmitted sometimes without a medium. The most obvious example is heat from the sun. The sun emits **radiation** that travels millions of miles through empty space in getting to the earth. Once here, most of the radiation passes through the earth's atmosphere, and is absorbed by the earth and objects on it. That way, heat is transferred from the sun to the earth, even though the radiation itself is not actually heat.

The radiation that transfers thermal energy is a form of what we call *electromagnetic waves*. In later chapters we'll say more about what these waves are. For now, we'll just notice that light, x rays, gamma rays, radio waves, and microwaves are also examples of electromagnetic waves. The only difference between them is in a property called *wavelength*. Figure 8.12 illustrates different wavelengths in a familiar type of wave. The wavelength is the distance from crest to crest, and is much longer for the ocean wave than for the waves in the pond. Electromagnetic radiation with wavelengths between about 10^{-7} m and 10^{-4} m transfer thermal energy; we call this *infrared radiation*. Infrared's neighbor on the short wavelength side is visible light, the radiation to which the eye is sensitive.

When you hold your hand near an ordinary incandescent light bulb, you feel heat. The radiant energy emitted includes not only visible light but infrared (heat) radiation as well. When you stand in front of an open fireplace, most of the heat you feel is from radiation. The heat from infrared radiation can relieve the pain of sprains, arthritis, bursitis, and various other aches.

All bodies radiate energy continually. This fact means that, if the temperature of a body does not drop, it must also continually absorb radiant energy. Some bodies are better emitters and absorbers than others. But we can show from the energy conservation principle that a good emitter is also a good absorber, and vice versa.

Suppose the fire goes out in the stove of Fig. 8.13, but the stove is still hot. (Imagine that the air has been pumped out of the room, so that we do not have to consider other forms of heat transfer.) The stove radiates heat to the walls, and the walls radiate to the stove. Since the stove is at a higher temperature than its surround-

Figure 8.12 The wavelength, the distance from crest to crest, is much longer for ocean waves than for waves in a small pond.

ings, it emits faster than it absorbs radiation, so that its temperature drops. Once the stove and walls are at the same temperature, the stove must emit and absorb radiation at the same rate; otherwise, its temperature would change. This condition must hold whether the stove is a good or a poor emitter of radiation. Since energy is conserved and the temperature neither rises nor drops, we conclude that a good emitter must also be a good absorber.

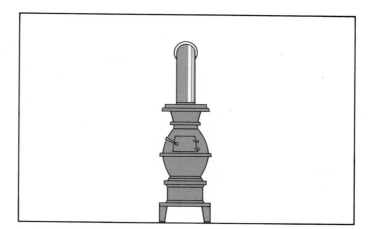

Figure 8.13   The stove emits radiation to, and absorbs radiation from, the walls, floor, and ceiling.

Influences on the Rate of Emission and Absorption of Radiation

Several factors influence the rate at which a body emits or absorbs radiation. We'll consider the main ones.

Temperature   The rate of radiation from a body increases greatly with increasing temperature. In fact, this rate is proportional to the fourth power of the *absolute* temperature. This means that a body at 600 K (327°C) radiates at about twice the rate of one at 500 K (227°C). Doubling the absolute temperature gives 16 times as much heat radiated. Any body at a temperature above absolute zero (0 K), which means *every* body, radiates heat. Depending upon the relative temperatures of a body and its surroundings, there may or may not be a *net* loss of heat.

Surface Condition   A smooth, shiny surface reflects much of the radiation that hits it. On the other hand, a rough surface effectively has more surface area that can absorb radiation—the reasoning here is the same as that of my college friends from West Virginia who claimed that, if West Virginia were flattened out, it would be bigger than Texas. In other words, a good absorber has a coarse-textured surface rather than a shiny one. From our earlier arguments, a good emitter has the same characteristics.

Surface Color   You may have noticed that a dark-colored car gets much hotter when left in the summer sun than a light-colored one;

a black asphalt driveway burns your feet worse in July than a concrete one; or various other evidences that dark objects get hotter when exposed to radiation. The natural conclusion to draw is that dark-colored objects are better absorbers than light-colored ones. However, that reverses the cause and effect. The object is dark-colored *because* it absorbs most of the radiation hitting it. A truly black object is one that absorbs all the radiation that hits it, thereby reflecting none. Physicists call such an object a *blackbody,* although a perfect one doesn't exist. The darker a body is, the closer it approximates a blackbody. By the reasoning we've used before, a dark-colored body is, therefore, one that is also a good emitter of radiation.

Question   Why will a black dog go outside and lie in the sun on a cold clear (even windy) day, rather than stay in its cozy insulated doghouse?

Temperature and Wavelength

All the radiation given off by a body is not of the same wavelength. The graph of Fig. 8.14 shows that the radiation is spread out over a broad range of wavelengths. The vertical axis shows the relative amount of radiation emitted at various wavelengths along the horizontal axis. The vertical wiggly lines symbolize the increasing wavelength to the right. The peak in this curve corresponds to the wavelength at which the most radiation is emitted, although some radiation has very short, and some very long, wavelengths.

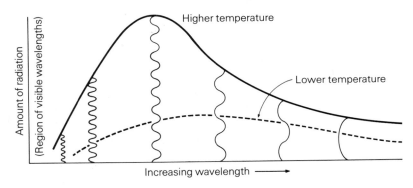

Figure 8.14   A graph of the amount of radiation emitted by a body versus wavelength of the emitted radiation. The separation between wiggles in the vertical lines indicates the variation in wavelength. The dashed line is for a lower-temperature body than the solid line.

As the body's temperature decreases, the peak in the graph shifts to the right (toward longer wavelengths), and the total quantity of radiation decreases. The dashed line shows the emission curve for a body at a lower temperature. This curve peaks at a longer wavelength, and its overall height is lower.

A portion of the radiation on the short wavelength side has wavelengths in the range of visible light. As the body's temperature is raised, more and more of the radiation given off is visible. At 500°C, a significant amount is in the long-wavelength part of the visible region, which corresponds to red light. You see a dull red glow. As the temperature is raised, not only is more heat given off but more light of shorter wavelengths is emitted, making the light whiter. At 900°C the body is bright red and at 1500°C, white hot. You can see this effect as the heating element of an electric range or portable heater gets hot.

Thermography   The infrared radiation given off by bodies can be a useful diagnostic tool in many areas. A device called an infrared scanner makes a *thermogram*, which is a picture based on the infrared radiation given off by a surface. Variations in temperature of small fractions of a C degree can be detected with sensitive scanners. An important application involves detecting cancers, which produce local "hot spots" in the skin temperature of as much as 1 or 2 C deg that can radiate at a 2 to 3% higher rate than normal (Fig. 8.15). Another use is the detection of areas of excessive heat loss from houses. The whole house is scanned on a cold day to detect spots where insulation is bad or where cracks exist. Electronic parts can be tested for flaws by looking for areas of excessive heat production.

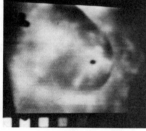

Figure 8.15   An infrared radiation thermogram can detect the slightly higher-than-normal temperature associated with breast cancer, even in its early stages. The light regions indicate the higher-temperature cancerous region.

Absorption of Radiation

We'll illustrate some of the important characteristics of the absorption of radiation by discussing some specific examples.

Transparency and the Greenhouse Effect   Ordinary window glass is transparent—that is, light passes through it. At least, most of it does. But some light is absorbed, and the fraction absorbed depends upon the wavelength of the radiation. Ordinary glass is also quite transparent to the short-wavelength infrared (heat) radiation emitted by very hot bodies. But glass absorbs most of the radiation from bodies at ordinary temperatures, because most of it has much longer wavelength.

In a glass greenhouse (Fig. 8.16), the short-wavelength radiation from the sun is transmitted easily, and absorbed by the contents of the greenhouse. The cooler objects inside emit mostly longer-wavelength rays that cannot pass through the glass, but are reflected back inside. The objects inside that are warmed by radiation conduct heat to the air confined inside the house, warming it. Convection distributes the heat throughout, but the relatively stationary air adjacent to the walls minimizes conduction to the outside.

The earth as a whole is a sort of greenhouse, housed not by glass but by the atmosphere. Most of the short-wavelength radiation of the sun is transmitted through the atmosphere and warms the earth. But much of the longer-wavelength rays reradiated by the earth do not get through the atmosphere. One of the potential hazards of air pollution is that polluted air is less transparent to radiation from the earth. This fact might at first seem to have a beneficial effect—a warmer climate for all. One possible consequence, however, is that the polar ice caps could start to melt, flooding all our coastal cities.

Figure 8.16 In a greenhouse, the short-wavelength radiation from the sun is transmitted, but the longer-wavelength radiation from the cooler objects inside is reflected.

Microwave Cooking   Water is transparent to visible light and short-wavelength infrared rays, but paper is not. For different wavelengths of electromagnetic radiation, different materials are transparent. In fact, for *microwave* radiation such as police radar, just the opposite is true—paper is transparent but water is not. The radiation that serves as police radar and the radiation that does the heating in microwave ovens is basically the same—electromagnetic radiation with wavelengths of a few centimeters.

In microwave ovens (Fig. 8.17), the radiation passes unobstructed through paper, plastic, or glass containers, but is ab-

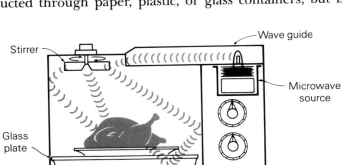

Figure 8.17 A schematic diagram of the operation of a microwave oven. Microwaves travel down the waveguide and reflect from the stirrer that slowly rotates to scatter waves in all directions. The waves reflect back and forth from the walls until absorbed by the food.

sorbed in water or food. The absorbed radiant energy is converted to heat, which raises the temperature of the food and causes cooking. Some of the radiation penetrates several centimeters into the food before absorption, thus distributing the heat production throughout the food and giving quicker cooking than could be accomplished by conventional means.

In addition, microwaves reflect from metal. They bounce around inside the oven unattenuated until absorbed by the food. This fact means that almost all the radiant energy goes into the food. A conventional oven, however, heats both the air inside and the oven walls, which in turn heat the outer surface of the food. Heat is slowly transmitted by conduction to the interior. A large fraction of the heat is wasted by leaking through the walls of the oven into the room.

8.5 Combinations of Heat Transfer Methods

When heat is transferred from one place to another, it usually moves by some combination of the three methods we have discussed. In all three methods, a major factor influencing the heat transfer rate is the temperature difference between the two regions. You can think of the temperature difference as the "driving effect" that causes heat to move from one place to another.

Isaac Newton, who probed into many different areas, noticed that the rate of heat loss from a body by all methods is approximately proportional to the temperature difference between the body and its surroundings. This fact is referred to as *Newton's law of cooling,* although it is not a fundamental law like Newton's laws of motion. It's only an approximate empirical law that holds under most circumstances.

Heating engineers, and others concerned with critical weather conditions, use a unit called a *degree-day.* The number of degree-days for a particular day is the difference between 18°C (65°F) and the average temperature for that particular 24-hour day. (If the average temperature is above 18°C, the number of degree-days is taken to be zero, since little or no heat would be needed on such a day.) This gives an estimate of the average temperature difference between the inside and outside of houses. For a given period of time, the total number of degree-days would be the sum of the values for each day. The degree-day is a measure of "coldness." When you hear a weather report something on the order of, "The weather in Slippery Rock was 40% colder this year than last," this statement means 40% more degree-days. Convince yourself that the number of Celsius degree-days is 5/9 the number of Fahrenheit degree-days for the same actual conditions.

Question   Based on Newton's law of cooling, what approximate relationship would you expect between the number of degree-days and the amount of fuel needed to heat your house (or apartment or room) during a particular season? (Not actual numbers, just proportionalities.)

To illustrate the way different methods of heat transfer can interact, we'll discuss two particular examples.

**The Thermos Bottle   A thermos bottle is designed to minimize heat transfer by any of the three methods. If the contents are hot, heat should be kept in; if they are cold, heat should be kept out.

Figure 8.18 shows how the thermos cuts down on each form of heat transmission. It consists of a double glass wall with silvered surfaces and vacuum between the walls. Conduction and convection cannot occur across the vacuum. Conduction can occur along the glass-and-silvered surface, but the distance there is quite long. The silvered surface cuts down on radiation emission and absorption. The stopper is made of a material that is a poor conductor. (The bottle is housed in a metal or plastic can for protection.)

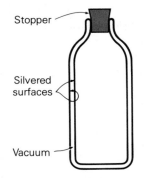

Stopper

Silvered
surfaces

Vacuum

Figure 8.18   A thermos bottle minimizes heat transfer by all three mechanisms.

**A Solar Water Heater   Radiant energy from the sun is free. All we have to do is find ways to absorb and use it. Today, the sun's energy is used in many ways—heating water, for example. A thermosiphonic (you can figure out that word by studying its roots) solar water heater operates by natural circulation with no moving parts or pumps required (Fig. 8.19). The bottom of the solar collector should be coated with a dull black material for best radiation absorption, and the top covered with a double glass pane for insulation and for taking advantage of the greenhouse effect. The water pipes in the collector should be a good conductor, like copper, to maximize conduction of heat into the water. The collector is located perpendicular to the direction of brightest sunlight, and below the level of the storage tank so that, once the water in the collector is hotter than that in the bottom of the tank, convection will bring cold water from the tank down into the collector, forcing hot water into the top of the tank.

Larger solar heating systems, such as might be used for heating a house, employ the same ideas but usually need a pump for more efficient circulation (forced convection) of the heated liquid. Air might be blown over pipes containing the heated liquid, and then

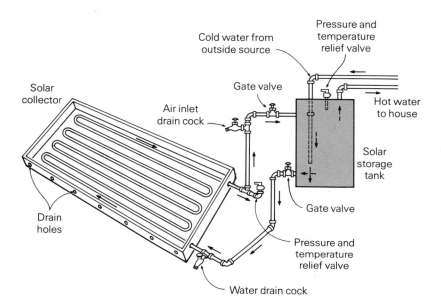

Figure 8.19 A thermosiphonic solar water heater.

into the interior of the house. Special coatings have been developed that strongly absorb visible and short-wavelength infrared radiation, but no long-wavelength infrared. That way they absorb the incoming, mostly short-wavelength radiation, but are poor emitters for wavelengths that dominate emission at the collector temperature.

Key Concepts

There are three mechanisms of heat transfer—*conduction, convection,* and *radiation.*

Conduction is the transfer of heat from molecule to molecule within a substance. The **thermal conductivity** is the property of a substance that tells how effective it is in conducting heat.

Convection is the transfer of heat by movement of all or part of a heated fluid. The circulation of fluid by means of a mechanical pump or fan is called forced convection. Conduction of heat from a body to an adjoining fluid is facilitated by convection (natural or forced), which brings more cool fluid into contact with the body.

Radiation is the transfer of thermal energy by electromagnetic waves. The rate of radiation of heat is determined by the temperature, color, and surface condition of the radiating body.

A body that is a good radiator is also a good absorber of radiation. All bodies radiate to, and absorb radiation from, their surroundings. A body at a higher temperature than its surroundings experiences a net loss of heat; one at a lower temperature, a net gain of heat. A body at the same temperature as its surroundings emits and absorbs radiation at the same rate.

The combined rate of heat transfer from a body to its surroundings by conduction, convection, and radiation is approximately proportional to the temperature difference between the body and its surroundings.

Questions

Conduction

1. Why are the handles of cooking utensils usually made of wood or plastic rather than metal?

2. When you pick up a cold hammer by the metal part, it *feels* much colder than when you pick it up by the wooden handle. Give two reasons why this is true.

3. Skis slide on snow that is not much colder than 0°C because friction melts the top layer of snow, letting the skis ride on water. Why do metal skis not slide as easily as wooden ones, even though they are smoother?

4. Queen Elizabeth is making tea to serve her guest, President Carter. The burners on her electric stove are all 19 cm in diameter. Why does 1 liter of water heat quicker on one of these burners if she heats it in a 19-cm-diameter tea kettle than if she uses a 14-cm-diameter one?

5. Discuss the electrical energy economy of using cooking utensils whose bottoms are larger than, or smaller than, the heating unit on which they are placed.

6. How much more heat will be lost through a window 3 m by 1 m than a window the same thickness but 2 m by 1 m, if the inside and outside surfaces of the glass are the same for both?

7. Aunt Jemima says that the pancake griddle is ready for cooking when "a small drop of water dances on it before evaporating." Why does the drop dance rather than quickly boil away?

Why does the drop evaporate quicker when the griddle is cooler?

8. Certain circus performers and physics teachers have been known to plunge their *wet* hands briefly into molten lead or iron without injury. Can you explain how this feat is possible?

9. When there is a period of extreme cold weather, underground pipes sometimes freeze *after* the weather has gotten warmer. Why?

10. If you put a mercury thermometer in something hot, the mercury moves up quickly at first, then slower and slower as it approaches the final reading. When you take it out, the same effect takes place—it goes down quickly at first, then slower and slower as it settles down. Why?

11. If the nurse tells you to keep the thermometer in your mouth for three minutes, is it okay to take it out and look at it in the middle, as long as it stays a total of three minutes?

Convection

12. True or false: Hot air always rises. Justify your answer.

13. Why does the wind at the beach change direction at night?

14. Only about 10% of the heat produced in an ordinary fireplace gets into the room. Why is a fireplace so inefficient?

Combinations of Conduction and Convection

15. The difference between conduction and convection is sometimes pointed out by comparing a bucket brigade with a group of people all carrying individual buckets of water. Which mode of heat transfer would you associate with the bucket brigade, and which with everybody having an individual bucket?

16. For a fireplace chimney, why can't you use a *bare* metal pipe running up the outside of your house?

17. When your skin temperature is warm, a large amount of blood flows through the small arteries near the outer layer of your skin. When the skin temperature drops below 37°C, your body compensates by constricting these blood vessels and limiting the flow of blood to the surface. How does this activity help to

keep the inner part of your body warm? (The skin and fatty tissue underneath are poor conductors.)

18. Charlie Brown turns a fan on in his house and goes away for a while, expecting the house to be cooler when he gets back. Is he fooling himself?

19. Baseboard hot water heaters are basically copper pipes with fins surrounded by a housing (see Fig. 8.20). What are the fins for?

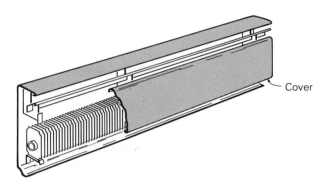

Cover

Figure 8.20 A hot-water baseboard heater.

20. Suppose you are comfortable when the temperature is 0°C and the wind is blowing at a speed of 40 km/h. If the wind stopped blowing, how much colder could it get without your being uncomfortable in the same clothing?

21. Some oven manufacturers claim that self-cleaning ovens "save energy" and cost less to operate in the long run because they have more insulation in the oven walls. Is this claim reasonable?

22. How does heavy snowfall help plant roots and small animals that live underground to survive the winter?

23. Why does the hair on animals "stand up" in winter? Why do birds *look* fatter in winter?

24. Justify the statement made in the text that, if you were a fish, you would be very glad that water expands as the temperature drops below 4°C, and expands even more when it freezes.

25. What was the temperature at noon yesterday in the deep underwater den of Nessie, the Loch Ness monster? What was it at midnight last night? Barring unexpected developments, what will it be in exactly six months?

Radiation

26. Many houses are heated with electric heaters imbedded in the plaster in the ceiling. By which of the three mechanisms does heat warm the room? Why do your feet get cold if you are sitting with them under a table in such a room?

27. Very roughly, the color of the visible radiation from bodies at different temperatures is given in the following tabulation:

Color	Approximate Temperature, °C
Incipient red	500
Dark red	700
Bright red	900
Yellowish red	1100
Incipient white	1300
White	1500

Turn the burner of an electric stove on "high" and estimate its temperature at various stages as it heats up.

28. It's likely that Joan Caucus' most dreaded job is polishing the silver coffee pot her mother gave her as a Christmas present. Aside from its appearance, is there a practical reason why this utensil will do its job better if kept shiny?

29. Explain why sprinkling soot over snow will cause it to melt more quickly.

30. If you want the snow to melt completely from your car as quickly as possible, should the car be a light or a dark color?

31. If you live in a warm sunny climate, should the roof of your house be light-colored or dark? Should you buy a light- or dark-colored car if you want it to be cool?

32. Can the temperature of the water in the collector of a solar water heater ever get higher than that of the air outside?

Combinations of Heat Transfer

33. The waiter brings apple pie and coffee to your date, who does not want to drink the coffee until after finishing the pie. Knowing that you are taking physics, your date asks you for this advice, "If I want my coffee to be as hot as possible when I drink it, should I add the cream and sugar now, or wait until I finish my pie?" What is your answer?

34. Why do you save fuel in winter by keeping your thermostat set low?

35. Do you save fuel in winter by turning your thermostat down at night and when you are away in the daytime? Why so or why not? Base your answer on the principle of conservation of energy.

Problems

1. A store window near the White House is a single sheet of glass 3 m by 4 m by 5 mm thick. If the air temperature inside is at 20°C and that outside at −10°C, heat would typically transfer through this window at a rate of about 0.50 kcal/s. For this rate, what is the actual temperature difference across the glass only?

Ans: 0.83 C deg

2. If Amy Carter throws a rock through the window of Problem 1 and the owner replaces it with a sheet of 1-cm-thick pine plywood, what will be the rate of heat transfer if the temperature difference across the plywood is the same as it was across the glass? Would you expect this temperature difference to stay the same?   **Ans:** 0.028 kcal/s

3. A copper rod 25 cm long and 2 cm in diameter has one end in boiling water and the other end in iced water. What is the rate of heat transfer down the rod? How much ice is melted per hour? (The rod and water are insulated so that no heat is transferred except along the length of the rod.)

Ans: 0.0116 kcal/s; 520 g

4. A styrofoam ice chest 60 cm by 30 cm by 40 cm has 2-cm-thick walls. If it is filled with ice and put in a 30°C room, what is the rate of heat loss through the walls, assuming the outside walls of the chest to be 5 C deg lower in temperature than the room? How much ice melts in 5 hours? (The thermal conductivity of styrofoam is 0.00001 kcal/m·s·C deg).

Ans: 0.0135 kcal/s; 3.0 kg

5. A wall 20 ft by 8 ft is insulated with R-11 insulation. Assuming this is the main resistance to heat flow, what is the rate of heat loss through the wall if the temperature difference is 70 F deg?

Ans: 1000 Btu/h

6. A room is heated by a steam radiator that gives up heat as steam condenses to water at a temperature of 100°C. An identical radiator contains hot water that heats a room as water cools from 90°C to 70°C. Compare the rate of heat transfer to the two rooms.   **Ans:** about 33% higher with steam

7. If the average temperature for each of seven days in Chicago is +2°C, −4°C, −16°C, −22°C, −20°C, −10°C, and −6°C, how many degree-days are there for this period?

Ans: 202 C degree-days

8. In Atlanta the average number of C degree-days is 1600, and in Cleveland it is 3400. Approximately how would the fuel consumption for heating a particular house in Cleveland compare with that for an identical house in Atlanta?

Ans: about 2.1 times as much in Cleveland

9. The net rate of heat loss by radiation from an object is given by

$$R = \epsilon \sigma A (T^4 - T_0{}^4),$$

where T_0 is the *absolute* temperature of the surroundings; T is the *absolute* temperature of the object; A is the area of the body; σ is a constant of value 5.67×10^{-8} W/m²·K⁴; and ϵ, the emmisivity, tells how effective the surface is as a radiator. The area of a human body is about 1.9 m² and ϵ is about 0.97. What is the rate of radiation of heat by such a body in a room at 20°C? The skin temperature is about 28°C. A person sitting still burns up about 100 kcal/h. What fraction of this energy is given off in radiation?   **Ans:** 88 W; 76%

Home Experiments

1. Check your answer to Question 11 by comparing your result from a continuous 3-min measurement of your body temperature with the result when you take the thermometer out once or twice during the measurement, but still have a total in-time of 3 minutes.

2. Using a candy thermometer or other thermometer that measures temperature variations as small as 1 C deg, check your answer to Question 33. Pour two identical cups of coffee and add a measured amount of cream and sugar to one. After about 10 min, add the same amounts to the other. Measure the temperature of each. You should see 2 or 3 C deg difference. To avoid using heat from your experimental coffee to warm the thermometer, place the thermometer in a third cup of hot coffee first and then quickly transfer it to the one being measured. Any stirring should be done with a small stick or other low-heat-capacity device. Except for the time difference, be careful to do the same thing to each cup of coffee.

3. Knowing that heat loss by all methods is approximately proportional to the temperature difference, you can now complete Home Experiment 3 of Chapter 6 by calculating how much money you save if you turn off your water heater for 8 hours each day. Let H, T_{on}, and T_{off} represent your measurements for heat loss, water temperature with the heater on, and that after the heater is off 8 hours. Then the energy you save by turning it off would be

$$H \times \frac{T_{on}}{T_{on} + T_{off}/2} - H.$$

Justify this equation, which assumes the average inside temperature while the heater is off equals the average of your two measurements. Knowing what you pay for energy, calculate the potential saving.

References for Further Reading

Anderson, B., *The Solar Home Book* (Cheshire Books, Harrisville, N.H., 1976). A beginner's book on solar energy.

Edelson, E., "Heat Pipes—New Ways to Transfer Energy," *Popular Science,* June 1974, p. 102. Shows how evaporation and condensation can be used as a very effective way to transfer heat.

Hodges, L., "Fuel Savings by Thermostat Setback: an Experimental Study," *The Physics Teacher,* Nov. 1977, p. 485. Experimental evidence that it does save fuel to set the thermostat back for part of the day.

Jones, T. H., "How to Figure the Best Places to Cut Heat Loss in Your House," *Popular Science,* Sept. 1975, p. 97. Shows how to calculate easily the rate of heat loss from a building.

Plumb, R. C., "Sliding Friction and Skiing," *Journal of Chemical Education,* Dec. 1972, p. 830. Describes evidence based on thermal conductivity for the importance of friction in the easy sliding of skis.

9

Using and Pumping Heat

At some time in your life, you've probably had a secret wish to get something for nothing. If that wish ever develops into an urge to try your hand with the dice, you might do well to remember what some people call the "Las Vegas laws":

1. You can't get something for nothing.

2. You can't even break even.

Using heat to do work is very much the same. As we discuss the first and second laws of thermodynamics and how they apply to engines that use heat to do work, you will see their similarity to the Las Vegas laws.*

Sometimes, rather than using heat to do work, we need to pump it from a cool place to a warmer one. That's what refrigerators, air conditioners, and less familiar devices *called* heat pumps do. We'll find that what happens in these devices is, in principle, just the reverse of what happens in an engine. We'll also find that for some of them, you get *more* than you pay for. As always, somebody has to pay the bill, just as when Mother takes you out to dinner. But here, we'll find that it's Mother Nature who foots the bill.

This chapter will draw on some of the ideas of the preceding three chapters. You need to recall: (1) that heat is internal energy added to, or removed from, a substance (Chapter 6); (2) that heat is required to change a substance from solid to liquid and from liquid to vapor, but heat is given up in the reverse changes (Chapter 7);

---

*The bit of folk wisdom incorporated in this terminology was passed on to me by Professor Jack M. Wilson.

and (3) that heat flows naturally from a warmer to a nearby cooler region, never the reverse.

Before we study the laws of thermodynamics, we need to understand how the temperature of a gas changes as it expands or is compressed.

9.1   **Expansion and Compression of Gases**

In Chapter 6 we discussed the expansion that usually accompanies an increase in temperature, and the contraction that usually accompanies a decrease. This effect is especially pronounced in gases, which expand or contract noticeably if kept at constant pressure and heated or cooled. (What happens to a balloon if you blow it up, tie it, and put it in a refrigerator?)

If we reverse cause and effect, you might expect that, by letting a gas such as air expand, it would become warmer. Instead, the opposite happens: A gas allowed to expand without adding outside heat gets cooler. For example, the air rushing out of a leaking tire is much cooler than it was when inside the tire. The air at high altitudes is cool because, as warm air rises, the pressure from other air above it decreases. The air expands, which cools it. Let's see why expansion of a gas causes it to cool.

The molecules of a gas have kinetic energy due to their continual motion. As gas molecules move, they collide with each other as well as with the walls of their container. When the gas is allowed to expand, it does work on the walls of its container as they move outward during this expansion. For example, as the piston in Fig. 9.1 moves to the right, the molecules of the gas exert a force on the piston in the direction of its motion—that is, they do work on the piston. The energy used in doing this work comes from the kinetic energy of the molecules, causing their average kinetic energy to decrease. A decrease in average molecular kinetic energy means a decrease in temperature. The expansion of a gas is, therefore, a cooling process.

Figure 9.1   The gas gives up energy as it does work on the piston moving to the right. The gas absorbs energy as the piston moving to the left does work on the gas.

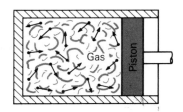

Compressing the gas has the opposite effect. To compress a gas into a smaller volume, you have to do work. We can compress the gas in the cylinder of Fig. 9.1 by moving the piston to the left. The gas molecules exert a force on the piston through their many collisions with it; from the action-reaction law, the piston has to exert an equal force back on the gas. As the piston moves to the left, this force does work on the gas. The work done increases the kinetic energy of the molecules, which means an increase in temperature.

In summary, *letting a gas expand causes it to get cooler; compressing a gas causes it to get warmer.* We can describe this behavior mathematically by the equation:

$$\frac{\text{pressure} \times \text{volume}}{\text{absolute temperature}} = \text{constant}$$

for a given mass of a gas. This equation, which we call the *ideal gas law,* is strictly correct only for an *ideal gas*—one whose molecules are volumeless points that interact only during collisions. Even though no such gas exists, most real gases at ordinary temperatures and pressures behave almost as an ideal gas would.

9.2   Conservation of Energy: the First Law of Thermodynamics

If you have a fixed quantity of matter, you can only treat it four ways from an energy viewpoint: (1) you can add heat to it; (2) you can remove heat from it; (3) you can do work on it; or (4) you can let it do work. Applying this statement to the gas of Fig. 9.1, you can add heat to, or remove heat from, the gas; you can do work on it by moving the piston to the left and compressing it; or you can let it do work by moving the piston to the right. Conservation of energy tells us that whatever energy or work you add to the substance must either stay there or come out. If it stays there, it increases the internal energy of the substance.

In summary,

net heat added = increase in internal energy + work done,

where "work done" means *by* the substance. We call this way of writing the energy conservation principle the **first law of thermodynamics.** It basically says that, as far as energy is concerned, you can't get something for nothing. Notice that if heat is removed, or internal energy decreases, or work is done *on* the substance, the particular quantity goes into this equation as a negative value.

We'll use this law in discussing engines that use heat to do work, and in discussing devices that pump heat from a cooler to a warmer region.

9.3   Engines That Use Heat

Many different types of **heat engines** are available that can do everything from power our cars to turn electrical generators. All of them do basically the same thing—add heat to a fluid that uses some of the heat to do mechanical work. To illustrate this process, we will describe the basic principles of an internal combustion engine, which is the type used in most automobiles as well as on most lawn mowers. We can then show how the same work and energy ideas apply to any heat engine.

The Internal Combustion Engine

An ordinary gasoline engine consists of one or more cylinders, such as those shown diagrammatically in Fig. 9.2(a). This figure

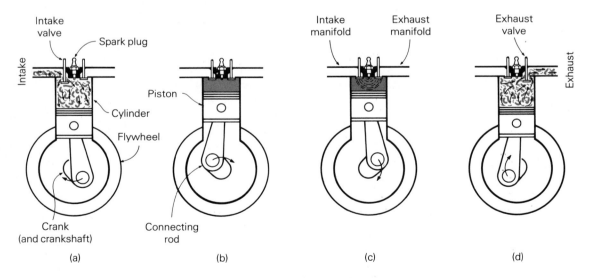

Figure 9.2   The four strokes of the four-stroke internal combustion engine. (a) Intake stroke: The intake valve opens, admitting fuel and air as the piston moves down. (b) Compression stroke: The rising piston compresses the fuel-air mixture. (c) Power stroke: The fuel burns, increasing the temperature and pressure. The high pressure pushes the piston down. (d) Exhaust stroke: The exhaust valve opens, letting the gases out as the piston rises.

shows the four strokes in a complete operating cycle of the engine. (Many small engines operate on a two-stroke cycle, but the basic ideas are the same.)

On the intake stroke (a), the intake valve opens, letting a mixture of fuel and air into the cylinder as the piston moves down. The intake valve closes, and this mixture is compressed during the compression stroke (b). A spark from the spark plug ignites the fuel, with the energy released during combustion causing a high rise in temperature. The resulting high pressure pushes the piston down for the power stroke (c). The exhaust valve then opens and the remaining gases are pushed out as the piston comes up during the exhaust stroke (d). The sequence then repeats.

Since only one of the four strokes is a power stroke, the engine has to be started by turning it with some external torque—either from a starter motor or a pull-rope, for example. A flywheel attached to the crankshaft stores some of the mechanical energy generated during the power stroke, and keeps the engine running until the next power stroke. The useful work output is done by whatever is attached to the rotating crankshaft. Automobile and other large engines usually have several cylinders of the type we've described.

The heat released during combustion either does mechanical work or is rejected as heat outside the engine. This rejected heat either goes out the exhaust pipe with the combustion products or is conducted through the cylinder walls. The heat conducted through the cylinder walls is either given up directly to the outside air or to circulating water that later gives up its heat to air passing through the radiator.

General Energy Considerations

We can show the relationship between the overall energy transfer and the work done in an internal combustion engine, or any heat engine, in a diagram such as Fig. 9.3. This diagram is purely symbolic and does not show the physical appearance of a real heat engine. The circle represents the engine. The "channel" on the left represents the heat input that, in the internal combustion engine, comes from burning fuel. The channel at the bottom represents the work output by the engine. The channel on the right represents the heat rejected to the outside.

At any point in the engine cycle, the internal energy of the fuel-air mixture is the same as at that point in the previous cycle. Thus, for a complete cycle, there is no net change in *internal*

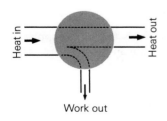

Figure 9.3 The energy and work transfer in a heat engine.

energy. The first law of thermodynamics then tells us that

$$\text{heat in} = \text{heat out} + \text{work out}.$$

The widths of the channels in Fig. 9.3 represent the amount of heat (or work) transfer in each region. The sum of the widths of the two output channels is equal to the width of the input channel. You might suspect that the width of the work channel is much too small compared with the heat output channel. Unfortunately, for most heat engines the widths on the channels in the figure are approximately in the correct proportions. Most of the input heat is rejected to the surroundings and does no useful work.

The term *thermal efficiency* rates the performance of a heat engine, and is defined as

$$\text{thermal efficiency} = \frac{\text{work out}}{\text{heat in}}.$$

This equation means that, the greater the work output for a given heat input, the higher the thermal efficiency. On the other hand, the more heat it takes to do some amount of work, the lower the thermal efficiency. Figure 9.3 shows that we can also write

$$\text{thermal efficiency} = \frac{\text{heat in} - \text{heat out}}{\text{heat in}}.$$

Example 9.1: It typically takes about 600 000 J of work from the engine of a compact car to move the car 1 km along the highway. If the car gets 14 km/liter of gasoline (33 mi/gal), what is the thermal efficiency of the engine? The heat is supplied by burning gasoline that has a heat of combustion of 36×10^6 J/liter. Basing our calculation on 14 km of travel:

$$\text{work out} = 60 \times 10^4 \text{ J/km} \times 14 \text{ km} = 84 \times 10^5 \text{ J};$$

$$\text{heat in} = 36 \times 10^6 \text{ J/liter} \times 1 \text{ liter} = 36 \times 10^6 \text{ J};$$

$$\text{thermal efficiency} = \frac{\text{work out}}{\text{heat in}} = \frac{84 \times 10^5 \text{ J}}{36 \times 10^6 \text{ J}}$$

$$= 0.23, \quad \text{or } 23\%$$

This efficiency is that of the engine alone. Multiply this by the approximately 40% combined efficiency of the transmission and rolling tires, and the overall efficiency is down to about 9%.

Exercise A power plant burns coal at a rate of 200 000 kg/h and generates 600 megawatts (MW) of electricity. The heat of combustion of coal is 7100 kcal/kg (8.2 kWh/kg). What is the overall thermal efficiency of the plant?

Ans: 36%

9.4 You Can't Break Even

The nineteenth-century French physicist Sadi Carnot studied the cycles a substance can be taken through in converting heat into work. He recognized that anytime you operate a heat engine, there must always be some heat rejected *as heat* from that engine. The **second law of thermodynamics** describes this effect. You may see this law stated many different ways, but the form of the law we're concerned with here is: *No actual or ideal engine that operates on a cycle can convert all the heat supplied to it into work.* In other words, the "heat out" channel in Fig. 9.3 must always be present. In any one expansion, all the heat can go to work. But for the complete cycle needed to *keep* an engine operating, heat must be rejected at some stage.

Carnot showed that for the most efficient possible engine you can have, the thermal efficiency is limited by the operating temperatures. If T_{hot} is the temperature at which heat is added during the cycle and T_{cold} is the temperature at which it is removed, then

$$\text{maximum thermal efficiency} \ = \ \frac{T_{hot} - T_{cold}}{T_{hot}} = 1 - \frac{T_{cold}}{T_{hot}} \ .$$

The higher you can make T_{hot} and the lower T_{cold}, the higher the theoretically possible efficiency.

A power plant using steam that enters its turbines at 873 K (600°C) and later condenses at 373 K (100°C) has a maximum theoretical efficiency of 1 — (373/873) or 57%. That makes the 37% efficiency of a modern power plant seem not quite as bad.

You may wonder why a car with a diesel engine typically gets higher gas mileage than a similar car with a gasoline engine. The main reason is that the fuel in the diesel burns at a higher temperature and therefore gives the engine a higher thermal efficiency. A diesel has no spark plugs. Instead, the air in the cylinder is compressed during the compression stroke to three or four times the pressure in the cylinder of a gasoline engine. This compression raises the temperature high enough for the fuel to spontaneously ignite when injected into the cylinders.

9.5 Pumping Heat

Heat flows naturally from a warmer to a nearby cooler region, just as water flows naturally downhill. However, not only can we pump water uphill by doing work on it, but we can pump heat from a

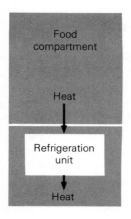

Figure 9.4 The refrigeration unit of a refrigerator pumps heat from the food compartment to the room.

lower to a higher temperature by doing work in just the right way. That's the purpose of any heat-pumping device, whether a freezer, refrigerator, air conditioner, or heat pump. Figure 9.4 illustrates this idea for a refrigerator. If its inside is to be cooled, and to stay cool when warm food is placed in the refrigerator, heat has to be removed by the refrigeration unit and dumped outside—usually at the bottom or the back.

The refrigerator accomplishes this feat by pumping a cold fluid through tubes inside the food compartment. This fluid absorbs heat from the contents of the compartment. The fluid is then pumped outside the compartment and compressed so that its temperature rises enough to give up heat to the warm outside air. We'll take a look at this process in more detail.

The Refrigeration Cycle

Figure 9.5 shows the main components of an ordinary electrically operated refrigerator. An air conditioner or heat pump has the same basic components, but the construction details differ.

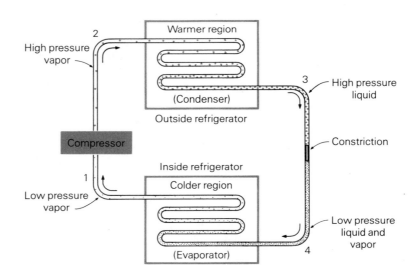

Figure 9.5 A schematic diagram for a compression refrigeration cycle.

The system contains coiled tubing that winds throughout the cold region inside the refrigerator. Another coil in the warm region outside the refrigeration compartment usually has fins attached to help dissipate heat. Tubing joins the two regions to form a complete loop. An electrically operated compressor pumps a fluid called *refrigerant* (usually freon) around the loop in the direction of the arrows. Opposite the compressor is a constriction that restricts the flow of refrigerant. There also may be a fan to blow air over one or both coils to speed heat transfer.

Let's follow the refrigerant through its cycle. At point 1, it has just passed through the cold region and is in the vapor state at low temperature and pressure. The compressor compresses this vapor to a high pressure, causing its temperature to rise.

At point 2 the refrigerant is still a vapor, but at high temperature and pressure. As it passes through the coils outside the refrigerator—the *condenser*—its temperature is higher than that of the surrounding air. Therefore, heat flows from the refrigerant to the air, causing the refrigerant to condense.

At point 3, having given up heat and condensed, the refrigerant is a liquid at roughly the same temperature and pressure as at point 2. The refrigerant is next forced through the small constriction where it expands into region 4, which is at much lower pressure. The constriction resists the flow, and keeps a large pressure difference between 3 and 4. The drop in pressure lowers the boiling point, causing some of the liquid to vaporize. The heat needed for vaporization comes from the liquid itself, causing the temperature to drop to a low value.

At point 4 the refrigerant is mostly liquid, but some is vapor, at low temperature and pressure. As the fluid passes through the cold inside region—the *evaporator*—it is at a lower temperature than the surrounding medium so that heat is absorbed into the fluid from the surroundings. This heat evaporates the refrigerant to the vapor state, bringing it back to point 1 at low temperature and pressure, where it continues through the same cycle.

When the refrigerant is in the high-pressure part of the cycle, its temperature is that of its boiling point at that pressure. Similarly, the low-pressure side is at the temperature of the boiling point of the refrigerant for *that* pressure. The two pressures are chosen to produce a refrigerant temperature several degrees higher than the surrounding medium in the warmer region, and several degrees lower than the surrounding medium in the colder region.

The overall result of the cycle is to pump heat from the colder to the warmer region. Keep in mind, however, that in the condenser heat flows of its own accord from the warm coils to the cooler surroundings. In the evaporator, heat flows of its own accord from the cold region to the colder coils.

General Energy Considerations

We can describe the overall energy and work transfer in a refrigeration cycle by a simple symbolic diagram. In Fig. 9.6, the circle

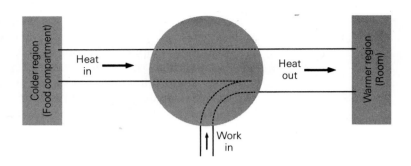

Figure 9.6   Energy and work transfer in a refrigerator.

represents the region where the refrigeration cycle operates. The region between the horizontal lines on the left symbolically represents a channel through which heat enters from the low-temperature region. Similarly, heat given up to the high-temperature region is represented by the channel on the right. The channel at the bottom represents the work done by the motor that runs the compressor.

The width of each of the channels is proportional to the amount of energy transferred in each process. Since there is no net change in internal energy for the complete cycle, the first law of thermodynamics tells us that

$$\text{heat in} + \text{work in} = \text{heat out.}$$

The widths of the channels shown indicate that the heat input is about 2.5 times the work, and the heat output is about 3.5 times the work. These are typical values for real refrigeration systems. In other words, about 2.5 times as much energy is removed in the form of heat as the amount of work done in the compressor. Similarly, 3.5 times as much heat is delivered to the high-temperature region as is used in running the compressor.

Let's compare the energy transfer in an air conditioner and in a heat pump (Fig. 9.7). Figure 9.7(a) for an air conditioner is similar to the diagram for a refrigerator. Heat is removed from the cool inside and dumped outside the house. The total heat rejected equals the sum of the heat removed and the work done.

In winter you are concerned with adding heat *to* the inside. A heat pump (Fig. 9.7-b) is really an air conditioner in reverse. In a sense, you air condition the outside. Heat is removed from the cold outside and deposited in the house. Since the heat deposited is the sum of the heat removed from the outside and the work done, you get more heat inside than the energy you use in running the heat pump. The same piece of equipment can serve as both an air con-

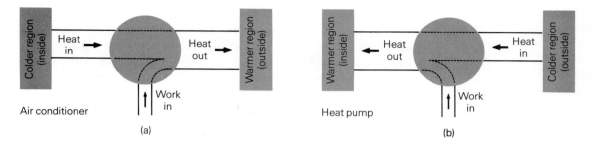

Figure 9.7 (a) A heat pump serves as an air conditioner in the summer. (b) In winter it is reversed, and pumps heat into the house from outside.

ditioner and a heat pump. A reversing valve reroutes the refrigerant so that the coils acting as an evaporator in summer become a condenser in winter.

Performance Ratings

Heating and cooling systems are rated in terms of how much heat they deliver or remove per unit of time. Most systems currently in use in the United States are rated in Btu/h, rather than the metric units of energy or power that we have used in this book. For example, a 6000-Btu/h air conditioner (for a room) removes an average of 6000 Btu (1500 kcal or 1.8 kWh) of heat from the room each hour of operation.

It is also useful to know how much work is done (that is, how much electrical energy is needed) to add or remove the heat. Rather than defining an efficiency, we use the term *coefficient of performance* to rate refrigeration systems. The coefficient of performance is the ratio of the useful heat transfer to the work needed to transfer the heat.

For a cooling system—refrigerator or air conditioner—

$$\text{coefficient of performance (cooling)} = \frac{\text{heat removed}}{\text{work done}}.$$

In a heat pump used for heating,

$$\text{coefficient of performance (heating)} = \frac{\text{heat added}}{\text{work done}}.$$

For an air conditioner or refrigerator, the coefficient of performance is typically about 2.0 to 2.6, meaning that about 2.3 times as much heat is removed as work is done.

If you look at the sales information provided for air conditioners (see, for example, the Sears, Roebuck catalog), you will notice that it usually refers to something called the *energy efficiency ratio* (EER). The EER gives exactly the same information as the coefficient of performance. But the way the units are used in the EER does about the same thing to a physicist as scraping a fingernail across the blackboard. The definition of EER is:

$$\text{EER} \quad = \quad \frac{\text{heat removed in BTU/h}}{\text{power to the compressor in watts}}.$$

An 8000-Btu/h air conditioner that uses 1000 W of electrical power has an EER of 8.

The EER presents problems for the physicist because its definition requires different types of units to be mixed in the same equation. By comparison, the coefficient of performance assumes that you use the same units for heat transfer as for work done. The heat transfer rate could be given in any *power* units, such as watts. Since 1 Btu/h is equivalent to 3.4 W, the 8000-Btu/h system removes heat at the rate of 8000/3.4 or 2400 W. For this air conditioner,

$$\text{coefficient of performance} \quad = \quad \frac{2400 \text{ W}}{1000 \text{ W}} = 2.4.$$

(Why can we use power here rather than energy?)

Air conditioner EERs usually run between 5 and 12, with a value of 10 or above being excellent. For low power consumption, you obviously need an air conditioner with a high EER. If the EER is not given directly, you can calculate it by finding the heat removal rate and the electrical power used from the nameplate or owner's manual.

Exercise   What is the EER of a 6000-Btu/h air conditioner that uses 1200 W of electrical power? What does it cost to operate this unit for 10 h if electrical energy costs 5¢ per kWh?   **Ans:** 5; 60¢

Exercise   What electrical power would be needed if this air conditioner had an EER of 10? What would it then cost for 10 h of operation?   **Ans:** 600 W; 30¢

The amount of work needed to pump a certain amount of heat depends upon the temperature difference through which the heat is being pumped. This is like saying it takes more work to carry 5 gallons of water up Mt. Rainier (elevation 4392 m) than up to Clingman's Dome (elevation 2024 m). For air conditioners and refrigerators, the temperature difference between warmer and colder regions usually doesn't change much. But for heat pumps that take heat from outside air for which the temperature varies widely, the coefficient of performance can vary appreciably with temperature change.

For an outside temperature of 10°C, a typical heat pump has a coefficient of performance around 3. This fact means that for every 1 kWh of electrical energy delivered to the heat pump, 3 kWh of heat are delivered to the house. You get 2 of the 3 kWh free! Compare this with direct electrical resistance heaters such as baseboard or ceiling heaters. For these heaters, the coefficient of performance is 1: You get only as much heat as you supply in electrical energy. But the coefficient of performance of heat pumps drops drastically with temperature. At about −10°C, most heat pumps need to be assisted by another heat source in order to keep up with the demand. In moderate climates where the temperature does not stay below −10°C for extended periods in winter, and where a central air conditioning system would be installed anyway, a heat pump can save both money and usable energy in the long run.

An interesting variation of the "air-to-air" heat pump we've been discussing is a water-to-air system that takes advantage of the large heat of fusion of water. In winter the heat pump removes heat from a large tank of water, turning it into a block of ice as the heat of fusion from the water is pumped into the house. In summer the process is reversed—heat is pumped from the house into the block of ice, turning it into water. Such an approach is obviously best suited for a climate where the amount of heat removed from a house in summer is about equal to the amount added in winter. This system has not yet been widely used.

Key Concepts

Allowing a gas to expand causes it to get cooler; compressing a gas causes it to get warmer.

The **first law of thermodynamics:**

net heat added = increase in internal energy + work done.

For a **heat engine,**

$$\text{heat in} = \text{heat out} + \text{work done};$$

$$\text{thermal efficiency} = \frac{\text{work out}}{\text{heat in}}.$$

The **second law of thermodynamics:** No actual or ideal heat engine that operates on a cycle can convert all the heat supplied to it into work.

$$\text{maximum thermal efficiency} = 1 - \frac{T_{\text{cold}}}{T_{\text{hot}}}.$$

For the **refrigeration cycle:**

heat in + work in = heat out;

$$\text{coefficient of performance (cooling)} = \frac{\text{heat removed}}{\text{work done}} ;$$

$$\text{coefficient of performance (heating)} = \frac{\text{heat added}}{\text{work done}} .$$

The energy efficiency ratio EER gives the same information as the coefficient of performance, but is defined in terms of a mixed set of units. The coefficient of performance decreases as the temperature difference between colder and warmer regions increases.

Questions

Expansion and Compression of Gases

1. Why is it usually cold on top of a tall mountain?

2. True or false: Since a gas expands as its temperature rises, if you allow a gas to expand, its temperature will rise.

3. Why does a tire valve get cool as you let the air out of the tire?

4. When you pump up a basketball with a hand-operated pump, the pump gets very warm. Is there anything causing this heating other than friction? If so, what? Why does a diesel engine not need spark plugs?

Heat Engines

5. Why do most automobiles have a radiator? Why can some get by without one?

6. Why do lawn mower engines have fins along the sides near the cylinder?

7. How does thermal efficiency compare with the efficiency in general as defined in Chapter 1?

8. Why does an automobile engine have a flywheel attached to its crankshaft?

Second Law of Thermodynamics

9. Could you run a ship by extracting heat from the water and

converting it to work for propelling the ship? Would it matter if the water were very hot?

10. Suppose a power plant has an overall efficiency of 37%. Does this mean that, if the plant were run properly, it could produce 2.7 times as much electrical energy for the same amount of fuel burned?

11. Power plants heat their steam to as high a temperature as the construction materials can withstand and still confine the steam. Why not just use more steam at a lower temperature, and that way have a safer plant?

Pumping Heat

12. Does heat ever flow naturally from a particular region to one at a higher temperature?

13. In the refrigeration cycle, how is it possible for the refrigerant to be at the boiling point in both the condenser and the evaporator?

14. Could a refrigerator or air conditioner be operated with water as a refrigerant?

15. Why does heat enter the refrigerant in the evaporator of a refrigerator? Why does heat leave the refrigerant in the condenser?

16. An enterprising freshman proposes to make a refrigerator in the following way: Draw room air into a compressor and compress it until it cools below the temperature inside the refrigerator. Pass this cold air through coils inside the food compartment to pick up heat inside. Then let it pass through a tiny opening and expand back to atmospheric pressure and be released into the air. What is wrong with this idea?

17. Can you cool your kitchen by leaving the refrigerator door open?

18. If you heat your house with electricity (resistive heating such as baseboard or ceiling heaters rather than a heat pump) does it cost you anything to operate your refrigerator or freezer in the winter?

19. A handy way to dry kids' tennis shoes after they've walked in water is to leave them on the floor just in front of the refrigerator. Would putting them in front of the TV work just as well?

20. What is wrong with calling the coefficient of performance an efficiency?

21. In a heat pump you get more heat energy out than you put in as electrical energy. Does this fact violate the principle of conservation of energy?

Problems

1. If 0.30 kcal of heat is added to the gas in the cylinder of Fig. 9.1, how much does the internal energy of the gas change if the piston does not move? If the piston then moves 20 cm to the right as the gas exerts an *average* 14 000-N force on it, what is the net change in internal energy for the two processes?

    Ans: 1260-J increase; 1540-J decrease

2. A particular gasoline engine has a thermal efficiency of 20%. How much work is done by the engine for each liter (0.74 kg) of fuel used? How much heat is released to the air? (The heat of combustion of gasoline is 11 500 kcal/kg.)

    Ans: 2.0 kWh; 7.9 kWh

3. What is the thermal efficiency of an engine that takes in 400 kcal of heat and does 500 000 J of work?   **Ans:** 30%

4. The steam that turns a particular power plant generator goes into the turbine at 500°C and is condensed at 100°C. How much would the ideal maximum thermal efficiency be increased if the steam entered at 600°C?   **Ans:** 5.5%

5. A 14 000-Btu/h air conditioner has an EER of 5.1. How much electrical power does it take to operate it? What does it cost per hour of operation if electrical energy costs 5¢ per kWh? What would be the power and cost if the EER were 10.1?

    Ans: 2750 W; 14¢; 1390 W; 7¢

6. If the 5.1-EER air conditioner of Problem 5 costs $270, and the 10.1-EER one costs $380, how long would you operate the higher-priced unit to recoup the additional purchase cost in lower operating expense?   **Ans:** 1600 h

7. What is the coefficient of performance of an air conditioner with an EER of 8.4?   **Ans:** 2.5

8. The *ton,* sometimes used in rating air conditioners, is a unit of

power equivalent to the rate of heat transfer needed to freeze 1 ton of water in 24 h. What is the ton's equivalent in watts?

Ans: 3.52 kW

9. A typical house in Ohio needs about 20 000 kWh (68 × 10⁶ Btu) of heat during a winter. If electrical baseboard heaters are used, how much electrical energy is needed to heat the house? If a heat pump with an average coefficient of performance of 2 is used, how much electrical energy is needed?

Ans: 20 000 kWh; 10 000 kWh

Home Experiments

1. Hold the ends of a rubber band in your hands, with the middle held lightly between your lips. Alternately stretch and then release the tension in the rubber band, keeping the ends in your hands and the middle between your lips. Notice the change in temperature that you feel in your lips. Which way, stretching or compressing, is the band a heat engine, and which way a heat pump? Interpret the results in terms of the first law of thermodynamics.

2. If you have access to an air conditioner, determine its EER from the information given on the nameplate or in the owner's manual.

3. Observe the heat given off at the bottom or back of a working refrigerator, the outside section of a window air conditioner, and/or the outside unit of a central air conditioning system.

References for Further Reading

Derwin, R., "Heat Pumps—Cheapest Cooling and Heating for your Home?" *Popular Science,* Sept. 1976, p. 92. Shows how the same heat pump is used for cooling and heating, and evaluates the economics of its use in various parts of the United States.

Fowler, J. M., *Energy and the Environment,* (McGraw-Hill Book Co., New York, 1975). See especially Chapter 4. Discusses efficiency of various types of heat engines and their uses in energy conversion.

Leff, H. S. and W. D. Teeters, "EER, COP, and the Second Law Efficiency for Air Conditioners," *American Journal of Physics,* Jan. 1978, p. 19. Compares actual coefficients of performance (or EERs) of air conditioners with their maximum values allowed by the second law of thermodynamics.

Walker, G., "The Stirling Engine," *Scientific American,* Aug. 1973, p. 80. A heat engine with low noise and low pollution.

Part III

Electricity and Magnetism

The experiments combine to prove that when a piece of metal (and the same may be true for all conducting matter) is passed either before a single pole, or between the opposite poles of a magnet, . . . electrical currents are produced across the metal transverse to the direction of motion.

M. Faraday

Michael Faraday, born at Newington, Surrey, England, in 1791, was the son of a blacksmith. After becoming an apprentice to a London bookbinder at age 14, Faraday became fascinated with electricity after by chance reading an article on the subject in an encyclopedia brought in for rebinding. His scientific career began as a chemist, and while working as an assistant to Sir Humphry Davy, he discovered benzene. However, his major contributions were in the field of electricity and magnetism. The law named in his honor expresses the relationship between changing magnetic fields and electric currents. Faraday is shown here with his wife at Christmas, 1821.

10

Electricity

When the electricity goes off, life as we know it in this country comes to a standstill. A power failure following a storm might be fun for a while—classes and work might be called off, or you might enjoy eating dinner by candlelight if you get it cooked before the storm. But a blackout quickly loses its appeal. Certainly, you miss the obvious: lights, electric ranges, TV, electric heating and air conditioning, and the hundreds of electrical appliances that both simplify and complicate our lives. But you might be in for a surprise if you have a gas-fired furnace that doesn't heat because the electrically operated valve doesn't open to let the gas in; or if you have to walk rather than ride the elevator to your high-rise apartment; or if your escape to the next town becomes impossible because your tank is empty and the gasoline pump at the station won't operate; or if you can't purchase snacks at the grocery store because the cash register is dead without electricity.

Not only are our personal lives greatly disrupted, but most industry is completely shut down without electrical power. Sudden lack of electricity can spawn massive social disorder as we saw, for example, in the rioting and looting during the blackout of New York City in the summer of 1977.

Our involvement with electricity doesn't stop with gadgets that we plug into a wall receptacle. Ever increasing in numbers are the battery-operated tools and toys—calculators, radios, cassette players, intercoms. Even though the electric car is still relatively rare, the ordinary gasoline-fueled automobile depends on electricity to start the engine, and to produce a spark thousands of times per minute in each cylinder of the engine. And where would we be without that electrical wonder, the telephone?

Even if we could live without all of these enjoyable electrical conveniences, we would still depend on electricity because that is what controls our own bodies. The messages sent back and forth between our brains and all parts of our bodies travel electrically. Electrical signals control even our heartbeat and breathing.

10.1   What Is Electricity?

Suppose you have an ordinary table lamp with a 60-watt bulb that you expect to use for reading. You would plug the lamp into a wall receptacle and then turn the switch on the lamp. If everything is working right, light will emanate immediately from the bulb. It *appears* that *something* is waiting inside the receptacle to get into the bulb and create light as soon as a path is provided through the cord, and a "gate" is opened by turning on the switch.

A similar thing happens when you use an electric shaver. When you plug it in and turn on the switch, you make a path for "something" waiting inside the receptable to get out and into the shaver to make it go.

There are a few general requirements for the working of electrical lights and appliances that you already know about from your own experience: (1) that this "something" needs a path—a cord—to get to the device, and that this path has to be continuous—if a wire breaks or comes off one of the screws in the plug, your TV doesn't work; (2) that it takes two paths going to the device, not just one—there are two wires in your hair drier cord: and (3) that this "something" waiting in the receptable to get into your electrical apparatus is choosy about what it travels through—it will go through the metal wire but not through the rubbery insulation outside the wire. You might have found, to your regret, that it will even travel through you.

Consider another familiar electrical device—a flashlight—that operates on batteries rather than from a power cord. A person who knows nothing about electricity still realizes that something in the battery gets into the bulb and makes it light when the switch is turned on. Again this "something" seems to be waiting, ready to get into the bulb as soon as it has a path that the switch somehow provides.

Let's look at how the flashlight works. Figure 10.1 is a diagram of the inside of a typical flashlight. When you push the switch on, the metal slide moves to the right, contacting the ring around the metal casing of the bulb. This contact provides a continuous *metal* path (dashed line) from the terminal labeled "+" on the right battery, through the filament of the bulb, through the metal slide on the switch, through the flashlight case, through the spring, and to

Figure 10.1 A flashlight bulb gives off light when there is a closed path from the positive terminal of the battery, through the filament, and back to the negative terminal (dashed line).

the end of the left battery labeled "−". The "−" end of the right battery contacts the "+" end of the left one. Evidently, the batteries cause the "something" to flow through this continuous path, but are unable to cause a flow when the switch is off (open), and the path is no longer continuous. This flow causes the bulb to produce light.

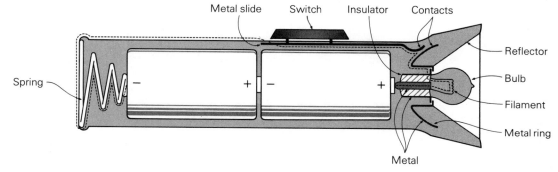

The desk lamp produces, or doesn't produce, light for the same reason. Figure 10.2 shows a simplified diagram of how the lamp works. When you move the switch so that there is a continuous path for our "something" to flow from the receptacle through the filament, the bulb gives off light.

To explain these effects we need to answer three questions.

1. What is this "something" that flows?

2. What makes it flow?

3. How does it do its job in the bulb, the motor, the heater, or whatever?

The answer to the first question is that what flows is a fundamental quantity called *electric charge.* To understand the properties of electric charge, and to answer the second question—"What makes it flow?"—we need to start by looking at the structure of the atoms that make up all matter. We must work from there toward a definition of a quantity called electrical *potential difference,* or often called *voltage.*

The third question is a broad one, and we will find the answers to it as we go through the next few chapters. In this chapter, we'll look at the basic ideas and quantities that we need in order to describe electricity. In Chapter 11, we'll find out how these principles are applied in some everyday ways. In Chapter 12 we will study magnetism, and its relationship to electricity. Chapter 13

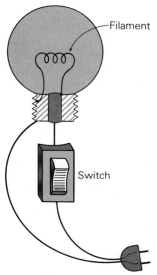

Figure 10.2 An ordinary light bulb produces light after it's plugged in only if the path is complete from one prong of the plug to the other.

takes an overall look at electrical energy: How it's produced from other sources, and how it's used to do specific kinds of jobs.

You may be getting the idea that all aspects of electricity deal with the *flow* of charges. Most practical uses of electricity do involve a flow of charge—called an *electric current*—and for that reason we will talk about currents as soon as we can in order to build on your everyday experiences. But we will consider more details of the important area of charges at rest—*electrostatics*—in Chapter 14.

In the next few sections we will lay the groundwork for electricity by defining the separate terms and ideas. By the time you get through the chapter, these should all fit together.

10.2   Atomic Structure and Electric Charge

All forms of matter—this book, the pencil with which you are supposed to be taking notes, the clothes you're wearing, as well as you and I—are made up of atoms. Your body consists of about a billion billion billion (10^{27}) atoms, which means that they are much too small to be seen with the eye or the most sensitive light microscope. In recent years electron microscopes have been developed that are sensitive enough to display individual atoms in a photograph. An example is shown in Fig. 10.3.

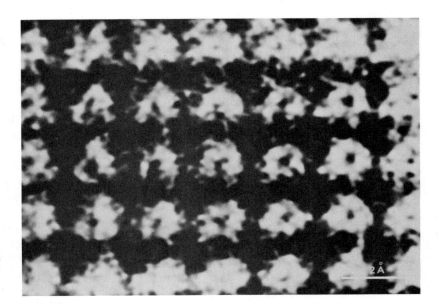

Figure 10.3 Photograph from electron microscope showing individual atoms in a gold crystal.

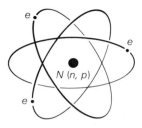

Figure 10.4 In the planetary model of the atom, electrons *(e)* orbit the nucleus *(N)*, which is made up of neutrons and protons.

No microscope, however, is capable as yet of showing the *internal* structure of an atom. For an understanding of the internal structure, we depend on indirect evidence that lets us develop models for the atom—representations that help us to visualize it in terms of more familiar concepts. A model can help to do that, even if it isn't correct in every respect.

We know that each atom has a nucleus made up of neutrons and protons at its center, and electrons orbiting about this nucleus. A model you have certainly seen of the atom is shown in Fig. 10.4. We might call this representation a *planetary* model because of its similarity to the planets and sun of our solar system. This simple model is not strictly correct, but it does help to give us a mental image of the atom. To describe the atom correctly, we use what is called *quantum mechanics,* which we will discuss in Chapter 18.

Aside from the need for quantum mechanics, there is another very misleading aspect of pictures like Fig. 10.4, which people so often use to describe atoms. *The relative size of the nucleus is greatly exaggerated.* If we draw the atom to scale, with the nucleus the size shown in this figure, the electron orbits would have to be about a quarter-mile in diameter. The atom is almost all empty space! (Roughly the same proportion of empty space as in our solar system.) Nevertheless, at least 99.95% of the mass of the atom is contained in its tiny nucleus.

Atoms differ greatly in complexity. The simplest one is the hydrogen atom whose nucleus of one proton has one electron in orbit around it. But atoms that exist in nature range in complexity on up to the uranium atom with 92 electrons orbiting a nucleus of 92 protons and 146 neutrons.

Electric Charge   A quantity called **electric charge** exists on every electron and every proton. We can't really say what charge *is;* that is, we can't define it in terms of something more fundamental. We can say only that it exists—in the same sense that mass, time, and length exist—and we can describe its properties.

First, charge exists in two types that, following the terminology used by Benjamin Franklin, we call *positive* and *negative*. **Positive** charge is the kind held by the proton, and **negative** charge the kind held by the electron. The *amount* of charge on the electron is exactly equal to the amount on the proton. The neutron has no net charge; we say it is electrically **neutral.**

The terms "positive" and "negative" are more than just identifying names. Just as positive and negative numbers balance each other, equal amounts of positive and negative charge balance each other. Since an atom in its normal state has the same number of protons and electrons, the total charge of such an atom is zero. As a *whole,* the atom is electrically neutral. A hydrogen atom, for example, with one proton and one electron has no *net* charge. If a neutral atom gains or loses an electron, it becomes negatively or positively charged. We call such a charged atom an **ion.**

Ordinary-sized objects contain many, many atoms and, therefore, many, many electrons and protons. When the total number of electrons and protons in an object is equal, the object is electrically neutral. We then say that such an object is **uncharged,** although what we mean is that it has the same amount of negative as positive charge. Because the charge on any object is made up of individual electron and proton charges, the total charge must be some whole number times the charge of the electron. In other words, charge is *quantized*—it exists in discrete bundles. These bundles are so small that, for ordinary purposes, the charge seems to be able to take any value. (If you have two truckloads of sand and are only interested in shovelsful, you don't care that the sand is quantized in individual grains.)

We usually use the symbol q (or Q) for charge and measure it in a unit called the **coulomb** (C), named in honor of Charles Augustin de Coulomb, an eighteenth-century pioneer experimenter on electric charges. The size of the coulomb is such that it takes 6.25×10^{18} electrons to give a charge of 1 C. Turning this around, the amount of charge on the electron (or proton) is 1.60×10^{-19} C. This amount of charge is very special in that it is the smallest quantity of charge known to exist on any object.* Physicists give this amount of charge its own symbol—the letter e.

Normally, your body is electrically neutral, and contains about 10^{28} electrons and protons. If you were to lose one electron out of every billion, you would have a net charge of a few coulombs.

Experiments show that the net amount of charge in any region where no charge enters or leaves remains constant. That is, no *net* charge can be created or destroyed; if a net positive charge appears in one place, a net negative charge must appear in another so that the total amount of charge does not change. This is a statement of the principle of **conservation of charge,** a law just as fundamental as the energy and momentum conservation principles.

Forces Between Charges   Another important property of electric charge is that any two charges exert a force on each other. If both

---

*Currently popular theories propose that particles such as protons and neutrons are made up of more elementary particles called *quarks,* which can have a charge of either $\pm \frac{1}{3}$ or $\pm \frac{2}{3}$ that of the electron. Some experiments have given evidence for quarks but, as of this date, no one has observed individual quarks.

DESY PROVIDES
EVIDENCE FOR FIFTH QUARK

HAMBURG, GERMANY, (January 15, 1979)—By stretching the energy limits of the DORIS storage ring to its fullest, experimenters at DESY in Hamburg have observed the upsilon and upsilon prime at roughly the masses reported in 1977 by Leon Lederman and his collaborators.

The new observations are generally taken as further evidence for the existence of a fifth quark, known as "bottom" whose charge is $-\frac{1}{3}e$. Knowing the exact mass difference between the Υ and Υ' gives further clues as to the nature of the force between quarks. The popular theory, quantum chromodynamics, assumes that the underlying force is the same for all quarks, regardless of their flavor (up, down, strange, charm, bottom, the predicted top, . . .).

The DORIS experiments provide an added fillip: Observations with PLUTO, DASP-2 and the NaI Lead-Glass Detector all show that outside the narrow energy region of the upsilon resonances, jets are visible. That is, the decay products tend to be emitted in two oppositely directed jets, a behavior expected from quarks. Although jetlike behavior had been observed earlier at SPEAR, the DORIS experiments show that, as expected for higher energy, the cones of the two jets become even narrower. At still higher energies, which will become available when CESR at Cornell, PETRA at DESY and PEP at SLAC start doing experiments, the study of jets is expected to provide a much better test of quantum chromodynamics.

DORIS results. Electron-positron storage rings are capable of giving cleaner and better resolved data on resonances like the ψ and Υ. So, accelerator specialists at DESY decided to push the total energy of the DORIS storage ring beyond its original total operating energy (8.6 GeV). By operating DORIS as a single-ring, single-bunch machine, the machine experts reached energies as high as 9.6 GeV total energy and finally more than 10 GeV. At 9.6 GeV, the PLUTO and DASP-2 groups confirmed the existence of the Υ at (9.46 ± 0.01) GeV; they obtained a partial width for decay into e^+e^- pairs of (1.2 ± 0.2) keV, ¼ the J/ψ partial width for decay to e^+e^- pairs.

Last July DESY machine experts added additional cavities to DORIS so that the total energy was raised to more than 10 GeV, allowing the search for the Υ', but the machine did not reach an energy high enough to see the possible third bump reported by Lederman and his collaborators, at 10.4 GeV, which would correspond to Υ''.

Planned experiments. For the time being, DORIS is out of action because it is being used as an injector for PETRA, which will soon be ready for experiments with 5–19 GeV electrons and 5–19 GeV positrons. The old PLUTO detector is already in place there, and two major detectors are being installed—the Mark J (a collaboration among the University of Aachen, DESY, MIT, NIKHEF in Amsterdam and the Institute for High-Energy Physics in Peking) and TASSO (a collaboration among Aachen, Bonn, DESY, Hamburg, Imperial College, London, Oxford, Rutherford, Weizmann Institute and Wisconsin). For a while the machine experts were having difficulty getting enough current in PETRA at the injection energy. PEP, the new storage ring at SLAC, is scheduled to operate this October in the same energy range as PETRA.

Late this spring or summer, DORIS will start looking for the Υ family, but by that time PETRA may be able to look for the Υ''. However, for low-lying states of Υ, lower-energy storage rings are preferable because the luminosity is higher at the needed energies. Thus CESR at Cornell, which will have up to 8 GeV in each beam, will be almost ideal. It is scheduled to be ready for experiments some time in late spring; CESR will have one multi-purpose detector, CLEO (patra), which will be operated by a collaboration among Cornell, Harvard, Rochester, Syracuse, Rutgers and Vanderbilt.

Editor's Note: These excerpts from the Jan. 1979 *Physics Today* give you the flavor of some of the terms used by particle physicists in their search for the fundamental building blocks of matter. The names in all capitals are acronyms for words that describe the nature of the particular machine.

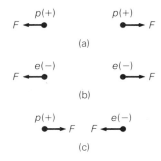

Figure 10.5 Like charges experience a force F repelling them from each other. Unlike charges experience a force of attraction.

charges are positive or both negative, then the force between them is repulsive—they push on each other. If one charge is positive and the other negative, the force is attractive—they pull on each other. Consider, for example, the two protons of Fig. 10.5(a). Each feels a force trying to push it away from the other. Similarly, two electrons (Fig. 10.5-b) feel a force trying to push them apart. On the other hand, an electron and a proton (Fig. 10.5-c) experience a force toward each other.

The force of attraction between the positive charge on the protons and the negative charge on the electrons holds the electrons in orbit around the nucleus of each atom. When atoms combine with each other to form molecules, the charges in the atoms rearrange themselves so that electrical forces hold the molecule together. The atoms or molecules that make up a solid are held in place by electrical forces. The charges arrange themselves so that when you try to pull the solid apart, the forces of attraction between unlike charges dominate, and the solid is held together. When you try to compress the solid, the forces between like charges dominate and prevent you from compressing it. There is some "give" in each direction and, if the applied force is strong enough, it can overcome these electrical forces holding the solid together, and either break it or crush it.

The situation just described means that, *aside from the gravitational force attracting you toward the earth, every force that you experience directly is electrical in nature.* (Gravitational forces from other bodies act on you, but are so weak that they're negligible.) Think about how sweeping this statement is, and try to reconcile it with the forces you've seen in action today.

Question   In ordinary table salt, sodium chloride, each chlorine atom takes one electron from a sodium atom. Does the sodium ion have a positive or negative charge? Does the chlorine ion have a positive or negative charge? What would you expect to be the nature of the force holding the two ions together in a sodium chloride molecule?

10.3   Coulomb's Law of Force Between Charges

In the late eighteenth century Charles Coulomb, the scientist for whom the unit of charge was named, found that the law governing the force between charges has the same form as that governing the gravitational force between masses. Figure 10.6 illustrates how this force varies with the size, sign, and distance of separation between two charges. As Coulomb found, the strength of the force is proportional to the amount of each charge, but is inversely proportional to the square of their distance of separation. Mathematically,

$$F \propto \frac{q_1 q_2}{d^2},$$

where q_1 and q_2 are the amounts of each charge, d is their distance of separation and, "$\propto$" means "proportional to." Putting in a proportionality constant, we have

$$F = k \frac{q_1 q_2}{d^2},$$

Figure 10.6 Coulomb's law describes the force between two charges. Each line (a through f) represents either a different pair of charges or a different separation. The length of the force arrows is proportional to the strength of the force ($1 \ \mu C = 10^{-6} C$).

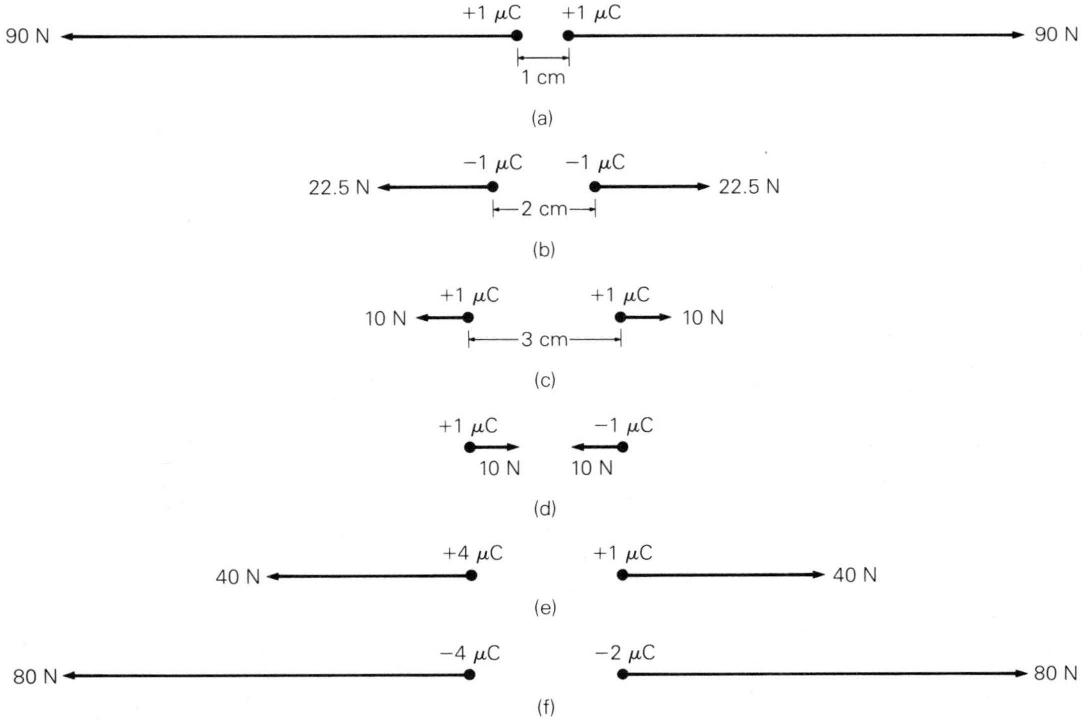

where k has the value $9.0 \times 10^9 \ N \cdot m^2/C^2$. For example, doubling either of the charges doubles the force, but doubling their separation reduces the force to ¼ its original strength, as Fig. 10.6 shows. Physicists call this law of force between charges *Coulomb's law.* You get a positive value for F if q_1 and q_2 are either both positive or both negative. If only one is negative, the force is negative. Thus, we interpret a negative force to mean attraction, a positive force repulsion. Newton's action-reaction law tells us that the force on each charge must be the same strength since they are an action-reaction pair. Coulomb's law is another of nature's "inverse-square" laws, in which a quantity varies as the inverse of the square of a distance.

Exercise   If you were standing an arm's length (about 1 m) from a friend and 1% of your 10^{28} electrons suddenly jumped from you to your friend, what force then would act on each of you? What is its direction? Compare this force with the weight on earth of an extremely dense rock with mass equal to that of the moon (7.4×10^{22} kg).

Ans: 2.3×10^{24} N attractive; electrical force 3 times as strong

---

10.4   Electric Fields

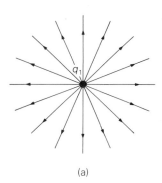

(a)

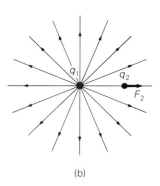

(b)

Figure 10.7   (a) The charge q_1 produces an electric field that surrounds it. (b) When the charge q_2 is in this field, a force acts on it.

Look again at a pair of the charges in Fig. 10.6. How can we explain the interaction between them? They are not physically in contact the way your hand is in contact with a door when you push on it. Yet one charge pushes or pulls on the other. The interaction is sometimes called "action-at-a-distance," but we usually picture it differently.

Consider one of the pairs of positive charges and call the one on the left q_1, the one on the right q_2. There is certainly something at the position of q_2 that causes the force on it. That something we call an **electric field.** We picture the interaction as follows: (1) an electric field results from, and surrounds, charge q_1; (2) when in this field, charge q_2 experiences a force.

We represent the fields surrounding q_1 by lines outward from the charge, as in Fig. 10.7(a). Figure 10.7(b) then shows the force on charge q_2 when it is placed in the field. The arrows on the field lines outward from the positive q_1 represent the direction of force on the positive charge q_2. If q_1 were negative, the force on q_2 would be toward q_1. That means the electric field is directed *toward* a negative charge, and *away* from a positive charge. If q_1 were positive but q_2 negative, the force on q_2 would again be toward q_1. But since the field of positive q_1 is outward, we conclude that the force on a negative charge is in the opposite direction of the field.

Similarly, the force on q_1 results from the electric field surrounding q_2.

Summary   (1) The electric field is outward from a positive charge, inward toward a negative charge. (2) A positive charge in an electric field experiences a force in the same direction as the field, a negative charge experiences a force in the opposite direction to the field. The wider spacing of the field lines with increasing distance from the charge causing them symbolizes that the field gets weaker with increasing distance from the charge.

Figure 10.7 shows only the field of one of the charges. The overall field is a combination of that due to both charges. Figure 10.8(a) shows the combined field of two equal positive charges. If both were negative the line shapes would be the same, but the arrows would head the other way. Figure 10.8(b) shows the combined field if the charges have opposite signs. Two equal but opposite charges separated by a small distance make up an electric *dipole.* When we study magnets, we'll define a magnetic dipole, and we'll want to look back at this figure then. A third charge in the vicinity of these would experience a force in the same or opposite direction as the combined field, depending on whether the charge is positive or negative.

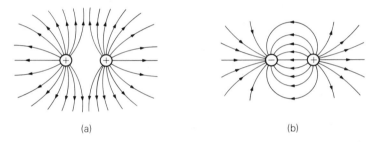

(a)   (b)

Figure 10.8   The combined field of (a) two positive charges; (b) a positive and a negative charge.

10.5   **Conductors and Insulators**

Any charge that's free to move, and is in an electric field, will move in response to that field. This may happen as charged particles move freely through space. An example would be the electrons that travel down a TV picture tube. (See Chapter 12 for more on this.) Usually, however, the movement or flow of charge takes place inside some material medium with charges that are free to move. We call such a material an electrical **conductor.** All materials conduct charge to some extent. Some are much better conductors—that is, the charge moves through them much more freely—than others. Those materials that are poor conductors are called **insulators.**

An extension cord consists of metal (usually copper) conductors surrounded by rubber or plastic insulator to prevent the flow of charge from one wire to the other or from a wire to your hand.

In general, metals are good conductors. Their atoms have a few (one, two, or three) electrons, called valence electrons, that are farther out from, and not as tightly bound to, their nuclei as the other electrons. In a solid metal these electrons are usually about the same distance from two or more nuclei. This makes them free to move throughout the metal as a sort of "electron gas". When an electric field exists across some part of the conductor, these (negatively charged) electrons move easily in the *opposite* direction of the field. Because of the specific ways the atoms are arranged in different metals, some conduct more easily than others. Silver, copper, and gold are the best conductors at ordinary temperatures, with aluminum not far behind.

Another important class of materials called *semiconductors* ought to be discussed here. This would be logical from the physics involved. But to avoid introducing any more new ideas now, we'll postpone discussing semiconductors until Chapter 13.

Ionic Conductors   Pure water is an insulator. Yet we are often cautioned not to be in water or have wet hands when using an electrical device. Are these statements contradictory?

The catch is in the word "pure." When certain salts are added to water, they form positive and negative ions that can move easily through the liquid in response to an electrical field. Ordinary tap water contains many impurities that form ions, which make it a good conductor.

There is one important difference in the way charge is conducted in an ionic solution and in a metal. In a metal, only electrons—negative charge carriers—move. In an ionic conductor, both negative and positive ions are free to carry the charge.

10.6   Electric Current

If we have some way of producing and maintaining an electric field in a conductor that has charges free to move in this field, we can have a continuous flow of charge. As we suggested in Sec. 10.1, a flow of charge is what enables most electrical devices to do their jobs. One device that can maintain an electric field is a battery (Fig. 10.9). In Chapter 13 we'll go into the details of how a battery uses chemical reactions to maintain one of its terminals—points for connection—more positively charged than the other. That charge difference gives an electric field outside the battery that is generally directed from the "positive" to the "negative" terminal.

Suppose we connect a wire from the negative terminal of a battery to one side of a bulb filament and a wire from the positive terminal to the other side of the filament, as shown in Fig. 10.10. The electric field acting on the charge of the free electrons in the wire causes them to move through the wire and filament in the direction shown.

Figure 10.9

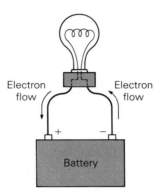

Figure 10.10   The electric field set up by the charges on the battery terminals causes a flow of electrons through the wire and bulb from the negative to positive terminal.

This flow of charge we call an electric **current.** The term current means the same here as it does when used elsewhere, such as water currents in a river, or air currents that we call wind. Sometimes we get careless and talk about a "flow of current." Since this description would literally mean a "flow of flow of charge," it's probably better to leave off the word "flow" when using the word "current."

More specifically, electric current means *the rate of flow of charge.* In Fig. 10.11, if charge is flowing in a section of the wire as shown by the arrows, the amount of charge passing through some surface (such as the plane shown) per unit of time is the current. We usually use the symbol I for current. Then,

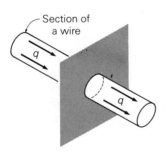

Figure 10.11   The current is the amount of charge passing through some area in each unit of time.

$$\text{current} = \frac{\text{charge}}{\text{time}},$$

or

$$I = \frac{q}{t},$$

where q is the amount of charge passing some point in time t.

Expressing the charge in coulombs, and the time in seconds, gives current in coulombs/second (C/s). We call this unit the **ampere:**

$$1 \text{ ampere (A)} = 1 \text{ coulomb/second(C/s)}.$$

In conversation, we usually shorten the word ampere to "amp." Sometimes the current itself is called "amperage."

Example 10.1 If a charge of 20 C enters and leaves the bulb filament in Fig. 10.10 in 1 min, what is the current?

Converting 1 min to seconds, we get

$$I = \frac{q}{t} = \frac{20 \text{ C}}{60 \text{ s}} = 0.33 \text{ C/s} = 0.33 \text{ A}.$$

Example 10.2 How many electrons pass a point in the filament of a flashlight bulb in 1 h if it carries a current of 0.20 A?

Since current is the charge flowing per unit of time—that is, the charge divided by the time—the total charge flowing is the current multiplied by the time:

$$q = I \times t = 0.20 \text{ A} \times 3600 \text{ s} = 720 \text{ C}.$$

Since 6.25×10^{18} electron charges make a coulomb,

$$\text{no. electrons} = 720 \text{ C} \times 6.25 \times 10^{18} \text{ electrons/C}$$

$$= 4.5 \times 10^{21} \text{ electrons}.$$

Direction of Current   I'm now going to ask you to twist your mind slightly out of shape and accept something that may at first seem quite illogical. It won't hurt so much after you get used to the idea.

The flow of negative charge in one direction is completely equivalent in external effect to the flow of positive charge in the opposite direction. If the charges moving to the right in Fig. 10.11 are electrons, this would have the same effect as *positively* charged electrons moving to the left. It is customary to refer to the direction of current as the direction of flow of *positive* charge. We call this *conventional current.* Then, if only electrons are flowing, as in a metal conductor, the conventional current is in the opposite direction to electron flow. In Fig. 10.10, even though electrons are flowing from right to left through the bulb, we say that the *current* is from *left* to *right*—from the positive to the negative battery terminals.

10.7   **Electrical Potential Energy and Potential Difference**

When we studied mechanics we found that we could learn a great deal about the motion of a body (or group of bodies) by studying

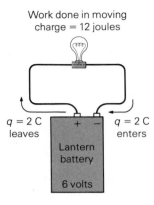

Work done in moving
charge = 12 joules

$q = 2$ C
leaves

$q = 2$ C
enters

Lantern
battery

6 volts

Figure 10.12 The electric field gives up energy in doing the work of moving the charge.

energy changes, without directly knowing anything about the forces on the bodies. We can do something similar in studying electricity. Knowing the exact electric field and the resulting forces on the charges in Fig. 10.10, for example, is extremely difficult. For practical purposes, we don't need to know them anyway; we only need to know the energy changes involved.

Potential Energy

Suppose an amount of charge q leaves the positive terminal of the battery in Fig. 10.12, and an equal amount of charge enters the negative terminal. (We'll talk in terms of movement of positive charge, even though negative charge actually moves in the opposite direction here.) The electric field does work in moving this amount of positive charge. This work must come from *some* form of energy. In this case, it comes from the potential energy the charge has in the field. *The change in electrical potential energy equals the work done in moving the charge.* Inside the battery, the same amount of charge moves from the negative to the positive terminal. The chemical energy of the battery does the work of moving the charge from negative to positive.

This flow of charge is similar in principle to a toy many young children enjoy (Fig. 10.13). The child acts as the "battery" that

Figure 10.13 The child is the "battery" that lifts the marble to the top of the zigzag track, where it tumbles to the floor as it loses potential energy.

increases the gravitational potential energy of the marble, which loses its potential energy in rolling down the track to the floor (where its anxious admirer quickly replenishes its supply). In the electrical case, the charge is "raised" to the higher potential energy by the battery, and then loses its potential energy in "falling" through the wire and bulb.

Potential Difference

The battery maintains its ability to move charge, whether or not it is actually moving any. In rating this "ability," we need a measure that is independent of the amount of charge being moved. The quantity we use is called *electrical potential difference*. We define the electrical **potential difference** as *the difference in potential energy per unit of charge moved*. That is,

$$\text{electrical potential difference} = \frac{\text{electrical potential energy difference}}{\text{charge moved}}.$$

If, for example, charge q in Fig. 10.12 is 2 C and the work done in moving it from "+" to "−" is 12 J, then the potential difference between battery terminals is:

$$\text{potential difference} = \frac{12\,\text{J}}{2\,\text{C}} = 6\,\text{J/C}.$$

If q were 6 C, the work needed would be 36 J and the potential difference still would be 6 J/C. In other words, the potential difference is independent of how much, if any, charge actually moves.

The unit of potential difference is the volt (abbreviated V). That is,

$$1 \text{ volt (V)} = 1 \text{ joule/coulomb (J/C)}.$$

Because this unit is universally used, potential difference often is called **voltage.** The symbol we use for voltage is V. (Voltage is one of the few quantities for which the symbol used for that quantity, and the abbreviation for its unit, are the same.)

What is the significance of the word "difference" in the term "potential difference"? To answer, let's compare electrical and gravitational energy. You know that when you say the gravitational potential energy of an apple is 3 J, you mean relative to some arbitrarily chosen zero of potential energy. You are really talking about a gravitational potential energy *difference* between the point in question and the point you've called zero. The same holds true of electrical potential energy and electrical potential. You can talk about the actual value of the potential at a point only if you assign it a value of zero somewhere.

Consider two examples. A flashlight battery "has" a voltage of 1.5 V, meaning that there is a potential difference of 1.5 V between the positive and negative terminals. We can say that the positive terminal is *at* 1.5 V only if we agree to say that the negative one is at 0 V.

A second example is that of commercial electricity. Ordinary light bulbs, for example, operate on a voltage of 110 V, meaning that the bulb needs a potential difference of 110 V across the filament to operate normally. We often consider the earth (ground) to be at 0 V. Then there is 110 V of potential difference between a 110-V power line and the earth. The term **ground,** when used in electricity, means whatever region is being defined as having *zero potential.* This region is often, *but not always,* the earth.

When discussing the voltage between two points, we often label the point at the higher voltage "+" and the one at the lower voltage "−", even though they may not be positive or negative relative to ground. Then a positive charge, if free, would try to move from + to −, a negative charge from − to +.

People use several different terms in talking about potential difference. We might refer to the voltage "across" the power lines, the voltage "supplied to" the TV, the voltage that a battery "has," or the voltage that a calculator "operates on." In every case, whatever the term used, when we talk about voltage we mean the potential difference *between* one point and another. If it's used concerning some electrically operated device, we usually mean the potential difference needed between the two connections on the device to make it operate normally.

10.8   **Electrical Circuits**

So far, we have been cutting out individual pieces of the electricity puzzle. Now let's try to put some of these pieces together and get an idea what the picture looks like—at least that part of the picture that shows currents.

Three basic ingredients are needed for the flow of electrical charge—a current—to do some practical job:

1. A source of potential difference.

2. A complete closed conducting path through which the charge can flow in response to this voltage.

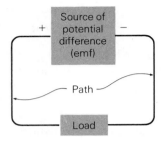

Figure 10.14 For a current to exist, we need a source of emf and a completely closed path.

3. A load—the device that does a job as the current passes through it.

Figure 10.14 shows these ingredients schematically. We'll look at each component separately.

Sources of Potential Difference—emf

Maintaining a steady current through a conductor requires some device for keeping a steady potential difference across the conductor. A potential difference produced by such a device historically has been called an "electromotive force." Since it isn't really a force at all, but a voltage, we now usually call it by its initials—**emf.** We've already talked about one source of emf, the battery. Generators and solar cells are among the others sources of emf that we'll consider in detail in later chapters.

Basically, a source of emf is a source of electrical energy. It converts some other form of energy—chemical energy in the case of a battery—into electrical energy. Being a particular potential difference, the emf of the source is, from the definition of potential difference, the work done on a unit of charge as it passes through the source:

$$\text{emf} = \frac{\text{work}}{\text{charge}}.$$

Turning this around, we see that if some known charge moves through a particular emf, the work done on it is:

$$\text{work} = \text{emf} \times \text{charge}.$$

Example 10.3 How much work is done by one of the 1.5-V flashlight batteries of Example 10.2 during 1 h of operation?

$$\text{work} = \text{emf} \times \text{charge}$$
$$= 1.5 \text{ V} \times 720 \text{ C} = 1100 \text{ J.}$$

This is about the amount of mechanical work a 60-kg (132-lb) person would perform in doing 6 push-ups.

The Complete Circuit

Suppose we have the situation shown in Fig. 10.15(a)—a circuit containing a battery, a light bulb, and a conducting path. The con-

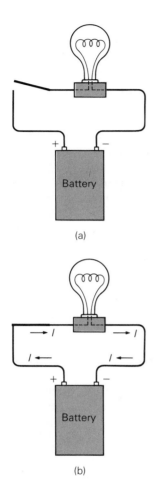

ducting path contains a gap across which no charge can flow. Even though the battery, the conductor, and the bulb filament all contain free electric charges, no current can exist anywhere in the circuit because it cannot get across the gap where there is no free charge to move. Repulsion between like charges doesn't let charge build up at some point in the circuit; if it cannot flow in one part of the circuit, it cannot flow anywhere.

On the other hand, for the situation of Fig. 10.15(b) there is no gap and the charge can move freely through all parts of the circuit. Once the circuit is continuous, and contains an emf for producing a potential difference in the circuit, an electric field propagates at nearly the speed of light throughout the circuit. This electric field acts on the free charge in the conductors, setting up a current almost instantaneously.

The situation is similar to the one shown in Fig. 10.16. Here, we have a pump and a pipe to carry water to a waterwheel, and then from the other side of the waterwheel back to the pump. The pump continually produces a difference in pressure. Figure 10.16 shows the higher pressure by a positive sign to the left of the pump. This means that the pump is trying to push water around the circuit in a clockwise direction, but when the valve is closed there is no path for the water to get through that part of the circuit. Water is still in the pipe on both sides of the valve, still in the paddlewheel, and still in the pump. The pump is still producing the pressure difference, but the water is unable to flow until the valve is opened. Once the valve is open, water flows throughout all parts of the circuit.

Figure 10.15   (a) Charge cannot flow when the conducting path has a gap. (b) Charge can flow when the circuit is complete.

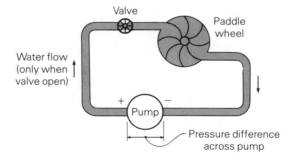

Figure 10.16   The flow of charge in a conducting circuit can be compared with the flow of water in a "water conductor."

If we compare this arrangement with our electrical circuit, we see that the pump producing the pressure difference is analogous to the battery that produces the electrical potential difference. Both are produced whether or not the circuit is complete. The flow of water in the pipe is like the flow of charge—the current—in the electrical circuit. The charge (water) is waiting to flow under the action of the potential (pressure) difference once the circuit is complete. The valve in the water circuit and the gap in the electrical circuit perform the same functions. Charge (water) cannot get across the gap (closed valve) in the conductor. Turning an electrical switch on means closing a gap in the circuit so that charge can flow. The closed switch is like the open valve in the water circuit.

This analogy breaks down in one place. Water flows because the water molecules on the high pressure side of the pump collide with those in front of them, pushing them as a locomotive would push a group of freight cars. But the electric charge is not bumped along in this manner. Instead, the electric field that propagates around the circuit acts on each charge, causing it to move.

The Load

The **load** in a circuit can be a light bulb, a motor, a heater, or any device that operates on electrical energy supplied by a current. A circuit may have more than one load, as we'll see in Chapter 11. Whatever the load, it uses the energy given to the charges set in motion by the electric field in the conductor.

One type of load is called a **resistance.** It acts much like a friction that impedes the flow of charge. Incandescent light bulbs and electric heaters are examples of loads that dissipate electrical energy by means of their resistance. We will consider resistance now, and other types of loads in later chapters.

Resistance and Ohm's Law

We have said that charge moves freely through a conductor whenever there is a potential difference across it. But we haven't really said what we mean by moving "freely." How "free" is free? We have talked so far as if the electrons that carry the current are sitting at rest until a potential difference causes them to move one way or another. Actually, due to their thermal motion, the electrons are jiggling around randomly at extremely high speeds. When a potential difference is put across the conductor, an overall drift in one direction or the other is superimposed on this random motion.

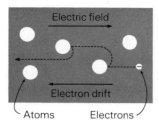

Figure 10.17 Electrons heat the conductor as they collide with its atoms, which get in the electrons' path.

When talking about currents, we are talking about this overall drift, and can usually ignore the random jiggling.

As electrons carrying a current move through a conductor, they collide with atoms or with other electrons that get in their paths (Fig. 10.17). When they do, they transfer some of the kinetic energy they had obtained from the electric field. This transferred energy increases the thermal energy of the atoms and the electrons collided with. Thus, the conductor is heated. (The process is somewhat akin to a burglar running from the police through a crowd of people. The burglar's motion is greatly impeded by the people and, every time he collides with one of them, he transfers some of his kinetic energy, which causes an increased jiggling of all the people in the vicinity.) Each time an electron loses kinetic energy by a collision, it is reaccelerated by the electric field.

We describe this interference with the electrical current by the term "resistance," which means just that: The atoms of the conductor get in the way and *resist* the flow or charge. Since the atoms of each different substance are put together differently, you might guess that each different substance would have a different resistance for the same-shaped conductor. If so, you would guess correctly.

Let's carry our guessing game a little further. How would you guess (educatedly, of course) that the voltage needed for a certain current in a conductor would change with its resistance? Since the voltage is the energy needed to drive each unit of charge through the conductor, you probably would expect the needed voltage to be higher, the higher the resistance. That's like saying it takes more work to push a wheelbarrow through tall dense weeds than over smooth grass. The higher resistance of the tall weeds means that more work is needed to push each wheelbarrow.

In fact, the potential difference needed is directly proportional to the resistance for a given current. So, we can write

$$\text{voltage} = \text{current} \times \text{resistance},$$

or
$$V = IR,$$

where the symbol R stands for resistance. We don't need a proportionality constant because the unit of resistance is defined so that this relationship holds as it is. The unit of resistance is the **ohm,** usually abbreviated Ω (the Greek omega).

George Ohm found by experimentation in 1826 that the resistance for most ordinary materials is almost constant, whatever the current and voltage. That's why we call the relationship $V = IR$ Ohm's law, and why the name ohm was given to the unit of resistance.

Example 10.4 What is the resistance of the flashlight bulb of Examples 10.2 and 10.3? We can rearrange the equation

$$V = IR$$

to give

$$R = \frac{V}{I}.$$

Since two batteries are used, the total voltage across the bulb is 3.0 V. For the current of 0.20 A,

$$R = \frac{3.0 \text{ V}}{0.20 \text{ A}} = 15 \text{ } \Omega.$$

When we talked about "good" conductors and "poor" conductors in Section 10.5, we were really talking about whether the resistance is low or high. A "good" conductor is one with low resistance. But even materials such as silver, copper, and gold, which are ordinarily the best possible conductors, have some resistance.

Any electrical energy lost by the current in a resistance is converted to heat. This heating effect can be useful. It is, for example, the way ordinary light bulbs produce their light. The current passing through the filament produces heat because of the resistance of the filament. As a result, the filament gets white hot. Almost all everyday electrical heating devices—ranges, toasters, hair driers, clothes driers, percolators, and so forth—generate their heat from this *resistive heating*.

Resistive heating also can be a nuisance. Since all conductors have some resistance at ordinary temperatures, there is always some loss of electrical energy to heat when current passes through conductors. In transmission lines, this represents a loss of usable energy. Even batteries and other sources of emf have resistance. This internal resistance of batteries usually increases with use.

Circuit Diagrams

To simplify drawing electrical circuits, people usually use a *schematic* diagram with standard symbols for electrical components. The conventional symbols used are as follows:

A battery or source of steady emf. The + and − signs are usually not written. You need to remember that the long line stands for the positive terminal.

A resistance.

An open switch.

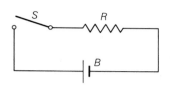

Figure 10.18   A schematic diagram of the circuit of Fig. 10.15, containing a battery B, a switch S, and a resistor R.

———•—•——— A closed switch.

——————— A conductor with zero resistance, or one whose resistance is small enough to be neglected.

Figure 10.18 is a schematic diagram, using these symbols, of the circuit of Fig. 10.15(a). The schematic diagram for 10.15(b) would be the same, except that the switch would be closed (on). For most purposes, the resistance of the wires leading to a device is so much less than the resistance of the operating device itself that the wire resistance can be neglected. For example, the resistance of a 100-W light bulb is about 120 Ω; that of 3 m of ordinary extension cord is about 0.06 Ω.

10.9   **Electrical Energy and Power**

If you look at the specifications stamped or printed on an electrical device, you usually find the operating voltage and either the current *or* the power requirements listed. If you are a consumer and purchaser of electrical energy, the power a device uses (the rate of energy use) has a direct impact on your pocketbook, since it determines the energy used in a given period of time. Our final task for this chapter is to find out how voltage, current, and power are related.

Remember, the potential difference, or voltage, between two points is the work done on each unit (coulomb) of charge in moving it from one of the points to the other. Therefore, the total work done, or electrical energy used, in moving some charge between these points is

work done = energy change = voltage × charge.

The **power** is the rate of doing work:

$$\text{power} = \frac{\text{work}}{\text{time}} = \frac{\text{voltage} \times \text{charge}}{\text{time}}.$$

But the charge flowing per unit of time is the current. Therefore,

power = voltage × current,

or $P = VI.$

If we know the power used during a certain time t we can obtain the total energy used during this time from

work done = energy used = power × time

$$= VIt.$$

When the current moves through a battery, the potential difference is the emf of the battery. The power is then being supplied by the battery at the rate $VI = \text{emf} \times I$, since V in this case is the emf. A resistance or some device that depends on electrical energy, uses it at the rate VI, where V is the voltage across the device carrying current I. The *power* used by some devices—typically lamps, toasters, electric blankets—is marked on them in watts or kilowatts. On others—typically motors, refrigerators, radios—the *current* rating is listed. Whichever is given, you can find the other by a simple calculation.

Example 10.5   What is the power delivered by the two flashlight batteries of Examples 10.2 through 10.4?

The voltage of the two batteries together is 3.0 V and the current delivered is 0.20 A. Then,

$$\text{power} = \text{voltage} \times \text{current}$$

$$= VI = 3.0 \text{ V} \times 0.20 \text{ A} = 0.60 \text{ W}.$$

Example 10.6   How much current passes through a 100-W light bulb that operates on a potential difference of 110 V?

Rearranging the power equation by dividing both sides by V gives

$$I = \frac{P}{V} = \frac{100 \text{ W}}{110 \text{ V}} = 0.91 \text{ A}.$$

A good rule of thumb for comparing different devices is to remember that a 100-watt bulb carries a current of *about* 1 amp.

Question   Fig. 10.18 is a circuit diagram for a flashlight. If R is the resistance of the bulb filament, how does the electrical power dissipated by the bulb filament compare with that delivered by the battery? How does it compare with the rate of use of chemical energy in the battery? How does it compare with the rate of heat plus light energy given off by the bulb?

Exercise   How much current passes through a 60-W bulb that operates on 110 V? At 5¢ per kilowatt-hour (kWh), how much does it cost to operate this bulb for 10 h?   **Ans:** 0.55 A; 3¢

10.10   **Summary**

Figure 10.19 summarizes the relationship between the various quantities we have defined. In the vicinity of an electric charge

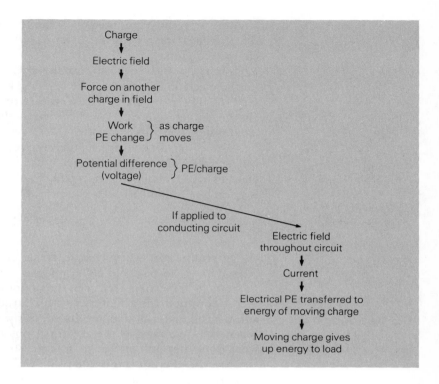

Figure 10.19   The cause and effect relationships between electrical quantities. PE stands for potential energy.

there is an electric field. This field exerts a force on any other charge in the field. If this charge moves, work is done, and the potential energy changes. The potential energy change per unit of charge is called the potential difference, or voltage.

To maintain a continuous flow of charge—a current—a device is needed that keeps a potential difference between its terminals. This particular potential difference is called an emf. If a continuous conducting path is connected across the terminals, an electric field will be set up through the conductor. This field acts on the charges, producing a current as the charges gain energy from the field. The moving charges then give up their energy to the load. If the load is a resistance, this energy transfer takes place as the moving charges collide with atoms in the resistor, transferring their energy to thermal energy (heat).

Key Concepts

Electric charge exists in two types: **positive** charge on protons, and **negative** charge on electrons. If a particle or body has equal amounts of positive and negative charge, we say it is **uncharged,** or that it is electrically **neutral.** An atom with a net positive or negative charge is an **ion.**

Charge is measured in **coulombs** (C). The amount of charge on the electron or proton—the smallest known charge—is 1.60×10^{-19} C. The principle of **conservation of charge** states that a net charge cannot be created or destroyed—the total charge in an isolated system remains constant.

A *force* exists between any two charges. Like charges repel each other; unlike charges attract each other. The forces on the separate charges are an action-reaction pair. The force between two charges is given by Coulomb's law:

$$F = k \ \frac{q_1 q_2}{d^2} .$$

Surrounding every charge is an **electric field.** A positive charge in an electric field experiences a force in the same direction as the field, a negative charge in the opposite direction.

A **conductor** is a material with charges free to move. An **insulator** is a material that conducts poorly. Electrical **potential difference,** or **voltage,** is the electrical potential energy change, or work done, per unit of charge. It is measured in **volts** (V).

Electric **current** is the rate of flow of charge. It is measured in **amperes** (A). A steady current requires: (1) a source of potential difference—an **emf;** and (2) a continuous conducting path across the potential difference. There is usually a **load** in the circuit—a device for using the energy associated with the current.

The **resistance,** current, and voltage across a conductor are related by: voltage = current × resistance; or $V = IR$. Resistance is measured in **ohms** (Ω). All electrical energy dissipated in a resistance is converted to heat.

Electrical **power** is given by: power = voltage × current; or $P = VI$. This is the power delivered *by* a source with emf V, or the power delivered *to* a load across which there is potential difference V. The total energy delivered or used in time t is power × time or VIt.

Questions

Atomic Structure and Charge

1. The distance from the center of a hydrogen atom to the most probable location of the electron is about 100 000 times the radius of its nucleus, which is a single proton. If you built a scale model of the hydrogen atom using a golf ball as the nucleus, how far away would you need to locate the electron?

2. If you and your chair are made of atoms, and atoms are mostly empty space, why don't you pass right through the chair when you sit on it?

3. What does it mean to say that an atom is uncharged? What does it mean to say that a baseball is uncharged?

4. A chemistry graduate student claims that he has an ionic solution in which each ion has a charge of 2.4×10^{-19} C. Is there a reason that you should doubt the accuracy of his claim?

5. Particle-physics theories postulate the existence of several kinds of quarks that have been given interesting names. These include: "Up" quarks with charge $+\frac{2}{3}e$; "down" quarks with charge $-\frac{1}{3}e$; "strange" quarks with charge $-\frac{1}{3}e$; "charmed" quarks with charge $+\frac{2}{3}e$; and "bottom" quarks with charge $-\frac{1}{3}e$. (Here e is the amount of charge on the electron.) In these theories, protons and neutrons are thought to be made of three quarks. What combinations of three up quarks and down quarks give the right charge to each of these "ordinary" particles? (The other quarks are needed for particles not quite so ordinary.)

6. Justify the statement in the chapter that, aside from the gravitational force attracting you and the objects around you toward the earth, every force that you experience directly is electrical in nature.

7. Arrows on the electric field lines tell you its direction. How can you tell from a diagram of the field what its relative strength is at different locations?

Conductors and Current

8. True or false: Insulators conduct absolutely no charge when a voltage is placed across them.

9. Name at least two purposes for the insulation on an extension cord.

10. Does the amount of current in a wire have anything to do with how fast the charges are moving?

11. If you connect a copper wire from the positive terminal of a battery to a bulb, and another wire from the bulb back to the

negative battery terminal, what is the direction of electron flow in the bulb? What is the direction of the conventional current?

Potential Energy and Potential Difference

12. When a mass gets closer to the earth, its gravitational potential energy always decreases. When a small charge gets closer to a large one, does its electrical potential energy always decrease?

13. When an electron moves from the negative terminal of a battery, through a light bulb, and to the positive terminal, does its potential energy increase or decrease?

14. Is an emf a force? Why do you suppose it was called "electromotive force" historically?

Circuits

15. In what ways is the analogy between opening a door to let people out of an auditorium and turning on a light switch a good comparison? In what ways is this comparison bad?

16. In what way does a water faucet perform the same type function as an electrical switch on the wall?

17. If a pump is used to pump water through a pipe that forms a closed circuit, the amount of water flowing is proportional to the pressure difference across the pump. In the analogous electrical circuit, which quantity—voltage or current—is analogous to pressure difference, and which to the rate of flow of water?

18. Why is it that on some sets of Christmas tree lights, if one bulb burns out, they all go out?

Resistance

19. Is there any similarity between the use of the word "resistance" in connection with electrical circuits, and its use in other areas? (The suspect offered no "resistance" when arrested by the police.)

20. The resistance of a metal conductor comes partly from the fact that electrons moving through it collide occasionally with atoms that get in the way. Since these atoms jiggle around

farther from their fixed positions as the temperature of the metal increases, how would you expect the resistance of a metal to change with temperature?

21. What is the resistance of a light switch that is off (open)?

22. A container of powdered carbon changes its electrical resistance as the pressure on the carbon changes. Can you speculate how this process is used in a telephone microphone to convert pressure variations in a sound wave to current variations in the telephone wire?

23. The internal resistance of a car battery increases with decreasing temperature, mostly because ions travel much slower through the battery solution when it's cold. What is one reason cars are much harder to start on a cold morning?

Energy and Power

24. In some European countries, the voltage on standard power lines is 220 V. What is the current supplied to a 100-W bulb operated on this voltage, if it takes 0.9 A to operate a bulb of the same power on 110 V?

25. A portable electrical heater has two settings: 1000 W and 1500 W. Is there any difference in the total amount of energy needed to keep a room's temperature near 20°C with this heater when you use its different settings?

26. Batteries are often rated in *ampere-hours*. Show that this rating, along with the rated voltage, tells you the total energy that can be supplied by the battery. Does it tell you anything about the number of months the battery will remain good?

Problems

1. How many electrons would you have to remove from a sheet of paper to give it a net charge of 50 μC*?   **Ans:** 3.1×10^{14}

2. What is the force of attraction between the electron and proton in a hydrogen atom when the electron is 53 pm* from the proton? Compare your result with the answer to Problem 3.

Ans: 8.2×10^{-8} N

---

*Metric prefixes are identified in Appendix II.

3. What is the force of repulsion from the electrical force between two protons 1.5 fm* apart in a nucleus? Compare your result with the answer to Problem 2. With the much stronger forces acting inside a nucleus than between the nucleus and electrons, do you get a hint as to why we can get much more energy from a kilogram of nuclear fuel than from a kilogram of coal?

Ans: 100 N (23 lb)

4. The mass of the earth is 6.0×10^{24} kg; that of the moon is 7.4×10^{22} kg. How much charge would have to be added to both earth and moon, in equal amounts, to overcome the force of the gravitational attraction? Assume there is one electron for every 4×10^{-27} kg of moon mass (slightly more than an average of one neutron and one proton per electron). What *fraction* of the moon's electrons would have to be removed to give this amount of charge?   **Ans:** 5.7×10^{13} C; 1.9×10^{-17}

5. A charge of 54 000 C enters the heating element of an electric heater in 1 h. What is the current through the filament?

Ans: 15 A

6. A 60-W light bulb carries a current of 0.55 A. How much charge leaves the filament in 8.0 h of operation? **Ans:** 16 000 C

7. A 100-Watt light bulb carries a current of about 0.91 A. How many electrons enter the filament per second? **Ans:** 5.7×10^{18}

8. If q in Fig. 10.20 is 4.0 μC, and 0.60 J of work are needed to move it from where it is to point P, what is the potential difference between these points?   **Ans:** 150 kV

9. What is the resistance of a 100-W bulb that operates on 110 V? What is it for one that operates on 220 V?   **Ans:** 120 Ω; 480 Ω

10. A hot plate with a resistance of 24 Ω is designed for a current of 9.1 A. What voltage should it be operated on?   **Ans:** 220 V

11. One wire of an ordinary lamp cord 25 m long has a resistance of about 0.51 Ω. What is the voltage drop in this wire if it carries a current of 3.0 A?   **Ans:** 1.5 V

12. A current of 100 mA through your body would probably stop your breathing. If you touched the two sides of a 110-V power line—one hand in contact with each wire—and the combined resistance of your body and the two points of contact were 1000 Ω, what would be the current through your body?

Ans: 110 mA

Figure 10.20   Charge q is moved to point P, closer to charge Q.

*Metric prefixes are identified in Appendix II.

13. A freezer is rated at 5.5 A on 120 V. What power is used by this freezer when operating on rated voltage?   **Ans:** 660 W

14. If the freezer of Problem 13 is actually running 20% of the time, how much energy does it use in a year? How much does it cost to operate at 5¢/kWh?   **Ans:** 1160 kWh; $58

15. How much electrical energy is used by a 100-W bulb operated for 24 h? How much water could you heat from 20°C to 100°C on this amount of energy?

    Ans: 2.4 kWh (2070 kcal); 26 liters

16. The resistance of the filament of an incandescent light bulb at room temperature is about 1/10 what it is at operating temperature. How does the power consumption the instant after the bulb is turned on compare with that after it reaches operating temperature?   **Ans:** 10 times as much initially

17. It takes about 0.02 s for an incandescent bulb to reach operating temperature. Assume the *average* power used during this time is five times what it is when operating normally (see Problem 16). For how long can you leave the light on and use less electrical energy than by turning it off and back on again at the end of the time?   **Ans:** 0.08 s

18. A 12-V car battery is rated at 120 ampere-hours. How much energy in kilowatt-hours and in joules can be supplied by this battery before it is discharged?   **Ans:** 1.44 kWh; 5.2 MJ

Home Experiments

1. Take a flashlight apart and examine the parts. Trace out the circuit, and see how it is completed and broken by turning the switch on and off. Flashlights often need a "whack" to make them come on. If you have one like that, try to find out why, and see if a judicious bending of some part will cure the problem.

2. Take the cover off and examine the switch mechanism, cord connections, and bulb contacts on a table lamp that uses an incandescent bulb. *Remember to unplug it first!* Follow the path of the current. If you have a lamp that doesn't work, try to locate its problem and fix it. If you can't find a broken wire or a loose connection between the wire and plug or switch, you may need a new switch and bulb socket from the hardware store. *Be certain that any part that carries a current is well insulated!*

References for Further Reading

Bueche, F., *Principles of Physics,* 3rd Ed. (McGraw-Hill Book Co., New York, 1977). An example of many good books that discuss basic electricity using techniques somewhat more mathematical than in this text, but nothing more advanced that algebra and trigonometry.

Dresner, S., "Superconductors," *Popular Science,* Oct. 1974. At a temperature near absolute zero, the resistance of some metals drops to *zero.*

Lubkin, G. B., *Physics Today,* July, 1977, p. 17. A report on one experimental search for quarks.

Resnick, R., and D. Halliday, *Physics,* 3rd Ed., (John Wiley & Sons, New York, 1977). If your mathematical abilities include calculus, you will find this book contains an excellent introduction to electricity.

11

Practical Electricity

Everybody knows that electricity is practical. *You* now know (after having read Chapter 10) the basic principles of electricity. But what do the practical and the basic have to do with each other? Now we want to see how basic electrical principles are applied to practical everyday uses.

We'll start by considering ways to combine more than one load in the same circuit. We'll then distinguish between alternating and direct current. Residential and automobile circuitry will provide down-to-earth examples of circuit principles. Resistive heating due to currents has many practical uses, but also can be a nuisance or a hazard—we'll discuss both aspects. Finally, we'll discuss an area important to everyone that uses electricity—shock, and how to avoid it.

One new idea introduced in this chapter is fundamental as well as practical. It's called superconductivity. We include this topic here because it has practical importance, and relates to other concepts in the chapter.

11.1 Combining Circuit Elements

You can plug several lamps into the same electrical outlet. Is there some special way these lamps have to be connected? What's inside that little adapter that lets you plug three lights into one outlet? Let's find out why loads have to be connected in a particular way in a circuit.

Resistances in Series   Compare the circuits of Fig. 11.1(a) and Fig. 11.1(b), where the only difference is an extra bulb in circuit (b). The two identical bulbs are in **series**—they are connected so that whatever current passes through one must pass through the other. How would you expect the current in circuit (b) to compare with that in circuit (a)? Since both resistances are impeding the same current, the resistance is twice as much in (b) as in (a). Since the emf is the same for both circuits, we know from Ohm's law ($V = IR$) that the current in (b) is ½ that in (a).

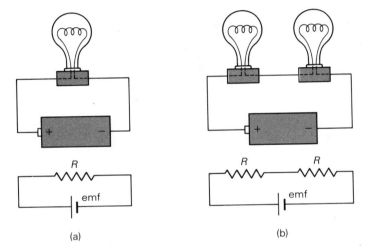

Figure 11.1   Actual and schematic diagrams of (a) a battery and one bulb; (b) a battery and two bulbs in series.

(a)   (b)

If the resistances are not identical, you just add all of them, whatever they are, to get the total resistance.

Example 11.1   If each bulb in Fig. 11.1 has a resistance of 200 Ω and the emf of the battery is 110 volts, what is the current in each circuit? What is the power used by each bulb?

First let's consider circuit (a). From Ohm's law, $V = IR$, we have

$$I = \frac{V}{R} = \frac{110\,\text{V}}{200\,\Omega} = 0.55\,\text{A}.$$

From $P = VI$, the power is

$$P = 110\,\text{V} \times 0.55\,\text{A} = 60\ \text{watts}.$$

Now for circuit (b), the resistance is twice that in (a), or 400 Ω. The current is

$$I = \frac{110\,\text{V}}{400\,\Omega} = 0.275\,\text{A}.$$

The power to *both* bulbs is then

$$P = 110 \text{ V} \times 0.275 \text{ A} = 30 \text{ W},$$

or ½ what it was to the one bulb of circuit (a). But since the bulbs are identical, the energy use is divided equally between them: 15 W to each bulb, or only ¼ the power to one bulb alone! Do you expect ½ as much light from two bulbs as from one? Evidently, we don't connect multiple light bulbs in series.

A Note of Caution about Light Bulbs as Resistors If your instructor decides to demonstrate the correctness of Example 11.1 by hooking up and operating the circuits of Fig. 11.1, it will probably look as if the light output of the bulbs is almost as predicted. But your eye is not a good judge of such matters. If a meter is hooked up to measure the current, don't be surprised if the current doesn't drop from (a) to (b) as much as expected. The reason is that the resistance of the bulb filament increases with increasing temperature. When the filament is at its normal operating temperature of roughly 2500°C, its resistance is about 10 times what it is at room temperature. Since the bulbs do not get as hot under the lower current of the series connection, the resistance of each bulb is less in the arrangement of circuit (b).

Furthermore, the power we are talking about is the *electrical* power, not the light output. Since the filaments do not get as hot, the fraction of the energy that goes into light is much less. The actual light output in Fig. 11.1(b) is therefore much less than ½ what it is in (a).

Resistances in Parallel In Fig. 11.2 we have added another bulb in **parallel** with the one of Fig. 11.1(a). Now the current I passing through the battery has two alternative paths between point x and point y. (A parallel connection does not mean that the paths are necessarily geometrically parallel.) Do you now expect more current out of the battery, or less, than for only the one bulb of Fig. 11.1(a)?

Consider an analogy. Suppose there are 50 000 people trying to get out of a stadium after a football game. If there is only one gate open, people will flow through the gate at a certain rate. If the stadium officials open another identical gate, people will pass through it at about the same rate. The total rate of flow of people from the stadium is about twice what it is with one gate open. People's urge to get out (their "emf") is about the same, whatever the number of gates open.

A similar effect occurs in electrical circuits. When we add path 2 in parallel, the same emf tries to push charge through the second

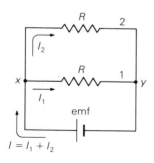

Figure 11.2 Two identical resistors in parallel need twice the current from the battery as one resistor alone.

path as through path 1. Since path 2 can carry the same current as path 1, the total current through the battery is twice as much as it would be with only one resistor.

Since more current passes through the battery with the two resistors in parallel than with one, the overall resistance of the combination is *less* than the resistance of one resistor by itself. If the two resistors have unequal resistance, the larger current flows through the lower resistance. (More people would leave a stadium through a larger, lower-resistance gate.) For more than two resistors, the current through each one is inversely proportional to the resistance, with the overall resistance of the combination always being lower than that of any one resistor.*

When resistances or other loads are connected in parallel across an emf, each one is independent of the other unless so much current is drawn from the source of emf that its voltage drops below normal. When we plug different devices into electrical outlets in a house, we are connecting them in parallel across the source of emf. That way, each one "sees" the full voltage supplied by the line and can operate as designed.

Exercise   Draw a schematic diagram showing a light bulb, a toaster, a percolator, and a TV connected in parallel across a source of emf.

Water Analogy   To keep all these circuit ideas straight in your mind, it sometimes helps to compare electrical circuits with the flow of water in a water circuit. Figure 11.3 shows the water version of the three circuits in Figs. 11.1 and 11.2. Recall that the current (the rate of flow of charge) is analogous to the rate of flow of water. The potential difference (voltage) of the battery is similar to the pressure difference produced across the pump that keeps the water flowing. The small wiggly pipes offer high resistance to water flow, whereas the large ones offer almost none. We'll assume all sections of wiggly pipe are the same, and therefore offer the same resistance to the water flow.

Let I represent the rate of flow of water in part (a). Then, in (b), the flow is $\frac{1}{2}I$ because the water has to go through two resistances. In (c), each resistance is connected independently across the

---

*The effective resistance R of a group of parallel resistors is (see Problem 3, this chapter):

$$\frac{1}{R} = \frac{1}{R_1} + \frac{1}{R_2} + \frac{1}{R_3} + \dots .$$

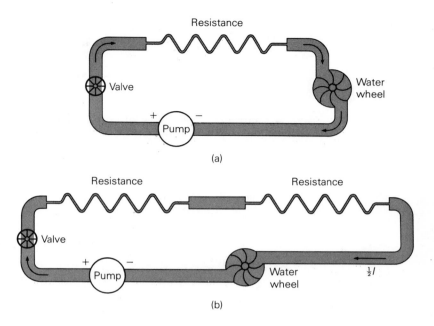

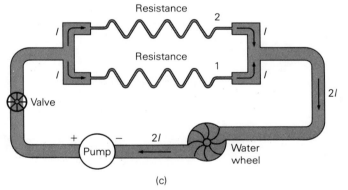

Figure 11.3 The current in an electrical circuit can be compared with the flow of water in a pipe. The small wiggly pipes offer resistance to water flow. All components are identical. If (a) I is the *water* current through the pump with one resistor, then (b) $\frac{1}{2}I$ is the current with two resistors in series, and (c) $2I$ is the current through the pump with two resistors in parallel.

pump, and each draws a flow of I. Then the total flow through the pump is $2I$.

11.2 Alternating Current

Batteries produce a steady emf in one direction. The current they cause when connected in a closed circuit is a steady current in one direction around the loop—what we call a **direct current,** or DC. But the emf produced by the generators at the power plant—and delivered to your room, your house, the place where you work, and your classroom—changes its direction 120 times per second. Therefore, the current it causes in your lights changes its direction 120 times per second—an example of what we call **alternating current,** or AC. We call such a current 60-hertz (Hz) AC, where 1 Hz is 1 cycle/s. There are 60 complete cycles each second, with current first in one direction, and then the other. It is fairly common to use terms like "AC current" or "DC voltage," even though their strict interpretation is not very logical. Most countries other than the United States and Canada use 50-Hz AC.

These changes in direction are not abrupt, but are smooth variations. Figure 11.4 is a graph of emf versus time for AC. The part of the curve above the horizontal axis stands for emf in one direction, that below the axis for emf in the opposite direction. We usually refer to this emf as positive in one direction, and negative in the other. The emf increases gradually to a maximum positive value, goes back to zero, then to a maximum negative value.

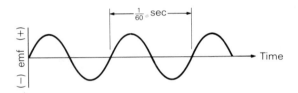

Figure 11.4   In 60-Hz alternating current, the emf and the current reverse their direction 120 times per second, varying smoothly with time as shown in this graph.

As the emf applied to a closed circuit varies according to the graph, the electric field throughout the circuit increases smoothly to a maximum in one direction, and then in the other. These changes in field propagate through the conductor at almost the speed of light. As a result, the current alternates smoothly at 60 cycles/s in proportion to the emf. A graph of current versus time would look just like Fig. 11.4, where positive current is in one direction and negative in the other.

The statements we have made about circuits and about combining resistances apply equally well whether we are talking about AC or DC. The details of the design of some pieces of equipment—for example, motors that run hair driers and fans, and electronic devices such as TV sets—depend upon whether the device in question is to operate on AC or DC. But light bulbs and electric heaters, which depend only on resistance, don't "care" whether they receive AC or DC.

NIAGARA PLANT FAVORS WESTINGHOUSE OVER EDISON

BUFFALO, N.Y., (November 8, 1891)—The Cataract Construction Company has decided to use alternating-current, rather than direct-current, generators for the large hydroelectric plant they are to build at Niagara Falls to supply electricity to Buffalo. Even though great controversy has surrounded this decision, if it proves successful it may set the pattern for any future plants that might be built.

Thomas Edison has strongly opposed the use of AC rather than DC for such a purpose—a highly understandable attitude since Edison Electric Light Company has invested vast sums of money in DC plants for many large cities. Edison's argument for DC is the following: Since electricity is used only for lighting, a need that exists just a few hours a day, the plant could store in batteries energy generated during the remainder of the day for use during peak-demand hours. No such storage device is available for AC-generated energy.

On the other hand, the Westinghouse Electric Company, under the leadership of George Westinghouse, has been building AC generating systems for about five years. Since any generator actually produces AC, Westinghouse's generators do not have the complicated devices called commutators that convert AC to DC at the generator output. Furthermore, Westinghouse has shown that, with AC, they can use a transformer to step up the voltage for transmission cross-country, and then step it down again on the user end. This voltage change, which is not possible with DC, drastically reduces losses to heat along the lines.

Only time will tell whether this decision was the correct one.

When we specify the amount of alternating current or voltage, we are giving the "effective" value—the AC that has the same heating effect as an equal amount of DC. It turns out that the peak (highest) value of the current or voltage during any one cycle is 1.41 times its effective value. For 110-V AC, the peak voltage is 155 V.

11.3   Residential and Automobile Electricity*

We can now see how the fundamentals of electrical circuits are applied in the distribution of electrical energy in the home and car. We will not learn any new physics in this section, but we will see how what we have already learned applies in some practical situations.

---

*Sections with the ** symbol are for illustration only, and contain no new principles of physics.

**Residential Circuitry

Figure 11.5 is a schematic diagram that shows the overall distribution scheme in a house. It is not intended to look like the real thing, but rather to show its workings. We'll discuss the parts of this diagram separately.

Voltages   Home electrical devices are designed to operate on a potential difference of either 110 V or 220 V.* Lights and low-power appliances—TVs, shavers, sewing machines, typewriters—use 110 V; high-power appliances—ranges, clothes driers, water heaters, many air conditioners—use 220 V. There is nothing sacred about these values—electrical industries in the United States and Canada have simply agreed to use them. Other countries use other voltages.

Three wires (shown in the upper right-hand corner of the diagram) run to each house from the main lines of the power plant. One of these wires, called the *neutral* one, is electrically connected to the earth (the one with the ground symbol $\perp$). This one's potential is defined as zero, or as electrical ground. Each of the other two wires is at 110 V relative to the neutral line. We often call these the "hot" lines, so named because they are at a voltage above that of ground. The two hot lines are at 220 V relative to each other. For AC this means that when one line is at its maximum positive voltage, the other is at its maximum negative voltage, giving an effective 220 V between them. (For DC this would mean that one line is at + 110 V and the other at − 110 V relative to the neutral line at 0 V.)

The Energy Meter   The incoming lines run to the energy meter, which is usually mounted on the side of the house (Fig. 11.6). The rotation of the dials on the meter is proportional to the energy used. What you pay for directly is the total energy (the number of kilowatt-hours). As we will see in Chapter 13, the rate at which you use energy (the power) at certain times of the day may also influence your bill. (You read the meter from left to right—the reading of each dial being the smaller of the two numbers that the pointer is between. Verify that the meter shown reads 31182.)

---

*Sometimes you see these specified as 115 V and 230 V, 120 V and 240 V, or 125 V and 250 V. Light bulbs, for example, are usually labelled 125 V. The "110-V" line is usually at a voltage somewhere between 110 and 120 V.

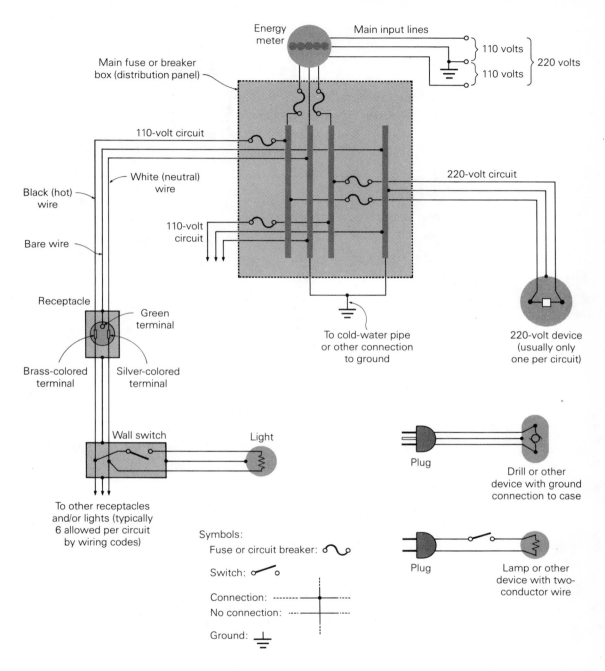

Figure 11.5   A schematic diagram of the overall distribution of electrical energy to a residence.

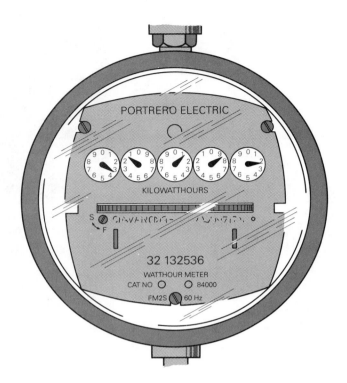

Figure 11.6   An electrical energy meter.

Distribution of Energy Throughout the House   Figure 11.5 shows schematically how wires are connected for distribution of energy throughout the house. We'll mention a few pertinent points.

1. The input lines are connected to large conductors (bus bars) inside the distribution panel (usually called the *fuse box* or breaker box—see Fig. 11.7). Cables are connected to these conductors for running through the walls to various parts of the house.

2. Each cable consisting of three wires runs from the breaker box to a small group of lights and receptacles (electrical outlets). This is called a *circuit*. A large appliance, such as a water heater or range, usually has its own private circuit.

3. The two insulated wires in the cable actually form the current-carrying loop. For a 110-V circuit, the two wires are hooked, one each, to a hot line and the neutral incoming line via the bus bars. A 220-V circuit is hooked to the two hot lines.

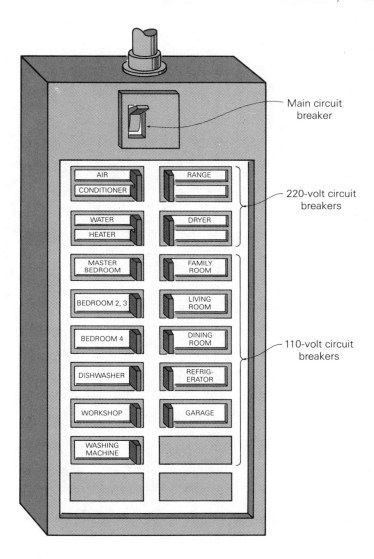

Figure 11.7 The service panel or circuit breaker box.

4. It's customary in this country to follow the color codes indicated in the diagram. In 110-V circuits, the wire with the black insulation is used for the hot wire; the one with the white insulation for the neutral wire. This code merely identifies which wire is which so that connections can be made properly.

5. The two current-carrying wires are connected to the terminals of a receptacle as shown in Fig. 11.8, again following conventional color codes.

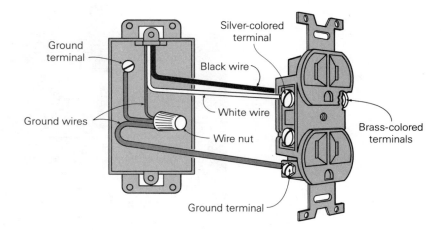

Figure 11.8 The connections to a receptacle.

6. Wall switches to lights are connected in the hot line. Figure 11.9 shows how the actual switch closes and opens a gap in the closed conducting path.

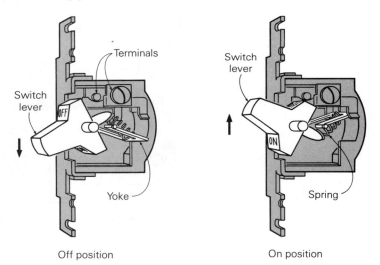

Off position

On position

Figure 11.9 A wall switch opens and closes the circuit when turned off or on, just as our schematic symbol shows.

7. When a wall switch is turned on (closed), a complete closed circuit is made from the power plant, through one of the supply lines, through the light, and back through the other line. This closed circuit lets current pass through the light bulb, "pushed" by the emf of the generator at the power plant. When other loads, such as the lamp in the lower right of Fig. 11.5, are plugged into a receptacle in the circuit, the connection is made in

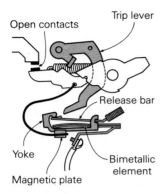

Open contacts — Trip lever

Release bar

Yoke

Magnetic plate — Bimetallic element

Figure 11.10 A circuit breaker has a bimetallic strip that is heated by the current in the circuit. When too much current passes through the strip, the temperature rise causes it to bend enough to open the switch.

parallel with the light or other devices plugged into another receptacle.

8. Excess current in the wires causes them to overheat. This "overload" can come from operating too many devices on the circuit or from faulty insulation, which allows the hot and neutral wires to come into contact (a "short circuit"). To guard against the dangers that may result, a fuse or circuit breaker is in series with each hot line. A fuse has a conductor through which the current in the circuit must pass. When the current exceeds that for which the fuse is rated, this conductor melts (the fuse blows), forming a gap that stops the current. The size of fuse is coordinated with the size of the wire in the circuit, which in turn is coordinated with the amount of current needed to operate the device or devices in the circuit.

Most houses now use circuit breakers rather than fuses. These operate by means of a bimetallic strip that opens the circuit when enough current is drawn to heat the strip and bend it enough to break contact (see Fig. 11.10). Rather than replacing the breaker as you would a fuse, you just reset it by throwing a switch *after finding and correcting the trouble that caused the excess current.*

Grounding The bare wire in the cable for each circuit is connected inside the distribution panel to a conductor hooked to a water pipe or something else in good contact with the earth. This wire is then connected to the terminal on each receptacle (Fig. 11.8) into which the odd prong on a three-pronged plug fits (Fig. 11.11).

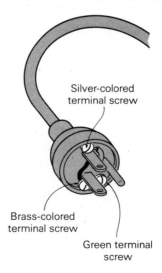

Silver-colored terminal screw

Brass-colored terminal screw

Green terminal screw

Figure 11.11 The wire attached to the third prong of the plug is connected to the housing of the appliance.

A device such as a drill, with this third prong on its connecting plug, has its third wire connected to its housing. Should the insulation inside the drill fail and the hot wire contact the casing, the casing will still remain at zero volts because it's in contact with the ground. Since you, the operator, normally would be at zero potential relative to the ground, there would be no voltage across your body, and therefore no reason for a current to flow through your body. Instead, the ground wire would carry the excess current from the "short" in the drill to ground.

Exercise   Analyze the diagram of Fig. 11.5 carefully, and convince yourself that all devices plugged into receptacles and all lights turned on with wall switches are connected in parallel across the main supply lines.

**Ground as Part of the Circuit

As we have said, electrical *ground* means whatever region we define to have *zero* potential. The symbol we use for electrical ground is ⏚. In home electrical circuits, ground is *the* ground—earth.

The conductor that serves as ground is often used as part of the circuit carrying the current. For example, when certain rural areas of the country first got commercial electrical power, they operated temporarily with only one wire from the power plant to the homes. The earth served as the other half of the conducting loop. As another example, electronic devices such as TV sets use their metal chassis as ground and as part of the conducting paths for currents within the sets.

In an automobile, only one wire runs to each of the electrical parts and accessories. The frame and body form the other part of the conducting path. The starter motor, for example, could be supplied with energy by two wires from the battery, shown as the heavy solid line containing the switch and the dashed line of Fig. 11.12(a). Instead, one terminal of the battery, usually the negative, and one terminal of each device are connected to the frame of the car. One wire containing the fuse and switch is then run from the positive battery terminal to the device.

Figure 11.12(b) shows the schematic diagram for the battery, the starter, and the tail light of the car. Each connection to ground implies a connection to a common conductor carrying current back to the negative terminal of the battery.

The wiring to the electrical parts of a car is divided into circuits

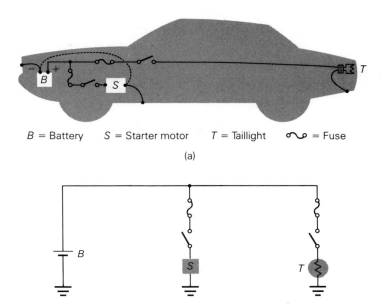

B = Battery   S = Starter motor   T = Taillight   ⌢⌣ = Fuse

(a)

Figure 11.12 An automobile uses the frame and body as part of the conducting circuit. The body is *ground* for the car. The pictorial representation in (a) is shown schematically in (b).

(b)

much as the wiring in a house. Each circuit has a separate breaker, and supplies current to a limited number of devices.

Question   What is ground on a flashlight? Does it carry the current? Draw a circuit diagram in the format of Fig. 11.12(b) for a flashlight.

11.4   Resistive Heating

How many ways can you think of that electrical energy is used to produce heat? We've already talked about the incandescent light bulb producing light as the filament gets white hot from current passing through it. Electric furnaces heat some of our houses, and most of our food is cooked on electric ranges. The small and large appliances that heat electrically are almost endless—toasters, electric blankets, water heaters, clothes and hair driers, steam and waffle irons, portable space heaters, and coffee makers, to name a few.

However, there are also situations in which electrical heating is a nuisance. Since all conductors have *some* resistance at normal temperatures, heat is produced as electrical energy is transmitted over power lines and through the wires in your house. This heat represents wasted electrical energy, and can be a hazard if too much is produced in a given area.

Rate of Heat Production

If we know the resistance of a conductor, and the current passing through it, we can calculate the rate at which heat is produced. We saw in Chapter 10 that power (rate of delivering energy) from a source of emf or to a load is given by

$$\text{power} = \text{voltage} \times \text{current},$$

or
$$P = VI.$$

If this power is delivered to a resistor, Ohm's law tells us the relationship between its resistance, the current through it, and the voltage difference across it:

$$\text{voltage} = \text{current} \times \text{resistance},$$

$$V = IR.$$

Using this for voltage in our power equation, we get

$$\text{power} = (\text{current} \times \text{resistance}) \times \text{current},$$

$$P_{\text{heat}} = (IR)I,$$

or
$$P_{\text{heat}} = I^2R.$$

This equation tells us the rate of production of heat in a conductor of resistance R carrying current I. The subscript "heat" on the P explicitly reminds you that this power all goes to heat production.

Example 11.2   A typical 15-cm-diameter surface unit on an electric range has a resistance of 43 Ω. What is the rate at which heat is produced in this unit, when operated at 220 V?

From $V = IR$, the current is given by

$$I = \frac{V}{R} = \frac{220 \text{ V}}{43 \text{ }\Omega} = 5.1 \text{ A}.$$

The heat production rate is then

$$P = I^2R = (5.1 \text{ A})^2 \times 43 \text{ }\Omega$$

$$= 1100 \text{ watts (W)}.$$

It might be interesting to find out how long it would take to boil away 1 liter (about a quart) of water, if all this heat goes into the boiling water.

Energy to boil water = (mass of water) × (heat of vaporization).

One liter of water has a mass of 1 kg, so

$$\text{energy} = 1 \text{ kg} \times 540 \text{ kcal/kg} \times 1.16 \text{ Wh/kcal}$$

$$= 630 \text{ watt-hours (Wh)}.$$

Since energy = power × time,

$$\text{time} = \frac{\text{energy}}{\text{power}} = \frac{630 \text{ Wh}}{1100 \text{ W}}$$

$$= 0.57 \text{ h} = 34 \text{ min}.$$

Most electrical cooking appliances produce their heat as current passes through a conducting heating element on the appliance, such as a surface unit on a range. An interesting exception is a commercially available "hot-dogger" (Fig. 11.13-b), named of course for what it cooks. As its schematic diagram (Fig. 11.13-a)

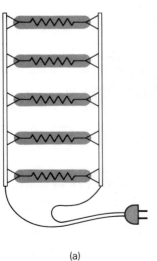

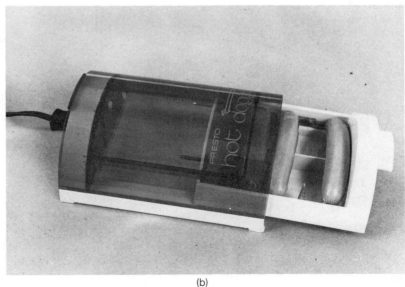

(a)

(b)

Figure 11.13   In a "hot dogger," the hot dog cooks from the heat produced as current passes directly through the hot dog. Each hot dog has a resistance in parallel with the others.

shows, the current passes directly through the hot dog. The resistance of a hot dog turns out to be just right to give the amount of heat needed to cook it in about 1 min when connected across a 110-V supply. Since the hot dogs are connected in parallel across the emf source, each one is independent of the other, and cooking time is independent of the number of hot dogs. The total current through the cord is, of course, proportional to the number of dogs.

Question   The wiring diagram of a popular portable electric room heater is shown in Fig. 11.14. Why does it use only six resistance elements for a high-heat setting but all eight elements for low heat?

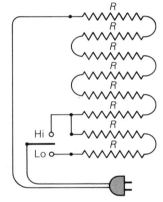

Figure 11.14   One type of portable electric heater gets its high-heat setting with six spans of resistance R in series. Adding two more in series gives the low-heat setting.

Exercise   If the portable heater of Fig. 11.14 operates on 120 V and the resistance of each segment is 1.5 Ω, what is the heat output for each setting?

Ans: 1600 W; 1200 W

Wire Sizes   Did you ever notice the lights dim briefly or the TV blink when a large appliance such as an air conditioner, or maybe a refrigerator, turned on? If so, you were seeing the effect of the resistance of the wires carrying the current.

Suppose you have an air conditioner and an incandescent lamp connected in parallel in the same circuit (Fig. 11.15), shown operating on a battery for simplicity. When the air conditioner first turns on, it needs a large current to get started. Since the wires have some resistance, there is a voltage drop given by $V = IR$ along the wires

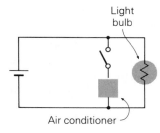

Light
bulb

Air conditioner

Figure 11.15 A circuit containing an air conditioner and a light bulb in parallel. Voltage drop along the wires due to current through their resistance causes lower-than-normal voltage across the devices being operated.

between the battery and air conditioner. This situation means that the voltage across the bulb is less than it would be with no current to the air conditioner. The bulb gets dimmer because, with less voltage across it, less current goes through it.

If a wire has resistance R and carries current I, heat is produced in the wire at a rate I^2R. This heat represents a loss of electrical power that does not reach the devices being operated. At the least, this is wasteful, but in extreme cases enough heat can be produced to cause a fire hazard.

For a given current in the wire, the only way to reduce heat production is to reduce the resistance, which can be done by increasing the wire sizes. The larger the cross-sectional area of the wire, the less its resistance. Compare this situation to the flow of water in a pipe: The larger the pipe, the more water it can carry for a given pressure difference between the ends of the pipe. Also, as you might expect, the resistance of a wire is proportional to its length. Thus, minimizing the length minimizes heat production.

If you notice the sizes of the wires used in high-current circuits, such as a clothes-drier circuit, you see that they are much larger than those in lighting circuits.

Question Why do refrigerators often come from the factory with a tag that warns you not to plug them in by way of a simple "dime-store" extension cord?

Loose Connections When you unplug your hair drier (or shaver, or drill, or back scratcher), do you pull on the plug or do you grab the cord and yank? If you do the latter, don't be surprised if after a few yanks the plug or cord starts to get hot when you use it. If your yanking loosens the connections to the prongs of the plug, or breaks some of the strands in the wire, you greatly increase the resistance to current through that point. The current to operate the device, passing through the increased resistance, produces heat—in extreme cases, enough to start a fire. Aside from the potential danger and electrical energy loss to heat, the voltage drop in this high resistance along the current path means that you get a lower voltage at the device and therefore less efficient operation.

Another place where the same effect can cause trouble: If your car starter won't turn the engine and yet the battery is good, corrosion may have formed an insulating layer that causes a bad connection between the battery and the cable connected to it. The resistance can become so high there that not enough current can get through to run the starter. Usually, however, enough current exists so that the connector gets extremely hot—hot enough to burn if you touch it—while you're trying to start the car.

Anytime the conducting path to or from an electrically operated device has too much resistance, either from bad connections at any point, or from too small or too long an extension cord or cable, energy will be lost to heat, and the voltage drop will reduce the efficiency of operation. A 10% drop in voltage lowers the output power of a motor by about 20%, and the light output of an incandescent bulb by about 30%.

Question   What would happen if you operated a 1500-W portable electric heater by plugging it into a small extension cord intended for use with lamps only?

Superconductors

A strange thing happens to some metals when their temperature is lowered to near absolute zero: Their electrical resistance drops to *zero*—no resistance at all. This circumstance means that a current set up in such a material would continue indefinitely with no additional input.

The resistance of all metals decreases as their temperature drops, mostly because less thermal agitation of the atoms causes less interference with the flow of charge. Ordinary good conductors, such as copper, continue to decrease gradually in resistance down to low temperature, as shown by the graph in Fig. 11.16. But for some metals—tin, for example—the resistance suddenly drops to zero as a certain temperature is reached. This transition temperature varies from metal to metal and, as the graph shows, it comes at 3.72 K for tin. We say that the material then becomes a *superconductor.*

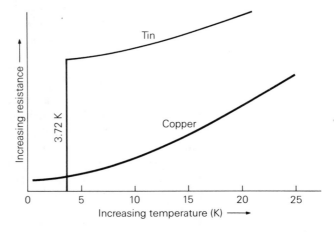

Figure 11.16 A graph of resistance versus temperature for two conductors. Tin is one of the metals that becomes a *superconductor*—its resistance drops to zero—at a few degrees above absolute zero of temperature.

You can easily imagine how a material with zero resistance could have many practical uses. The transfer of electrical energy over wires of zero resistance would eliminate the loss of electrical energy due to resistive heating. Unfortunately, no material known at the present time is a superconductor at temperatures much *above* $-250°C$. Cooling a long transmission line to this temperature is not a trivial matter. For applications such as electromagnets that require large currents in a fairly small region of space, superconductors are already being used. Some interesting applications are discussed in the references at the end of the chapter.

11.5 **Shock**

Everything your body does is controlled electrically. Therefore, it shouldn't be surprising that an electric current forced through your body by an outside potential difference would greatly disrupt your body functions. We'll look at some of the conditions for causing such a current, which we call an electrical *shock*.

First of all, notice that it's the *current* that gets you, *not* the voltage. The amount of voltage is of no *direct* consequence other than the fact that it takes some voltage across your body to drive the current through.

Let's illustrate by a specific example of how you can be shocked. Suppose your roommate has an electric toothbrush plugged into a receptacle in the bathroom, and accidentally closes the medicine cabinet door on the cord. Suppose, in the process, a metal part of the door cuts through the insulation and makes contact with the wire at 110 V. You come along later to open the medicine cabinet. In so doing you touch its metal frame. As Fig. 11.17(a) illustrates, as long as you are standing on the dry bathroom rug, or your body is insulated from ground in some way, nothing happens. Your body is not part of a complete closed circuit. However, if (as in Fig. 11.17-b) while you are touching the medicine cabinet, you reach down to turn on the water, *zap!* Your body provides a path between the 110-V line touching the medicine cabinet and the grounded water pipe at 0 V. Current can then pass through your body, giving you a potentially serious shock.

Any time your body forms a path across a potential difference, it can carry a current that gives you a shock. In Fig. 11.17(a), your body is at 110 V, but no harm is done because there is no potential difference across your body, and therefore no current through it. In Fig. 11.17(b), the potential difference produces a current. Figure 11.18 shows *living* evidence that the voltage doesn't kill you if you provide no path *across* a potential difference.

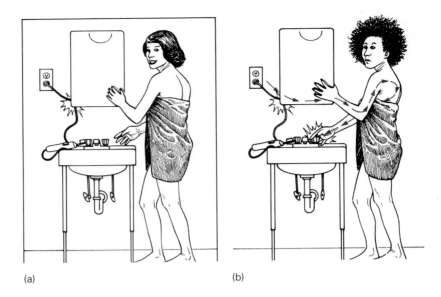

Figure 11.17 (a) Your *whole* body is at 110 V relative to ground. There is no potential difference across it, and therefore no current through it—no shock. (b) There is 110 V across your body, giving a current that follows the path shown by the arrows.

Why do some people receive only a small jolt, but others get electrocuted by contact with a 110-V line? The main difference is the resistance between the body and the conductors at the points of contact. Figure 11.19 is a schematic diagram of the situation of Fig. 11.17(b), where, for simplicity, the source of potential difference is shown as a battery. Here R_B stands for the resistance of the body; R_L and R_R are the resistances of the contact points at the left and right hands. Any current passing through your body has to pass through all three resistances. If any one is high, the current will be low. The body contains many ions, and is a pretty good conductor. The current, therefore, is determined mostly by the resistance at the contact points.

Fortunately, dry skin has a resistance high enough that the current from contact across 110 V is usually too small to be deadly to an adult, although certainly not comfortable! However, if the skin is wet from perspiration or water, contact is much better, and the resistance can be low enough that the current may be fatal. This is why shock hazards are worse in bathrooms and laundry rooms. A person in one of these rooms is often in contact with water that is, in turn, in contact with grounded water pipes. Water can easily seep into the inside of appliances picked up with wet hands, thereby providing a path for the current.

Figure 11.19 The schematic circuit diagram for Fig. 11.17. R_B is the resistance of the body; R_L and R_R are the resistances of the contacts made at the left and right hand. In Fig. 11.17(a), R_R is infinite, so there is no current. The ground is, in effect, a conductor shown by the dashed line.

Figure 11.18 Danny Matthews of Irving, Texas sits on a live 138 000-V power line, where he landed during a parasail accident. He received only burns on his hands and feet, probably from the surge of charge that initially brought his body up to 138 kV.

Question   If you are listening to the radio in the bathtub, and the radio falls into the tub with you while it's plugged in, what should you do? Remember the water is usually grounded by the connecting pipe, and to get a shock your body has to provide a path across a potential difference. What would happen to you if you just stayed in the tub with the plugged in radio? Discuss the consequences of the various possibilities for action.

For most people, it takes a current larger than about 1 mA (1 mA = 0.001 A) to be felt. At about 16 mA, the current stimulates the muscles enough that you have no control over them and cannot let go of the conductor. Between 25 and 100 mA breathing stops, and between 100 and 200 mA the heart stops pumping and goes into uncoordinated contractions (ventricular fibrillation). At higher currents the heart completely stops, irreversible damage may be done to the nervous sytem, and serious burns can result from the heating effect of the current.

If a hospital patient has wires connected to internal electrodes in the heart, currents as low as 0.01 to 0.1 mA in these wires can cause ventricular fibrillation. Leakage currents this large exist in many types of ordinary equipment. Even though extreme precautions are now usually taken to avoid even small currents around patients with electrodes implanted internally, many accidental electrocutions are alleged to occur each year among these patients.

Key Concepts

When two or more circuit elements are connected in *series,* whatever current passes through one must pass through them all. The total resistance of resistors in series is the sum of the individual resistances.

When circuit elements are connected in *parallel,* the current divides so that the sum of the currents to the individual components equals the current leading to the parallel combination. In parallel resistors, the current to any branch is inversely proportional to its resistance.

In 60-Hz *alternating current* (AC), the current and voltage smoothly reverse directions 120 times per second. The effective current and voltage are 0.707 times their peak values.

Practical electrical devices are connected in parallel across the power lines. The grounded body of an electrical device can safely form part of the electrical circuit.

The electrical power converted to heat is given by:

$$\text{power} = (\text{current})^2 \times \text{resistance},$$

or $$P = I^2 R.$$

Shock is caused by *current* passing *through* your body, not directly by the potential of the body. This current exists when your body forms a path *across* a potential difference.

Questions

Combinations of Circuit Elements

1. If you're hooking up two baseboard electric heaters and you make a mistake and wire them in series when they should be in parallel, how will the current through the heaters compare with what they need for rated heat output? Assume the resistance to be independent of temperature.

2. A 150-W bulb is connected to a 110-V power line.

   a. How would the current supplied by the power line change if two such bulbs were connected in parallel?

   b. How would the current through each bulb in parallel compare with that through the one bulb?

   c. How would the voltage across each bulb in the two-parallel-bulb case compare with that in the one-bulb case?

   d. If we had two such bulbs in series, how would the current supplied to each compare with the current with one bulb? How does the current supplied by the power line compare with that through each bulb? Assume the resistance of each bulb stays constant with changing current.

   e. Compare the total power supplied to one bulb, two bulbs in parallel, and two bulbs in series, assuming the resistance of each bulb is the same in each case.

3. When your car battery "dies" and you start the car by hooking jumper cables to a friend's battery, you should connect them as shown in Fig. 11.20. Draw the schematic diagram for the cir-

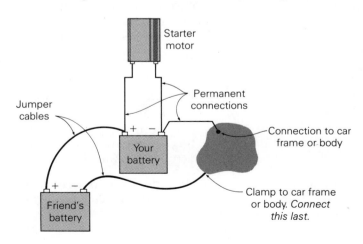

Figure 11.20 When hooking up jumper cables to your car battery, you hook positive to positive and negative to negative. Why? (Connect the cable used for negative-to-negative *last* to some point on the body, frame, or engine so that any spark produced will not cause an explosion of the gases released in the battery.)

cuit. Are the two batteries in parallel or in series? Why do you hook them "positive-to-positive" and what would happen if you hooked them "negative-to-positive"?

Alternating Current

4. Why are there 120 reversals of current direction per second in 60-Hz alternating current rather than just 60?

5. What is the average value of the voltage across a 110-V, 60-Hz power line? What is the average value of the current from this line to a 600-W hair drier? Why do we rate AC in "effective" values rather than average values? (The answer to the first two questions is *zero*.)

Residential and Automobile Electricity

6. Does the "electricity meter" on a house register energy or power?

7. In house wiring, why is the wire with the black insulation used for the "hot" wire?

8. What is the safety advantage of connecting a wall switch in the "hot" line?

9. What are the purposes of a fuse and a circuit breaker? Does one of these have an advantage over another? What determines the "size" fuse or circuit breaker that's needed?

10. When you plug a radio into one receptacle and a desk lamp into another receptacle in your room, are you connecting them in parallel, in series, or in some other fashion? Does it matter if these two receptacles happen to be on different house circuits?

11. What determines the size wire needed for a circuit or appliance?

12. Suppose you have two 2200-W hot plates, one that operates on 110 V and one that operates on 220 V. If you planned to operate nothing else on the circuits used for each hot plate, how would the size fuse or circuit breaker needed for the 110-V plate compare with that needed for the 220-V one?

13. Why do large power-using devices (ranges, water heaters, clothes driers, etc.) usually operate on 220 V rather than 110 V?

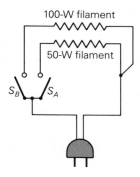

Figure 11.21 Circuit diagram for a three-way bulb and switch that gives three light levels corresponding to 50, 100 and 150 W.

14. Draw a circuit diagram for a DC power supply having three output wires arranged so that one wire is grounded, so that there is 110 V between two pairs of wires, and so that there is 220 V between the other pair.

15. A ground-fault interrupter (GFI) is a special circuit breaker that quickly turns off a circuit's current if the current in the circuit's hot line exceeds that in the neutral line. The GFI is especially recommended for circuits in high shock-risk areas, such as bathrooms. Why does the GFI provide added safety from serious shocks?

16. The circuit diagram for a three-way bulb and its accompanying switch is shown in Fig. 11.21. The bulb has two filaments and the switch has four positions that give 0 W, 50 W, 100 W, and 150 W of power used. The four switch positions correspond to: both switches open; S_A only closed; S_B only closed; and both S_A and S_B closed. Explain how this gives three light levels. Examine such a bulb and socket and you will see an extra contact point on the bulb base and in the socket, not found on ordinary bulbs and sockets. Why do you have to turn the switch two notches to get a change when you use an ordinary bulb in a three-way socket?

17. Figure 11.22 is a circuit diagram for a "three-way" switch that lets you control one light with two switches. What are the two switch-setting combinations that turn the light on? What are the two that turn it off?

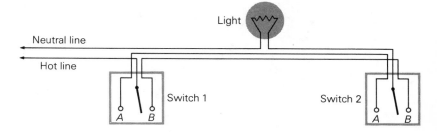

Figure 11.22 Circuit diagram for a "three-way" switch.

18. Does the ground wire hooked to the casing of an electric drill prevent current from passing through the casing?

19. Why does the tail light of your car need only one wire running to it rather than two, as in a table lamp used in the house?

20. Why do the lights sometimes dim when a room air conditioner or other large appliance comes on?

Resistive Heating

21. Which has a thicker filament, a 100-W or a 200-W incandescent bulb? (Assume the length of the filament to be the same in each.)

22. A portable electric heater for operation on 110 V has two heating elements: one 16-Ω and one 12-Ω element. Of the four alternatives—the 16-Ω element alone, the 12-Ω element alone, both in series, both in parallel—which way would you get the most heat? Which way would you get the least heat?

23. If you connect two identical hot plates in series, will you get a higher or lower rate of heat production from the two together than from one by itself? If the resistance of the heating elements does not change with temperature, how much higher or lower is the rate?

24. Why does the plug on an extension cord sometimes become hot if the connection between wire and prongs is loose?

25. Why does a hot dog cook when you pass a current through it?

Shock

26. Does wearing thick rubber-soled shoes assure you of not being shocked?

27. If you grab a high-voltage power line, you will not necessarily be seriously shocked. Why? Why aren't birds killed when they land on a high-voltage line?

28. Why is an automobile a safe place to be during a lightning storm?

29. If you're in an automobile accident and a power line falls across the car, what are the safe and unsafe alternatives for action? In particular, what's wrong with *stepping* out of the car? How should you get out?

Problems

1. Resistances of 10 Ω, 20 Ω, and 30 Ω are connected in series across a 120-V emf. What is the current through the 20-Ω resistor?   **Ans:** 2.0 A

2. Resistances of 10 Ω, 20 Ω, and 30 Ω are connected in parallel across a 120-V emf. What is the current through each resistance, and the current through the battery? What is the effective resistance of the parallel combination ? Hint: Knowing the battery current and emf, you can calculate the resistance it "sees."   **Ans:** 12 A; 6.0 A; 4.0 A; 22 A; 5.4 Ω

3. Show that for several resistors R_1, R_2, R_3, ... in parallel, the effective resistance R of the combination is given by

$$\frac{1}{R} = \frac{1}{R_1} + \frac{1}{R_2} + \frac{1}{R_3} + \ldots$$

Remember that $V = V_1 = V_2 = V_3$, etc., that $I = I_1 + I_2 = I_3 + \ldots$; and that $I = V/R$ for each resistor, as well as the combination.

4. Calculate the overall resistance in each arrangement of Question 2. Calculate the actual value of the current, voltage, or power as required by the various parts of the question.

5. What is the peak voltage in 220-V AC?

6. Calculate the rate of heat production in a 40-Ω resistor when connected to a 220-V line.   **Ans:** 1200 W

7. In Problem 6, calculate the total heat produced when two such resistors are connected in series. In parallel.

   Ans: 600 W; 2400 W

8. A resistance of 10 Ω is connected across a 100-V power supply. How much heat is produced in the resistor? How much heat is produced if the resistance is doubled?   **Ans:** 1000 W; 500 W

9. For each of the alternatives of Question 22, what is the rate of heat production?   **Ans:** 760 W; 1000 W; 430 W; 1800 W

10. The tungsten filament of an incandescent light bulb has a resistance at operating temperature of about 10 times what it has at room temperature. How does the rate of heat production when the bulb is first turned on compare with that a small fraction of a second later, when it reaches operating temperature?   **Ans:** 10 times as high initially

11. How does the *diameter* of the wire needed to supply current to a 5000-W range that operates on 220 V compare with the *diameter* of the wire needed for a 5000-W range that operates on 110 V? Hint: The cross-sectional area of the wire should be pro-

portional to the current to keep the amount of current through each unit of area the same.

Ans: $\sqrt{2}$ times as big for 110 V

12. If R_L and R_R in Fig. 11.19 add up to about 10^5 Ω when you grab the two sides of a 110-V power line, what will be the approximate current through your body? Is this current likely to hurt you badly? If the body resistance is 500 Ω, approximately how low would the contact resistances need to be to cause ventricular fibrillation?   **Ans:** 1.1 mA; 600 Ω

Home Experiments

1. After being sure that the "power" is off to that particular circuit, take the cover off a wall switch and a receptacle, and examine the connections to each. Notice the color coding. (Just look, *don't tamper,* unless you've had some wiring experience.)

2. With two wires, paper clips, or other conductors in contact with the terminals of a 1.5-V flashlight battery, touch them to your tongue. Notice the distinct taste. You won't get the taste without the wires contacting the battery terminals. Does this experiment give you a hint that taste is an electrical phenomenon? Try a 9-V calculator battery; it tingles a little, but the taste is still there. Why don't you feel that tingle by putting your finger across the terminals? (Some people test batteries in this way—by "tasting" the terminals.)

References for Further Reading

Aidley, D. J., *The Physiology of Excitable Cells* (Cambridge University Press, Cambridge, Eng., 1971). A detailed but readable treatment of the electrical aspects of nerve and muscle cells.

Blatt, F. J., "Nerve Impulses in Plants," *The Physics Teacher,* Nov. 1974, p. 455. Electricity plays a role in plants as well as animals.

Dalziel, C. F., "Electric Shock Hazard," *IEEE Spectrum,* Feb. 1972, p. 41. Discusses the effect of various current levels on the body, special precautions needed in hospitals, and equipment for reducing shock hazard.

Dresner, S., "Superconductors," *Popular Science,* Oct. 1974, p. 70. Some of the fundamental aspects of superconductors.

Jensen, K., "How to Troubleshoot Electrical-Accessory Failures," *Popular Science,* Aug. 1977, p. 124. Discusses wiring in automobiles and how to locate failures therein.

Stevens, C. F., *Neurophysiology: A Primer,* (John Wiley & Sons, New York, 1966). A good introduction to electricity in biological systems.

Swartz, B. B., and S. Foner, "Large-Scale Applications of Superconductivity," *Physics Today,* July 1977, p. 34.

Time-Life Books, *How Things Work in Your Home (and What to do When They Don't),* (Time-Life Books, New York, 1975). Contains a very good discussion of the layout and principles of home electrical wiring. Also gives a well-illustrated description of most of the ordinary electrical appliances that produce heat.

12

Electromagnetism

Almost everybody has played with magnets at some time. They're fun as toys and they're interesting, but are they useful? Magnets hold cabinet and refrigerators doors closed; magnetized screwdrivers are handy for putting screws in inaccessible places; compasses, really just small magnets, assist in navigation; magnets sometimes do specialized jobs, as in medicine (Fig. 12.1); but are magnets *widely* used? Do they actually influence your life very much? The answer is yes *only if* you do such unusual things as listen to a record, turn on a tape recorder, start your car, shave with an electric shaver, dry your hair with an electric hair dryer, wash your clothes in a washing machine, beat your eggs with an electric mixer, ring a doorbell, listen to a radio or watch a TV, talk on a telephone, read under an electric light, eat food cooked on an electric range, or go into an air conditioned building.

Figure 12.1 As this news article shows, people sometimes use magnets in surprising ways.

SWALLOWED OBJECTS REMOVED WITH MAGNETS

CHAPEL HILL, N.C. (AP)—A potholder magnet inserted in the end of a rubber tube is an ideal instrument for the removal of bobby pins, needles or nails from the stomach, according to two local radiologists.

Drs. George M. Himadi and Gary L. Fischer of North Carolina Memorial Hospital suggest that this easily assembled device can be used for the removal of elongated foreign bodies frequently swallowed by children. And the technique often makes surgery unnecessary, the radiologists say.

The physicians recently used the magnet taken from a common kitchen potholder to remove a bobby pin from a child's stomach. The tube was passed through the child's mouth into the stomach where rapid contact was made with the foreign body under fluoroscopic control. The pin was removed easily within two minutes.

As we learn the fundamentals of magnetism in this chapter, we will see what it has to do with such ordinary everyday activities. And we will see that we are not really changing subjects from the last chapter, that magnetism is actually part of electricity.

We'll start this chapter by discussing permanent magnets, and describe their effects in terms of magnetic fields. We'll see that these ideas apply not only to magnets of ordinary size, but to magnets ranging in size from that of the earth down to that of the atom. We'll then show that all magnetism results from electric currents, even that of permanent magnets. We then will turn our viewpoint around, and examine how magnetic fields exert forces on moving charges. This will equip us to describe how motors and generators work. We'll find that a changing magnetic field can produce an emf, a fact that is the basis for the functioning of transformers. We'll close the chapter by discussing electromagnetic waves, in particular those sent out by radio and TV stations.

12.1 Permanent Magnets

The most commonplace magnets are small bar or horseshoe-shaped pieces of iron that attract nails and other objects made of iron or steel. These are called permanent magnets because they hold their magnetism indefinitely. If you stick one in a bucket of nails, the nails will cling only to the ends (Fig. 12.2). That is, the force attracting the nails is concentrated at the ends of the magnet. We call these regions of force concentration magnetic *poles*.

You probably already know that the two poles on each magnet are of different types that we call *north* and *south* poles. And, as Fig. 12.3 shows, like poles—two north poles or two south poles—repel each other. Unlike poles—a north and a south pole—attract each other. Compare this effect with what we said about forces between like and unlike charges. There is one important difference: The individual charges can exist by themselves, but magnetic poles always come in pairs. On every magnet there is a north and a south pole. If you cut the bar magnet into two pieces, you always end up with two magnets, each of which has a north and a south pole. No matter how many times you cut each piece, each part always has a north and a south pole. You cannot isolate the poles.

Magnetic Fields

In our discussion of forces between charges, we described the interaction in terms of the electric field surrounding any charge. A

Figure 12.2 Only the poles of a magnet attract iron objects.

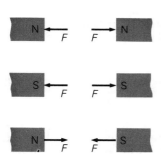

Figure 12.3 Like poles repel each other; unlike poles attract each other. N stands for north pole, S for south pole. *F* is the force on each pole due to the other one nearby.

second charge placed in the electric field of the first experiences a force because of the field. We can describe magnetic interactions the same way. Surrounding any magnetic pole there is a **magnetic field.** If we place another magnetic pole in this field, there will be a force on the second pole. Just as we did for the electric field, we represent a magnetic field by lines that tell us two things about the field: (1) the direction of the line represents the direction of the field; and (2) the stronger the field, the more closely the lines are spaced. (We'll use the symbol *B* to represent the magnetic field.)

Since magnetic poles always exist in pairs, we never have nice straight field lines extending outward from the poles as we do for an isolated electric charge. Instead, the magnetic field lines curve around, as illustrated in Fig. 12.4 for a straight bar magnet. For magnets of different shapes, the field lines take on different configurations, but in a general north-to-south direction.

Figure 12.4 The field lines surrounding a bar magnet.

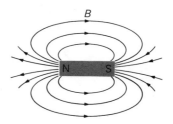

The direction of the magnetic field is the direction of the force on a north pole placed in the field (Fig. 12.5-a). The north pole of the right-hand magnet feels a force to the left—is attracted to the center one—because the field there points to the left. The north pole of the left-hand one experiences a force to the left—is repelled by the center one—because the field *there* points to the left. Figure 12.5(b) illustrates that the force on a south pole is in the opposite direction of the field at that point. The (unshown) forces on the center magnet result from the fields of the outer two magnets.

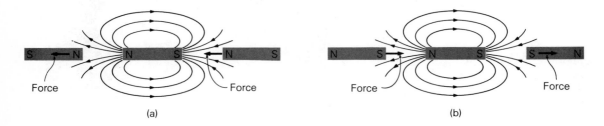

(a)   (b)

Figure 12.5   (a) A north pole experiences a force in the direction of the magnetic field. (b) A south pole experiences a force in the opposite direction.

The Earth as a Magnet

A compass needle aligns itself nearly in a geographic north-south direction. The same end of the needle that points north will point toward the *south* pole of a bar magnet. That's because the earth has a magnetic field similar to that of a bar magnet and oriented as shown in Fig. 12.6. The axis of the magnetic field is at a slight angle to the earth's axis of rotation, which determines geographic north and south.

The names "north" and "south" for magnetic poles came about because suspended magnets aim themselves toward the north and south poles of the earth. Since a north magnetic pole is attracted to a south magnetic pole, the earth's *geographic north* pole is a *magnetic south* pole.

Question   Is the end of a compass needle that points north a magnetic north or a magnetic south pole?

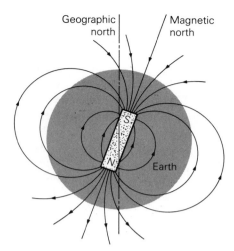

Figure 12.6 The earth's magnetic field is very much the same as if it were a large bar magnet whose axis is tilted slightly relative to the earth's axis of rotation.

Figure 12.7 In a permanent magnet, the atomic magnets are aligned with their north poles (shown by arrowheads) predominantly in one direction.

Atomic Magnetism

We mentioned that if you break a magnet in half, you do not get two isolated poles, but two magnets, each of which has a north and a south pole—each is a *dipole* (two poles). This effect will continue, no matter how many times you break the magnet, until you get to the individual atoms. Each atom of a permanent magnet is a tiny magnetic dipole.

Figure 12.7 shows how you can get a bar magnet with only two poles from many small "bar" magnets. The arrows represent atomic magnets, with the arrowhead standing for the north pole of each small magnet. All the atomic magnets are aligned in the same direction. In the interior, north and south poles are adjacent and cancel the outside effect of each other. But on one end there is an exposed set of north poles; on the other end an exposed set of south poles. The overall result is a net north pole at one end, and a south pole at the other. In an unmagnetized material, the atomic magnets are arranged randomly so that there are no overall magnetic poles.

In iron, the magnetism of each atom is so strong that the atoms group themselves into clusters called *domains,* with all the atoms in a domain pointing in the same direction. In an unmagnetized piece of iron, the domain directions are oriented at random, giving no overall preferred orientation and no overall outside magnetic effect (Fig. 12.8). Even though each cluster contains many atoms, the domains are still extraordinarily tiny by ordinary standards, and we can think of them as small single dipoles.

Figure 12.8 In iron, all the atoms in a given domain have their magnetic dipoles pointing in the same direction. When the material is unmagnetized, the domains are oriented randomly.

Figure 12.9 (a) The nail is unmagnetized, and the domains are oriented randomly. (b) The magnetic field of the magnet causes a rotation of the dipoles, giving a net south pole at the end closest to the magnet.

If we put a piece of iron in a magnetic field, the atomic north poles experience a force in the direction of the field, the south poles a force opposite to the field. As a result, the dipoles shift their orientation more nearly in line with the field. The stronger the field, the better the alignment. In reltively pure iron, called "soft" iron, the domains regain most of their random orientation when removed from the outside magnetic field. But in steel and in certain other alloys, they cannot shift so easily and the material stays magnetized even without an outside field. The amount of magnetism retained depends on the nature of the material.

Let's see why unmagnetized iron is attracted to *either* pole of a magnet. The nail of Fig. 12.9(a) has its atomic dipoles oriented at random when there is no outside field. When the north pole of a magnet comes near, as in Fig. 12.9(b), the magnetic field causes the dipoles to rotate so that a net south pole is at the end of the nail nearest the magnet; a north pole is farther away at the other end. The nail is then attracted to the magnet. If a south pole were brought near the nail, the dipoles would rotate so that the end of the nail toward the magnet becomes a north pole, again giving attraction. The alignment is usually not as complete as implied by the figure.

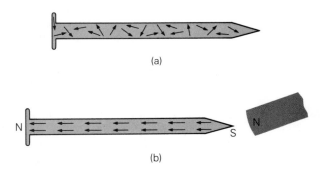

(a)

(b)

Question   Explain why the nail is attracted to the magnet if one of the poles is brought near the middle of the nail.

12.2   The Magnetic Field of a Current

In 1820 Hans Christian Oersted reported the discovery of an effect that makes possible the widespread use of electricity that we enjoy today. He discovered the connection between electricity and magnetism. That connection makes it possible to generate electricity in the huge quantities used by our society.

Even though he had been searching for a link between electricity and magnetism for some time, Oersted apparently made the actual discovery during one of his lectures. Having a battery and compass on his lecture table for other demonstrations, he noticed that when he ran a current through a circuit, the nearby compass needle deflected. On closer examination, he found that if he ran a current-carrying wire directly above the compass, with the wire perpendicular to the needle, as in Fig. 12.10(a), nothing happened to the compass when he closed the switch (b). But if the wire were first parallel to the needle (c), the needle would swing around perpendicular to the wire when he turned the current on (d). This discovery illustrated that a *magnetic field exists in the region surrounding every electric current.*

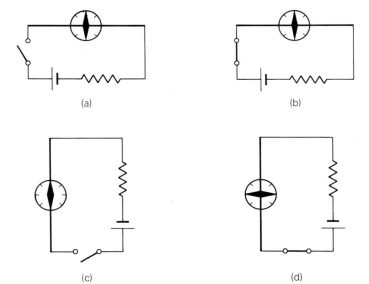

Figure 12.10   (a) If the wire is perpendicular to the compass needle, then (b) nothing happens when the switch is closed. (c) If the wire is first parallel to the needle then (d) it turns until it is perpendicular when the current comes on.

To better understand this field, imagine a wire perpendicular to and through the page, as shown in Fig. 12.11, with a current upward in the wire (out of the paper). If you place compasses around the wire, they will point in the directions shown: The north pole of each points counterclockwise around the wire, the south pole points clockwise. Apparently the field around the wire is in a direction tangent to a circle centered on the wire. Thus, the field lines are concentric circles around the wire (Fig. 12.12).

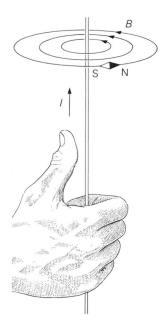

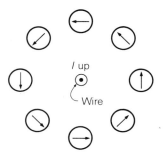

Figure 12.11   Compasses around a straight current-carrying wire (current up through the paper) point in a direction tangent to a circle around the wire. The arrows represent the north poles of the compass needles.

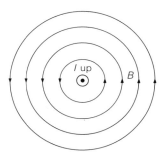

Figure 12.13   Right-hand rule: Imagine that you grab the wire with your right hand so that your thumb is in the direction of the current; your fingers then point in the direction of the magnetic field.

Figure 12.12   The magnetic field lines around a straight current are concentric circles around the wire. The field is everywhere tangent to the circles, and is in the direction of the arrows.

We can remember the direction of the field (whether clockwise or counterclockwise) by what is called a *right-hand rule:* Imagine that you grab the wire with your *right* hand so that your thumb, extended along the wire, points in the direction of the current (Fig. 12.13). Your fingers then point in the direction of the field around the wire. Your hand, of course, has nothing to do with the field. This rule is merely a way to remember its direction.

The Field around Curved Conductors

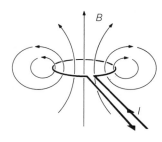

Figure 12.14   The magnetic field around a circular loop of current.

Suppose that, instead of being straight, the wire carrying the current is bent into a circular loop, as shown in Fig. 12.14. If you use the right-hand rule near any segment of the wire, you can find the direction of the magnetic field there. For the current direction shown, the field is upward everywhere *through* the loop; it is downward everywhere outside in the plane of the loop. The field lines curve around as shown in the figure.

Exercise   Use the right-hand rule to explain the direction of the field lines near the wire of Fig. 12.14.

If we have several *loosely-wound* turns of wire, the field from the various turns combines to give an overall shape as in Fig. 12.15. A straight coil of many *closely-spaced* turns, called a *solenoid,* gives field lines outside the solenoid that are similar to Fig. 12.16(a). Notice that these field lines are very much like those for a bar magnet (Fig. 12.16-b). This similarity means that a solenoid carrying a current acts as a bar magnet. There are some useful differences. You can vary, or turn on and off, the magnetism of the solenoid merely by varying, or turning on and off, the current. Also, you can reverse the north and south poles of the solenoid by reversing the current's direction. Compare the magnetic field outside these magnetic dipoles with the electric field surrounding the electric dipole of Fig. 10.8(b).

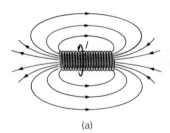

(a)

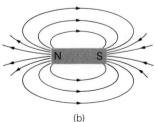

(b)

Figure 12.16   (a) The field outside a solenoid is identical to (b) the field outside a bar magnet.

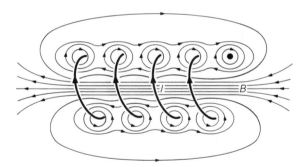

Figure 12.15   The field of several turns of current.

Solenoids with Iron Cores—Electromagnets

If we put a material medium inside the windings of a solenoid, the magnetic field inside and outside is changed due to the action of the field on the atoms that themselves act as little magnets. If this material, called a core, is iron, it becomes strongly magnetized. As a result, the field outside the solenoid is much stronger than it is without the core. If the core is soft iron, it demagnetizes almost as soon as the current is turned off.

We call such an arrangement—coils of wire wound around soft iron—an *electromagnet*. Electromagnets have hundreds of practical applications. We'll mention a few.

**Applications of Electromagnets   The reason electromagnets are so useful is that their fields can be varied easily by controlling the current to their coils. Industry uses many kinds of electromagnets, such as the large ones that pick up scrap iron and release it at will. But you and I use many electromagnets in our everyday routines.

Door chimes, for example, are a simple application of electromagnets. Figure 12.17 shows their basic components. When you push the button, it closes a circuit that passes current through the solenoid. This draws the core into the solenoid, making the nonmagnetic core extension hit the tone bar on the right (bing!). When you release the button, the magnetic field vanishes so that the spring pushes the core against the tone bar on the left (bong!). Most door chimes have more solenoids than this, but the basic principle is the same.

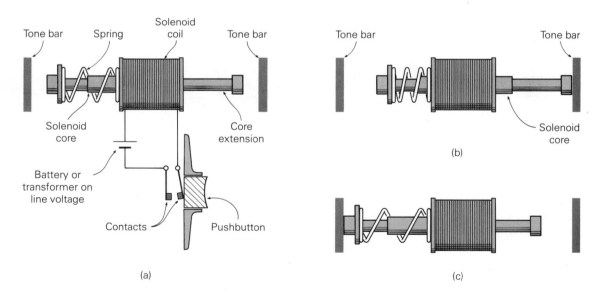

Figure 12.17   (a) Door chimes are electromagnets that pull the core into the solenoid when you close the switch (push the button) thus (b) causing the core extension to hit the tone bar on the right. When you release the button, the spring pushes the core back to the left (c), hitting the tone bar on the left.

Relays are solenoids that operate switches in a second electrical circuit. A current through the solenoid of Fig. 12.18 attracts the armature on the right, which swings over, closing the contacts. This allows current from the large battery, or other source of emf, to pass through the device being operated. Why not just put the switch S in the circuit on the right and avoid the solenoid? This arrangement is sometimes appropriate—for example, in a table lamp—but for many applications the switch used cannot carry enough current to operate the device.

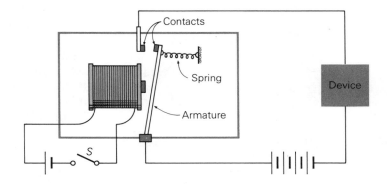

Figure 12.18   A relay: When the switch S in the low-current circuit is closed, the electromagnet closes the switch in the high-current circuit.

For example, we discussed in Chapter 6 how thermostats turn heating systems on and off. It isn't practical to have the bimetallic strip of the thermostat carry the full current needed to operate a furnace. The thermostat is used as the switch S in Fig. 12.18. This low-voltage circuit then operates the relay that closes the main circuit, which sends a much larger current to the furnace.

You can find solenoids and relays in all types of useful devices: they turn on air conditioners when the thermostat calls for more heat removal; they operate the mechanisms in tape recorders; they release the toast when it's done in a pop-up toaster; they control the filling and mixing of water in washing machines, to give the selected water temperature; they connect your phone to your friend's when you dial her number.

How would your life be different without loudspeakers? Everything from sound systems for listening to music to telephone receivers would vanish without them. Most speakers use electromagnets as their main component, and are typically arranged as in Fig. 12.19. The purpose of a loudspeaker is to convert variations in electrical current to vibrations in the air molecules. The current fed to the loudspeaker passes through the coil shown in the diagram. The force of attraction between coil and permanent magnet is proportional to the current in the coil. The coil, therefore, oscillates in

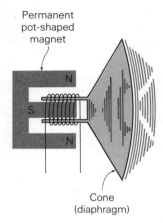

Permanent
pot-shaped
magnet

Cone
(diaphragm)

Figure 12.19   In a loudspeaker the coil vibrates in response to the force between the coil and the magnet. This force varies in proportion to the current through the coil.

proportion to the current. The resulting vibrations in the cone set up vibrations in the air that have the same variation with time as the current in the coil.

Superconducting Magnets   Large electromagnets use thousands of amperes of current in their windings. Even if the windings have very low resistance, you can easily estimate the huge amount of electrical energy lost to heat in such magnets. (Remember the rate of heat production is I^2R.) If the resistance of the windings can be made *zero,* this loss of electrical energy is eliminated.

As we discussed in Chapter 11, some materials do become superconducting—have zero resistance—at temperatures near absolute zero. Magnets with superconducting coils can carry very high currents and can routinely produce fields six or seven times those conveniently reached with conventional magnets. The big drawback is that the coils have to be kept at a temperature near absolute zero: no easy feat in itself. The usual way of doing this is to surround the coils with liquid helium at a temperature of $-270°C$ (3 K).

The Source of Permanent Magnetism

We made the statement in the opening remarks of this chapter that all magnetism results from electric currents. Yet a permanent magnet needs no wires connecting it to a source of emf to make it work. There are two ways in which electric currents establish the magnetic fields of permanent magnets. Both of these are related to the electrons within the atoms of the magnet. The nuclei of the atoms also contribute a small amount, but this is negligible for most purposes.

As you know, an electron has a negative charge. In the planetary model of the atom (a useful approximation for many purposes), the electron is pictured as orbiting the nucleus much as a planet orbits the sun (Fig. 12.20-a). The movement of this negative charge in orbit makes up a current along the path of the orbit (in the opposite direction to the movement of the electron). This current sets up a magnetic field in the direction shown just as would the current in a loop of wire.

The second and larger contribution to the atom's magnetism comes about because each electron acts as if it were spinning about an axis through its center. Imagine it to be a spinning charged sphere as shown in Fig. 12.20(b). Then each little chunk of charge moves in a circle and forms a current loop in the opposite direction to its direction of motion. Each of these current loops contributes to a magnetic field in the direction shown.

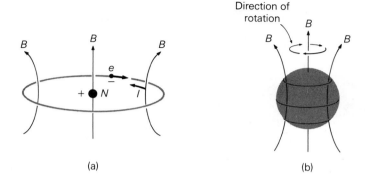

Figure 12.20   (a) An electron e orbiting the nucleus N of an atom constitutes a current I that causes a magnetic field B. (b) If the electron is assumed to be a spinning charged sphere, each bit of moving charge is a current that contributes to the overall magnetic field.

These two effects of the atomic electrons give each atom its magnetism, and cause it to act as a tiny bar magnet. The combined effect of all these atomic magnets gives the permanent magnet its magnetism. The reason iron can be so strongly magnetized is that several electrons in each iron atom have their spin axes pointing in the same direction, making each atom strongly magnetized. The description in terms of planetary motion and spinning electrons is only an approximation. The correct description, as it is now understood, is quite complicated and requires the use of quantum mechanics, a topic we'll discuss in a later chapter.

12.3   Moving Charges in a Magnetic Field

We saw in Chapter 10 that a net charge, either stationary or moving, gives rise to an electric field. Furthermore, another charge put in such an electric field experiences a force on it caused by the field. So far in this chapter we have seen that currents—*moving* charges—give rise to *magnetic* fields. Comparing effects in electric and magnetic fields, you might ask whether or not a second charge moving in a magnetic field would experience a force due to this field. The answer to that question is *yes*, provided its motion is in the right direction. Just as a *stationary* charge produces no *magnetic* field, a *stationary* charge in a magnetic field experiences no force.

Consider the situation shown in Fig. 12.21, where a positively charged particle moves between the poles of a C-shaped magnet. The magnetic field is down (north to south) and the particle moves in the direction shown by the arrows on its path. You might expect

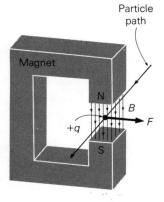

Particle path

Magnet

N

B

+q

F

S

Figure 12.21 A positive charge +q moving in the direction shown through a magnetic field B feels a force directed to the right that is perpendicular both to the charge's direction of motion and to B.

a force to be in the direction of the magnetic field *B*. Instead, the force is perpendicular to both the magnetic field and the direction of motion of the charge: to the right in this case.

The force has the following characteristics:

1. The force is proportional to the amount of charge.

2. The force is proportional to the speed of the charge. (This implies *no* force on a stationary charge.)

3. The force is proportional to the strength of the magnetic field.

4. For a given charge, speed, and magnetic field, the force is strongest when the particle motion is perpendicular to the field. It decreases in strength as the angle between its path and **B** decreases, becoming zero when the motion is parallel to the field.

5. The force is directed along a line perpendicular to both the particle direction and the magnetic field.

Item 3 in this list concerns the *strength* of the magnetic field. The term usually used to express this strength has the unfortunate name *magnetic induction*. (The symbol *B* we've been using is the one normally used for magnetic induction.) The SI unit of magnetic induction is the tesla (T). To give you an idea of the size of the tesla, here are a few examples. The field very near a magnetic cabinet latch is typically of the order of about 1/100 to 1/10 T. The earth's average magnetic field at its surface in the United States is about 0.000055 T. Large electromagnets usually have fields no stronger than about 2 T.*

You might see magnetic fields expressed in the unit gauss (G), where 1 G = 0.0001 T.

Item 5 doesn't tell us which way along the line the force acts. We find the direction by the right-hand rule shown in Fig. 12.22. Imagine that you hold your right hand so that its thumb points in the direction of motion of a *positive* charge, and your extended fingers in the direction of the magnetic field. Then the force is in the direction of your palm: the direction you would push with your hand. The force on a *negative* charge is in the *opposite* direction to this. Again, remember that your right hand has nothing to do with the force. It merely gives you a way to remember the direction.

---

*Quantitatively, when the particle's velocity v is perpendicular to B, the force is given by $F = qvB$. This equation shows that the tesla is that magnetic induction which would cause a force of 1 N on a charge of 1 C moving perpendicular to the field with a speed of 1 m/s.

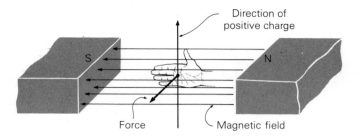

Figure 12.22   The right-hand rule for the force on a positive charge: If the thumb points in the direction of motion of the charge, and the fingers in the direction of the magnetic field, the force is in the direction you would push with your palm.

If you are wondering *why* the force is in this particular direction, that's not a question we can answer. We can say only that that's the way nature is. Here, as always, the laws of physics describe *what* happens but not *why* it happens.

** The TV Picture Tube and Forces on Moving Charges

By now you're probably saying, "Big deal! Who, besides some ivory-tower physicists, cares if there's a force on a charged particle moving through a magnetic field?" Well, *you* do if you enjoy watching TV.

Figure 12.23(a) shows the overall structure of a TV picture tube. Electrons emitted from the "electron gun" (Fig. 12.23-b) travel along the tube until they strike the inner surface of the screen that's coated with a phosphor. The phosphor gives off light (we'll see why in Chapter 16) at any point hit by electrons. If the electron beam could be swept quickly back and forth as well as up and down over the screen, the entire screen would give off light. That's where our force on particles moving in a magnetic field comes in.

Figure 12.23(c) shows the coils (solenoids) that are above and below, as well as on the left and right, of the tube neck. Current in the vertical coils gives rise to a vertical magnetic field that causes a horizontal force on the electron beam, bending its path to the right or left. Current in the horizontal coils gives a field that bends the beam up or down. For example, an upward vertical field bends the (negatively charged) electron beam to the left in the diagram (to the right facing the front of the screen). A field to the right bends the beam upward. Varying the current in the magnet coils in just the right time sequence makes the beam trace out from left to right the top line on the simulated screen, return quickly along

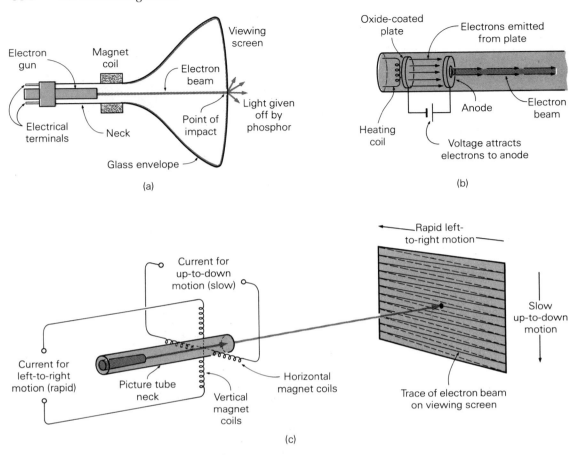

Figure 12.23 (a) The general layout of a TV picture tube, as if you take a slice through its middle. (b) The "electron gun." Electrons given off by the heated oxide coating are accelerated by the voltage between this surface and the anode. The anode has a hole in it that lets electrons pass through to the neck of the tube. (c) Varying the current through the magnet coils varies the magnetic field, causing it to sweep the electron beam back and forth, up and down along fixed lines on the screen.

the dashed path, trace out the second line, and so on down the screen. In TV sets in North America, 525 such lines are traced out across the screen every 1/30 second. That's so fast that your eyes see a uniformly bright screen.

But a solid white TV screen doesn't make a very interesting picture. Suppose you want a TV picture of a black circle on a white piece of paper. Figure 12.24 shows the lines traced out by the electron beam on the screen. (On a real screen these lines are much closer together.) By turning the beam off just as it gets to the edge of the circle in each sweep, and back on at the other edge, we get a

Figure 12.24   We get a picture of a dark circle on a light background if the beam intensity in the TV picture goes to zero just as the beam gets to one edge of the circle and comes back to full intensity at the other edge. The horizontal lines represent the traces of the electron beam sweep. On a real screen, the lines are so close together that you don't see individual lines.

dark circle on a light background. We can get pictures *even more* exciting than this by varying the intensity of the electron beam to give the right shade of grey at just the right spot on the screen. (We're assuming black and white TV.) To produce a moving picture, each scan of the screen depicts the proper distribution of light intensities at 1/30-s intervals.

The detailed methods of getting the variations in beam intensity exactly synchronized with the beam sweep to produce a real picture are interesting engineering techniques discussed in some of the references at the end of this chapter. In Section 16.3 we'll talk about how to get color TV.

Question   Would you expect the picture of a TV to shift when you rotate the TV from facing north to facing east? If so, which way? If you expect a shift, but don't see it when you try, why do you think you don't see it?

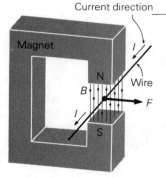

Current direction

Figure 12.25   If the moving charge in a magnetic field is a current *I* confined in a wire, the entire wire feels a force *F* in the direction shown.

The Force on a Current-Carrying Conductor

To get a force on a charged particle moving through a magnetic field, that particle does not have to be free; it can be inside a material medium. In fact, the moving charges can be those carrying a current through a conductor.

Figure 12.25 shows the magnet of Fig. 12.21 with the moving free charge replaced by a current-carrying conductor. A current *I* in the direction of the arrows gives a force to the right on each moving charge. Since the charges cannot get out of the conductor, *the whole wire feels a force in the direction of F.*

From the factors that determine the force on a moving *free* charge, it is probably reasonable to you that the force on the wire is proportional to:

1. the current;

2. the length of wire in the magnetic field;

3. the strength of the magnetic field.

As with the free charge, the force is a maximum when I and B are perpendicular, and decreases to zero when they are parallel. The direction of F is perpendicular to both I and B, and given by the same right-hand rule as for the free charge. The direction of the current is the direction of motion of positive charge.*

As we will see in the next section, the forces on moving charges and on currents form the basis for both electric motors and generators.

12.4   Motors and Generators

Motors

Electric motors are the most widely used devices we have (aside from our hands) for doing mechanical work. They operate not only large industrial machines, but also home appliances such as refrigerators, washing machines, and alarm clocks as well as hand-held devices such as drills, mixers, and shavers. If a device runs on electricity and hums or whines when it operates, it probably has an electric motor in it.

Suppose we have a wire carrying a current I to the right in the magnetic field of Fig. 12.26. (To have a current in the wire, it of course has to be part of a complete circuit. We show only the part of the circuit in and near the magnetic field.) There is a force on the wire directed toward you out of the page. If the wire is free to move, it will move toward you because of this force.

Now suppose that, instead of a single wire, we have a rectangular loop in the magnetic field. Assume the plane of the loop to be in the plane of the page, with the current clockwise around the loop, as shown in Fig. 12.27. The current in the upper wire is in the same direction as in the straight wire we talked about. It, therefore, feels a force directed *out* of the page. The ends of the loop, where the current is moving up or down, feel no force since the current is

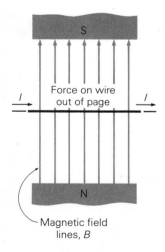

Figure 12.26   The force on the current-carrying wire is directed out of the page.

*When a length L of conductor perpendicular to a uniform magnetic field B carries a current I, the equation $F = BIL$ gives the magnitude of the force F on the conductor.

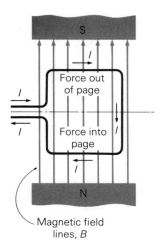

Figure 12.27 The force on the top of the loop is out of the page; that on the bottom is into the page. These two forces tend to rotate the loop about the dashed line.

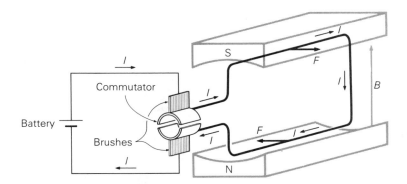

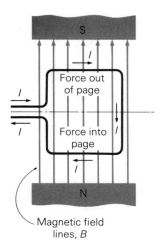

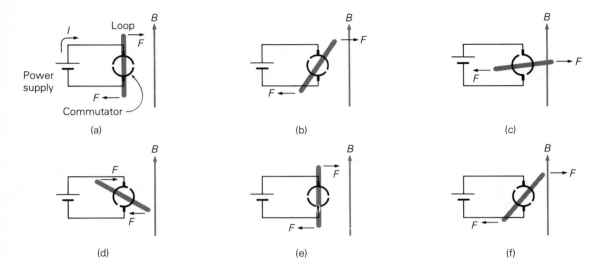

Figure 12.29   An edge view of the current-carrying loop at various orientations. The commutator sides interchange brushes when the coil is horizontal, thereby keeping the force to the right on whichever side of the coil is highest.

Question   Except when the plane of the loop is parallel to the magnetic field, there is a force on each end of the loop. Why do these forces not affect the rotation of the loop?

For a motor that runs on AC, the current reversal is handled by the AC power supply. Each side of the loop connects to a continuous *slip-ring*, as in Fig. 12.30. The speed synchronizes with the frequency of the voltage reversal so that the loop passes the orientation perpendicular to the field just as the AC reverses direction.

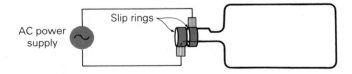

Figure 12.30   A motor for AC has slip-rings instead of a commutator. Each side of the loop is connected to a continuous ring that slides on one of the brushes, thereby maintaining contact between one side of the power supply and the same side of the loop. The AC power supply handles current reversal.

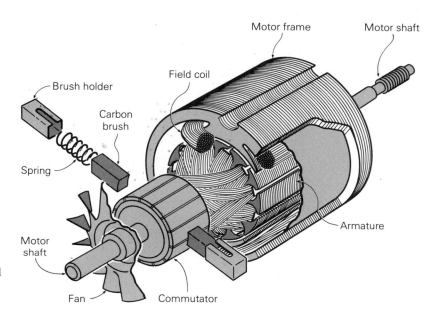

Figure 12.31 The components of a practical electric motor (or generator).

Practical electric motors have several rotating loops wound on an iron core—called the *armature*—in order to increase the torque, and to give a smoother rotation. Also the magnet providing the field is usually an electromagnet fed by the current supplied to the motor (see Fig. 12.31).

Energy Interchanges in a Motor An electric motor uses electrical energy to do mechanical work. This happens as the electric field set up by the power supply transfers energy to the charge to move it through the motor windings. This current in the armature experiences a force that moves the armature windings, thereby doing work on them as they rotate. In turn, the device hooked to the armature shaft does work on whatever it's designed to move or turn: a clipper through your hair, a fan blade that ventilates your room, a compressor in your air conditioner, or a saw that slices logs into boards for your house or desk.

Generators

Most electricity for home and industrial use is generated in devices made almost exactly like electric motors. To see how these *generators* work, let's go back and consider the single wire perpendicular to a magnetic field.

Suppose the wire of Fig. 12.26 does *not* have a current passing through it. Instead, imagine that you grab the wire and quickly pull it toward you (straight out of the page). The charges in the wire would then experience a force perpendicular to both the motion and to the field—the positive charges a force to the left, the negative charges a force to the right. If this wire were then part of a complete conducting circuit, current would flow to the left in the wire. (Remember that negative charge moving to the right is a *conventional* current to the left.)

Figure 12.32 is like Fig. 12.28, with the battery replaced by a light bulb. If you grab this coil and rotate it clockwise as viewed from the commutator end, the force on the free charges causes a current to the left in the top of the loop, and to the right in the bottom. The same amount of current passes through the bulb from top to bottom, as shown. Just as the plane of the loop is perpendicular to the field, the brushes switch contact to the other halves of the commutator, thus keeping the current in the bulb always in the same direction.

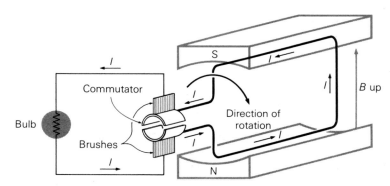

Figure 12.32 The basic DC generator. If the coil is rotated about its axis in the direction shown, positive charge will be pushed to the left in the top and to the right in the bottom of the loop, giving a current throughout the circuit in the direction shown. The commutator keeps the current always in the same direction in the outside circuit.

If slip-rings such as those of Fig. 12.30 are used instead of a commutator, the current through the bulb is first in one direction and then the other, depending on which part of the loop is at the top. We then have an AC generator, sometimes called an *alternator*. Alternators are used for charging the batteries in most automobiles. Since a battery can be charged only with DC, the AC alternator output is converted to DC in devices called *diodes*, which allow current through them in one direction but not in the other.

The same device can act as both a motor and a generator, depending on what is input. Figure 12.31, therefore, also shows the internal construction of a typical generator. In a motor, electric current is input causing the armature to rotate and do mechanical work. In a generator, the armature is mechanically rotated by an outside source of work, thereby generating current through the attached circuits. The generator, then, uses mechanical work to generate electrical energy. In power plants, the mechanical work of turning the generator armature is usually done by a turbine operated by steam produced in a boiler or nuclear reactor. Hydroelectric plants use water turbines to turn the generator armature.

As in a motor, *electro*magnets usually supply the magnetic field of a generator. *Some* of the current supplied by the generator is fed back into these *field coils* to produce the necessary magnetic field.

Question   Does this last statement imply that the generator is running itself? Once a generator is operating, could you use some of the current from the generator to run a motor that, in turn, runs the generator, thereby eliminating the need for a turbine or other outside source of mechanical work? (The principle of conservation of energy will provide the answer.)

12.5   Electromagnetic Induction

Suppose you have the items shown in Fig. 12.33(a): a permanent bar magnet, and a solenoid with the ends of the wires connected to a sensitive current-measuring meter. If the magnet and solenoid are not moving relative to each other, there is no current in the coil—the meter reads 0.

Now suppose you quickly shove the magnet into the coil as in part (b). During the movement you will find the meter registering a current in the direction shown. Once the magnet is at rest inside the coil (c), the current stops. If you pull the magnet out (d), a current will register in the opposite direction, but only during the motion. While the magnet and coil are moving relative to each other, a current exists. When there is no relative motion, there is no current, no matter where the magnet is. Furthermore, the faster the relative motion, the higher the current.

The current is induced during the motion because the magnetic field through the coil is changing. Anytime the magnetic field in *any* circuit changes, an emf is induced. If it's a closed conducting circuit, a current results from this emf. Michael Faraday discovered this effect in 1831. The law that now bears his name states that *the*

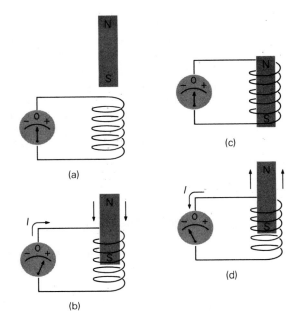

Figure 12.33 (a) and (c) When the magnet is stationary relative to the coil, there is no current. (b) When the magnet is being inserted into the solenoid, there is a current in one direction. (d) Removing the magnet induces a current in the other direction.

emf induced in a circuit is proportional to the rate of change of the magnetic field through the circuit. This process we call **electromagnetic induction.** If the changing field is within a coil of many turns, the same emf is induced in each turn. Therefore, *the total emf is proportional to the number of turns of the coil through which the field is changing.*

The change in magnetic field *through* a circuit can result not only from a change in the field itself, but from a movement or rotation of the circuit so that it intercepts more or less of the field. For example, when the armature loop of a generator is in the orientation of Fig. 12.34(a), it intercepts a maximum number of field lines. When it's oriented as in (b), it intercepts *no* lines. As the loop rotates, the amount of field intercepted continually changes. Instead of the way we described the generator's action in the last section, we could explain it in terms of changing magnetic fields. You get the same result from either analysis.

Figure 12.35 shows another way a changing magnetic field can induce a current. When the switch is closed in the left-hand circuit, the current sets up a magnetic field inside its coil. Since this magnetic field is confined mostly to the iron ring, almost the same magnetic field passes through the coil on the right. Immediately after the switch is closed, the field is changing and a current exists in the meter. Once the current is steady, the field is not changing, and

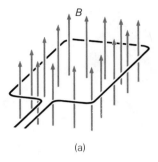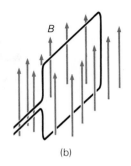

(a)   (b)

Figure 12.34 The number of magnetic field lines that pass through a generator loop changes as the loop rotates.

there is no current on the right. When the switch is opened, the current exists in the opposite direction while the field is decreasing. Faraday used an arrangement such that shown in Fig. 12.37 in his original discovery of electromagnetic induction.

An interesting application of induced currents may be seen in magnetic levitation for use in high-speed trains (Fig. 12.36). The idea is that, as the train moves along, the strong magnetic field of a *superconducting* magnet on board the train gives a *changing* field in the track below. This induces a circulating current in the track. The magnetic field of this circulating current repels the magnet inside the train, lifting it off the track. The only friction is then due to air resistance on the train, which rides along, suspended on the magnetic field. The train is propelled by the interaction of the field of the on-board magnet with that due to an alternating current fed successively into short sections of the track. As the current moves from one section of the track to the next, the train moves along at the same speed—much as a surfer is propelled along by a wave on the ocean.

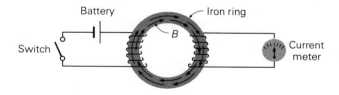

Figure 12.35 A current exists in the meter immediately after the switch is closed or opened, but not while the current in the left-hand circuit has a steady value.

Figure 12.36 A Japanese experimental train that uses magnetic levitation produced by superconducting magnets on board the train.

12.6 The Transformer

The main reason alternating current is so widely used instead of direct current is that AC can be run through a *transformer,* and the voltage stepped up or down easily. Let's see how this is done, and why it's useful.

Figure 12.37 is a slight variation of Fig. 12.35. An alternating emf replaces the battery and switch; a resistor replaces the meter; the iron core has a different shape, but most of the field is still confined inside it. An alternating emf and, therefore, an alternating current exist in the left-hand coil—called the *primary* winding. Since the current is continually changing, the magnetic field is continually changing. This change induces a continual alternating emf in the right-hand coil—the *secondary* winding.

If the primary and secondary coils have the same number of turns, the emf at the secondary equals that input to the primary. Since the emf induced is proportional to the number of turns, the

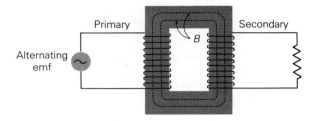

Figure 12.37 A transformer consists of two coils, a primary and a secondary, wound on the same iron core.

voltage output at the secondary can be changed by changing the number of secondary turns. The relationship is:

$$\frac{\text{voltage at secondary}}{\text{voltage at primary}} = \frac{\text{number of turns in secondary}}{\text{number of turns in primary}}$$

If N stands for the number of turns and V for the voltage, this proportionality can be written in symbols as

$$\frac{V_s}{V_p} = \frac{N_s}{N_p}.$$

(The subscripts indicate "secondary" and "primary.")

For example, if the input voltage to a transformer is 110 V AC, the primary winding has 1000 turns, and the secondary has 4000 turns, what is the output voltage? Multiplying both sides of the above equation by V_p, we have

$$V_s = \frac{N_s}{N_p} \times V_p.$$

Therefore,

$$V_s = \frac{4000}{1000} \times 110 \text{ V} = 440 \text{ V}.$$

Exercise   The heating element for the electron emitter in the electron gun of a TV picture tube operates on 6.3 V. Design a transformer to step down normal line voltage (110 V) to this value.

Ans: For example, $N_p = 10\ 000$ turns; $N_s = 570$ turns

With this ability to step up the voltage in a transformer by merely having more secondary than primary turns, are we getting something for nothing? The principle of conservation of energy tells us *no*. From this principle, we know that the total energy input to a transformer per unit of time—that is, the power input—must equal the total power output. Some electrical energy is converted to heat by the currents within the transformer, but this fraction is small in a well-designed transformer. Therefore, to a good approximation we can assume the electrical power ouput equals the electrical power input. Thus,

$$\text{power from secondary} = \text{power to primary,}$$

or
$$V_s I_s = V_p I_p,$$

where we have used the fact that the electrical power is the product of the voltage V and current I. Dividing both sides by V_s and I_p, and remembering that V is proportional to N, we get

$$\frac{I_s}{I_p} = \frac{V_p}{V_s} = \frac{N_p}{N_s}.$$

The current is *inversely* proportional to the voltage and, therefore, inversely proportional to the number of turns. *When the voltage is stepped up, the current is stepped down proportionally.*

The biggest advantage to using AC, and the reason for its widespread use, is the ability to step the voltage up to high values for transmission over long distances, and to step it down to whatever voltage is needed. For the same power transmission, the current is stepped down and then up proportionally. Since the power dissipated as heat in a conductor is proportional to the current squared (I^2), this transformation greatly reduces the loss of electrical energy to heat along the transmission line. Some typical voltages between generator and consumer are shown in Fig. 12.38.

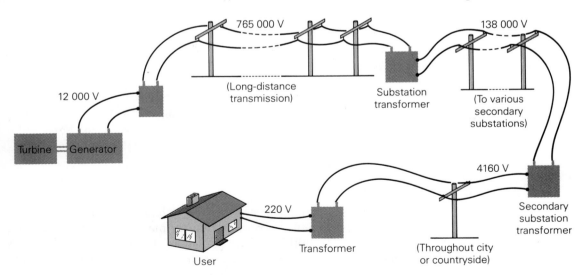

Figure 12.38   Typical voltages at various stages in the distribution of power from power station generator to user.

Example 12.1   Suppose 550 W of power are to be transmitted a distance of 100 km over 1.2-cm-diameter copper wire that has a total resistance for this length of 16 Ω. Compare the electrical power lost to heat if the transmission voltage is 110 V with that lost at 110 000 V.

Recall that

$$power = voltage \times current,$$

$$P = VI,$$

and that the rate of heat generation is the current squared multiplied by the resistance, or

$$P_{heat} = I^2R.$$

At 110 V:

$$I = \frac{P}{V} = \frac{550\ W}{110\ V} = 5.0\ A;$$

$$P_{heat} = I^2R = (5.0\ A)^2(16\ \Omega) = 400\ W.$$

At 110 kV:

$$I = \frac{P}{V} = \frac{550\ W}{110\ 000\ V} = 0.0050\ A;$$

$$P_{heat} = I^2R = (0.0050\ A)^2\ (16\ \Omega) = 0.00040\ W.$$

For transmission at 110 V, 400/550 or 73% of the power is lost along the way as heat. At 110 kV, the heat loss is only 0.000073%.

Notice also that the voltage drop, given by $V = IR$, in the transmission line for 110 V at the source would be

$$V = (5\ A)(16\ \Omega) = 80\ V.$$

There would be only 110−80 V or 30 V potential difference across the line at the user's end. At the higher voltage, you can show that the voltage drop is negligible.

Exercise   Show that the transmission of as much as 1100 W at 110 V would be impossible over the line described in Example 12.1, but that 1100 kW could be transmitted at 110 kV with a loss of only 1600 W or 0.14%.

Question   Why are heavy power users in the home (electric ranges, water heaters, clothes driers, etc.) usually designed to operate on 220 V rather than 110 V?

The ignition coil of an automobile is just a transformer with typically about 200 turns in its primary winding, and about 20 000 turns in its secondary. The primary is connected in series with the battery and a switch—the breaker points—located inside the distributor. Just before each spark is needed, a rotating cam separates the breaker points, cutting off the current to the primary. The rapid change in field induces a high voltage (about 20 000 V, or more) in the secondary winding that is connected by cables to the appropriate spark plug to initiate the spark.

Transformers do lots of jobs around the house. For example, they step down the 110-V line voltage for operating door bells, toy electric trains and racing cars, calculator battery chargers, and furnace thermostat circuits, to name a few.

12.7 Electromagnetic Waves

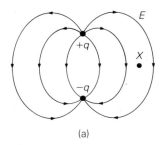

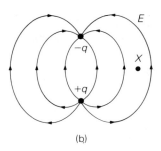

Figure 12.39 The electric field lines E around an electric dipole when (a) the upper charge is positive; and (b) when it is negative.

In Chapter 8 we described the transfer of thermal energy from one body to another by electromagnetic waves of the type called *infrared* radiation. Many other forms of electromagnetic radiation are very important to our lives. These include radio waves, light, x rays, and gamma rays. All of these are the same stuff; they differ only in wavelength (distance from crest to crest) and in the way they are produced. We'll examine the production of radio waves—electromagnetic waves that carry broadcast signals from radio and TV stations to our homes.

Consider the electric *dipole*—a positive and an equal negative charge separated by some distance—shown in Fig. 12.39, along with its accompanying electric field. In part (a), the upper charge is positive and the lower negative. In part (b), the charges are reversed. The electric field at point X is therefore downward in Fig. 14.39(a) and upward in Fig. 14.39(b). If the charges in the dipole could somehow be made to change continuously from positive to negative, the electric field at point X would continually alternate between up and down.

One way to get such an *oscillating dipole* is to connect two metal rods (see Fig. 12.40) to an alternating emf. As the emf changes sign, the ends of the rod are alternately charged positively and negatively. The field at point X then alternately changes from up (as shown) to down. The change is gradual from zero to maximum up, back to zero, to maximum down, and so on.

The field extends from the charges out to large distances. But the *change* in the field takes a finite time to propagate outward. In vacuum, it propagates at a speed that is the same anywhere in the universe: *the speed of light*. Thus, at an instant when the field at point X is up, that a long distance away at point Y may be down. At both points the field is continuously changing from up to down and back.

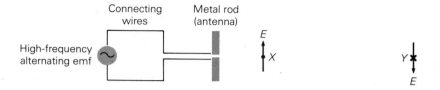

Figure 12.40   When an alternating emf is applied to two metal rods as shown, they form an antenna that radiates field variations outward. If, at some instant, the electric field is upward at *X*, it could be downward a long distance away at *Y*. The magnetic field would be outward (·) at *X* and inward (×) at *Y*.

We get a *wave* in the electric field that moves along as shown in Fig. 12.41(a). This graph shows the electric field at an instant at various distances from the oscillating dipole: the *antenna*. This up and down variation in electric field moves outward from the antenna at the speed of light (3.0×10^8 m/s, or 186 000 miles/s).

Look again at Fig. 12.40, and notice that when charge is flowing in the antenna so that the top is becoming negative and the bottom positive, a net downward current exists in the antenna. This situation requires a *magnetic* field at nearby point *X* directed *out* of

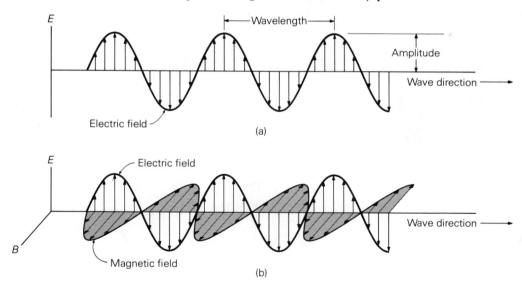

Figure 12.41   (a) The wave in the electric field that is produced by an oscillating dipole. (b) In an electromagnetic wave, the electric and magnetic field variations are in phase, but directed at right angles to each other.

the page, at a right angle to the electric field. As the charge oscillates up and down, the antenna current is first up and then down, meaning that the magnetic field at X alternates between in and out. This magnetic field change also propagates outward at the speed of light, so that when the field is out (indicated by the ·) at X it may be in (indicated by the ×) some distance away at Y. The variations in electric and magnetic field are in phase, but directed at right angles to each other. This overall **electromagnetic** *wave* at any instant is shown in Fig. 12.41(b).

The electromagnetic wave propagates out from the transmitting antenna. When it impinges on a receiving antenna, the variations in electric and magnetic field induce current variations in direct proportion to the field variations, which in turn are in direct proportion to current variations in the transmitting antenna.

We use the term *frequency* to describe the number of oscillation cycles per second and measure it in the unit hertz (Hz) where 1 Hz = 1 cycle/s. AM radio stations broadcast in the frequency range 550 to 1700 kHz,* while FM stations use the frequency range 88 to 108 MHz.† TV stations broadcast between 59 and 890 MHz. You may have noticed that you can get the sound from Channel 6 on your FM radio, since its frequency is within the FM band.

The broadcast frequencies we have mentioned are called the *carrier* frequencies. The sound or picture information is "carried" along with this wave by superimposing on it a variation in amplitude or frequency. For example, in AM (amplitude modulated) waves, the carrier wave of Fig. 12.42(a) is varied in

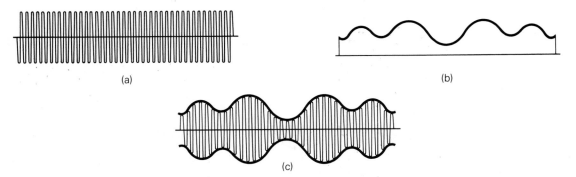

(a)

(b)

(c)

Figure 12.42   (a) The carrier wave, modulated in amplitude by (b) a lower-frequency wave, gives (c) an overall AM wave.

---

*Remember, the prefix k (kilo) means 10^3.

†The prefix M (mega) means 10^6.

amplitude according to the waveform of Fig. 12.42(b), which contains the sound information. The resulting wave takes the form of Fig. 12.42 (c). In FM (frequency modulated) waves, the frequency rather than the amplitude is varied. In TV broadcasts, the sound is transmitted by FM and the picture by AM, both on the same carrier wave.

The Electromagnetic Spectrum

Any form of electromagnetic radiation is a wave of electric and magnetic fields that propagate at the speed of light. The various forms differ in frequency and in wavelength—the distance from crest to crest, as indicated in Fig. 12.41.

Figure 12.43 shows the approximate frequency and wavelength range for each type radiation in the entire **electromagnetic spectrum.** The boundaries between types are not well-defined, and there is overlap between the various regions. The figure also shows the specific regions in which particular devices operate.

All the radiation from microwaves up to the longest wavelengths (lowest frequencies) come under the general category usually called *radio* waves, which are all produced electrically in some way. Radiations from x rays through infrared are all pro-

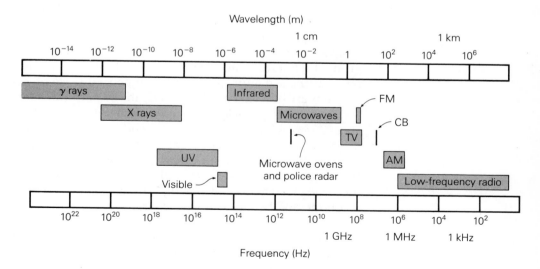

Figure 12.43   The electromagnetic spectrum, indicating the approximate ranges of the various forms of radiation.

duced within individual atoms or molecules, as we will see in Chapter 18. Gamma rays are emitted from the nuclei of atoms as the neutrons and protons inside rearrange themselves. (This is a form of radioactivity that we'll consider in Chapter 18).

As you look at the spectrum in Fig. 12.43, be careful to notice the vast range of wavelengths shown. Between any two vertical lines, the wavelengths increase (or decrease) by a multiplication factor of 100. The ratio of the longest to the shortest wavelength shown is about the same as the ratio of the size of the earth's orbit around the sun to the size of a single proton in the nucleus of an atom. Viewed in this perspective, you see that visible light is just a "speck" in the vast electromagnetic spectrum.

Key Concepts

Permanent magnets always exist with two *poles,* a north pole and a south pole. Two north poles or two south poles repel each other; a north and a south pole attract each other. Surrounding a magnet is a **magnetic field** whose direction is the direction of the force on a north pole.

A magnetic field surrounds every current. Around a long straight wire carrying a current, the field is directed tangent to a circle concentric with the wire. *Solenoids* have fields similar to those of bar magnets. A solenoid wound on an iron core forms an electromagnet.

The atoms of a magnetic substance act as tiny bar magnets because of the electron spin and orbital motion. In iron the atoms group themselves into *domains* in which all the atomic magnets point in the same direction. In a magnetized piece of iron all the domains are oriented predominantly in the same direction.

A charged particle moving through a magnetic field experiences a *force* whose direction is perpendicular to both the field and direction of motion. The strength of the force is proportional to the charge, the particle speed, and the magnetic field strength; the force increases from zero when the field and particle direction are the same to a maximum when they are perpendicular. A current in a conductor—being a moving charge—experiences a force with the same characteristics.

An electric *motor* uses electrical energy to do mechanical work by means of the forces on the two sides of a current-carrying loop in a magnetic field. A *generator* uses mechanical work to produce electrical energy. Its basis is the force on the free charges inside a conducting loop as it rotates in a magnetic field.

A changing magnetic field through a circuit induces an emf proportional to the rate of change of magnetic field. This effect is called *electromagnetic induction.* In a transformer,

$$\frac{\text{output voltage}}{\text{input voltage}} = \frac{\text{input current}}{\text{output current}} = \frac{\text{number of secondary windings}}{\text{number of primary windings}}.$$

Electromagnetic waves are oscillating electric and magnetic fields that propagate out from their source at the speed of light. They include x rays, light, microwaves, and radio waves.

Questions

Permanent Magnets

1. What are the similarities and what are the differences in the forces between magnetic poles, and those between electric charges?

2. Suppose you were locked inside a room made of glued-together wood. You have a bar magnet and an identical-looking piece of iron, but nothing else. The guard will let you out *only* if you toss out the magnet and keep the iron bar. How can you be sure your choice will not condemn you for life?

3. Why can't you create a single isolated magnetic pole by breaking a bar magnet in half?

4. Why is a nail attracted to *either* pole of a magnet, but another magnet attracted only to one of the poles?

Magnetic Fields of Currents

5. A magnetic field surrounds a current-carrying wire. Is there also an electric field around such a wire? Why not?

6. Would it matter if the battery in Fig. 12.17(a) were reversed? How about the small one in Fig. 12.18?

7. Design an addition to the door chime of Fig. 12.17 that has another solenoid for the back door—one that "bings" but doesn't "bong."

8. Figure 12.44 is a diagram of a doorbell that rings continuously while the button is pushed. Explain its operation. Notice that as the clapper moves to the left, the electrical circuit is broken.

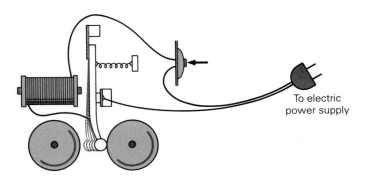

Figure 12.44 Why does the bell ring continuously when the button is pushed?

9. Justify the statement that all magnetism, even that of permanent magnets, results from electric currents.

Moving Charges in a Magnetic Field

10. If the charged particle in Fig. 12.21 is an electron, how does the force compare with that on a proton moving at the same speed?

11. A charge of 0.5 C is at rest in a 2-T magnetic field. What is the force on the charge due to the magnetic field?

12. If there were a magnetic field parallel to this page and directed from the bottom to the top of the page, what would be the direction of the force on a negatively charged chlorine ion if it were moving toward the page from above, along a line perpendicular to the page? What if it were moving from top to bottom of the page? What if it were moving from the lower-left to the upper-right corner?

13. Describe the relative orientation of a combination magnetic and electric field, and particle direction that would give no force on an alpha particle (a helium nucleus) moving through the fields. Would it make a difference if the charge were negative?

14. If you have a Ping-Pong ball with a net positive charge on it, how could you swing a bar magnet in the vicinity of the ball and make it jump without actually hitting it? How would the ball move in relation to the magnet and its motion?

15. When a charged particle enters a uniform magnetic field perpendicularly, the force is always perpendicular to its direction of motion. What kind of path will the particle follow in the

field? Will the particle be speeded up or slowed down by the field?

16. Two parallel wires carrying a current in the same direction are attracted to each other. Why?

17. If a TV set were placed between the poles of a large magnet as shown in Fig. 12.45, what would be the effects on the picture?

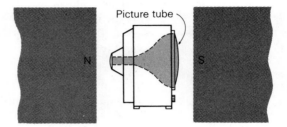

Figure 12.45 What happens to the picture on this TV?

18. Two electron beams are shot from side-by-side electron guns so that the beams are parallel. What is the direction of the electrical force between the electrons in the two beams? What is the direction of the magnetic force? Would you expect the beams to converge or diverge? If that depends on some unspecified variable, what is the variable?

19. One straight current-carrying wire lies across another so that they are perpendicular to each other. What is the nature of the forces on each wire?

Motors and Generators

20. If you run a large current through a horseshoe lying on a table, what would you expect to happen to it? (Remember the earth's magnetic field.)

21. The basis for electromagnetic meters that measure currents is similar to that of motors. Figure 12.28 would have to be changed little to show their operation. Instead of the brushes and commutator, each terminal is connected permanently to one side of the loop. The loop has a spring attached to it that opposes its rotation and lets it rotate only a limited amount. The angle of rotation is proportional to the torque applied to the loop. Explain why the angle of rotation is proportional to the current through the loop.

22. How could you use the same device as both a motor and a generator?

23. True or false: If you take a ring off your finger and spin it on the tabletop, there will be a current in it. (Remember the earth's magnetic field.)

24. Po-Dunk Power and Light Company supplies electrical power to city of Po-Dunk Holler. How does the mechanical work used to turn Po-Dunk Power's generator in any one hour compare with the combined work done (or energy expended) by all the electric lights, tooth brushes, can openers, sawmills, aircraft assembly plants, and the like in Po-Dunk Holler during that hour?

25. When you turn on an electric light, it's a little harder to turn the armature of the generator supplying your power. Explain this fact on the basis of the energy conservation principle.

26. Imagine the generator of Fig. 12.32 connects to the bulb through a set of slip-rings (as in Fig. 12.30). Now, draw a graph of bulb current versus time. What is the orientation of the loop relative to the magnetic field when the current is a maximum and when it is zero?

Changing Magnetic Fields and Electromagnetic Waves

27. If the bar magnet in Fig. 12.33 were dropped to fall straight through the loop, describe the deflection of the meter at various points during the fall.

28. Why isn't commercial electricity DC rather than AC?

29. What is the difference between AM and FM radio waves?

30. Why can you get the sound from a Channel 6 TV station on an FM radio?

Problems

1.* A baseball that has a net charge of 3.0×10^{-5} C is thrown with a speed of 40 m/s into the 13-T field of a superconducting

---

*Solutions for Problems 1, 2, 3, and 4 depend upon the equations in the footnotes of Sec. 12.3.

magnet. If the ball moves perpendicular to the field, what is the force on the ball? If its mass is 0.15 kg, what is the acceleration of the ball due to this force?

Ans: 0.0156 N; 0.10 m/s² (0.01 g)

2.* An electron in a TV picture tube is moving with a speed of 3.0 × 10⁷ m/s in a direction perpendicular to the earth's magnetic field of strength 5.5 × 10⁻⁵ T. What is the magnetic force on this electron? What is its acceleration due to this force? (The electron mass is 9.1 × 10⁻³¹ kg.)

Ans: 2.6 × 10⁻¹⁶ N; 2.9 × 10¹⁴ m/s² (3 × 10¹³ g)

3.* A straight wire carrying a current of 5.0 A passes through 1.5 m of a 2.0-T magnetic field. If the current direction is perpendicular to the magnetic field, what is the magnetic force on the wire? **Ans:** 15 N

4.* Suppose the loop of Fig. 12.28 is a square 20 cm on a side, and carries a current of 3.0 A. If the magnetic field strength is 0.90 T, what is the torque on the loop when its plane is parallel to the magnetic field? **Ans:** 0.11 N·m

5. A DC motor is 90% efficient and operates on 110 V. It has an output work rate of 250 W (⅓ horsepower). What is the current to the motor when operating at full power? **Ans:** 2.5 A

6. A generator is 98.0% efficient and has an output voltage of 6.00 kV. What must be the power provided by the turbine turning the generator if the load draws 20.0 A of current?

Ans: 122 kW

7. If the transformer that steps the voltage down from 4160 V to 220 V in Fig. 12.38 has 20 000 turns in its primary winding, how many turns are in its secondary? **Ans:** 1060 turns

8. Suppose 100 A of current at 220 V are supplied to the user in Fig. 12.38. Neglecting losses, what current is needed in each part of the transmission line to give this current?

Ans: from user end, 5.3 A; 0.16 A; 0.029 A; 1.8 A

9. Design a transformer to operate on 120 V and provide the 12 000-V power needeed for a neon sign. **Ans:** $N_s/N_p = 100$

10. An automobile ignition coil typically has about 200 turns in its primary winding and 20 000 turns in its secondary. If you hooked the primary of such a coil to 110-V AC, what would be the output voltage of the secondary? **Ans:** 11 kV

Home Experiments

1. Make a homemade compass. Magnetize a steel sewing needle by rubbing it on one pole of a magnet. Rub always in the same direction, and move the needle away from the magnet on the return for the next stroke. Test your efforts, and make the compass, by floating the needle horizontally on a cork in a cup of water. It should align itself in a north-south direction.

2. Hold a strong magnet near the front of a working *black and white* TV and notice what happens. Do *not* do this to a color set; you may magnetize the metal mask in back of the screen!

3. Swing a finger back and forth quickly in front of a lighted TV screen. You see the finger only every 1/60 s as the beam sweeps down the screen. (The beam sweeps across every other line each 1/60 s, so that it covers all lines every 1/30 s.) Estimate how fast you're moving your finger by the spacing between the points where you see it.

References for Further Reading

Aldridge, B. G. and G. S. Waldman, *Automobile Ignition Systems* (a Physics of Technology Module, McGraw-Hill Book Co., New York, 1975). Discusses electromagnetism by focusing on the automobile ignition system.

Bobeck, A. H. and H. E. D. Scovil, "Magnetic Bubbles," *Scientific American,* June 1971, p. 78. Discusses the way in which magnetic domains are formed into bubbles for use in computer memory.

Fink, D. K. and D. M. Lutyens, *The Physics of Television* (Anchor Books, Doubleday & Co., Garden City, N.Y., 1960). Very good discussion of the physics of television broadcasting and receiving. Written at the level of this text, and includes discussion of the modulation of the carrier wave.

Gould, J. L., J. L. Kirschvink and K. S. Deffeyes, "Bees Have Magnetic Remanence," *Science,* Sept. 15, 1978, p. 1026. Magnetic materials within honey bees let them orient to the earth's magnetic field.

Mee, C. D., *The Physics of Magnetic Recording* (North-Holland Publishing Co., Amsterdam, 1968). A technical discussion of the techniques and components of magnetic tape recording devices.

"Ole, Amigo," *Time,* August 29, 1977, p. 32. An interesting example of the use of electromagnetic radiation in controlling ants, mice, gophers, and other pests.

Schwartz, B. B. and S. Foner, "Large-scale Applications of Superconductivity," *Physics Today,* July 1977, p. 34. Applications of superconducting magnets.

The Way Things Work, an Illustrated Encyclopedia of Technology (Simon and Schuster, New York, Vol. 1, 1967, Vol. 2, 1971), or an equivalent publication, *How Things Work, The Universal Encyclopedia of Machines* (George Allen & Unwin Ltd.,

London, Vol. 1, 1967, Vol. 2, 1971). Discusses the physical principles and construction details of many electromagnetic devices.

Time-Life Books, *How Things Work in Your Home (and what to do when they don't)* (Time-Life Books, New York, 1975). Well-illustrated discussion of the construction and principles of electric motors, and how they are used in appliances.

13

Electrical Energy

Electrical energy is energy in its most convenient form for everyday use. Some of us use natural gas, oil, coal, or wood for jobs that are always done in the same place—heating our houses and water, or cooking our food. But try to think of a design for a portable hair drier, clothes iron, coffee pot, or automatic food mixer that doesn't use electricity.

In this chapter we'll look at the two ends of the electrical energy spectrum: its sources and its uses. Concerning sources, we'll consider the overall scheme for electrical energy production in a power plant, and we'll discuss the mechanism that converts chemical to electrical energy in a battery. We'll also discuss solar cells and other sources of electrical energy that are becoming more important as our fuel supplies dwindle. To talk about solar cells, we will need to describe the fundamental properties of a class of materials called *semiconductors*.

In examining electrical energy uses, we'll compare the energy requirements for different kinds of jobs. This section will tie together electrical and other forms of energy that we've discussed throughout the book, and will provide some practical insights into conserving usable energy. Finally, we'll investigate the interaction between energy and power as they influence the cost of supplying electrical energy.

13.1 Sources of Electrical Energy

We've mentioned electrical energy sources at various places in this book. We'll now discuss them in more detail.

Electrical Power Plants

In the previous chapter we described the basic physical laws that apply to using mechanical work to produce electrical energy in a generator. We'll now look at what else we need besides the generator to get electrical energy from a power plant.

First notice that the amount of work done in rotating the generator armature coils must be at least as much as the energy supplied to the devices using the generator's current. For example, if ten 50-kW motors operate on the current from the generator, work must be done at a rate of *at least* 500 kW in turning the generator. What supplies this work?

Most power plants use steam turbines to turn their generators. A simplified block diagram of the overall processes in a steam plant that burns fossil fuel—coal, oil, or natural gas—is shown in Fig. 13.1. We say "simplified" diagram because a real power plant has many auxiliary components along with each of the basic ones shown here. The result of the process is the conversion of chemical energy in the fuel to electrical energy, with the intermediate forms, thermal energy and mechanical work, along the way.

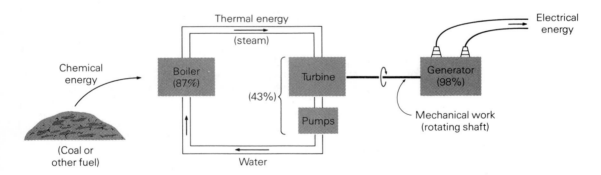

Figure 13.1   A simplified block diagram of a fossel fuel power plant. Efficiencies of each stage are given in parentheses. The energy to run circulating pumps and other accessories is included as part of the energy "loss" in the turbine.

The diagram shows typical efficiencies of the devices performing the various processes. For each 100 kWh of energy entering the boiler as chemical energy in the fuel, 87% (or 87 kWh) are passed on to the steam. Of these, 43% (or 37 kWh) do useful work in turning the generator. The generator is quite efficient, converting 98% of this work (or 36 kWh) to electrical energy. A typical overall efficiency is thus 36%. The most efficient modern fossil fuel plants have an overall efficiency of about 39%. This efficiency does not include the loss, typically about 8%, of electrical energy to heat in the transmission line between power plant and consumer.

For *nuclear* power plants the block diagram looks the same. The boiler is then a nuclear reactor that uses uranium or plutonium as a fuel. The nuclear energy released in the reaction produces heat that generates steam. Because of the need to insure no leakage of radioactive material, boilers in nuclear plants are run at lower temperatures to increase the structural strength of the materials, and lower pressures to decrease the chance of rupture. This lower steam temperature and pressure reduce the turbine efficiency so that the overall efficiency is typically about 33%.

All the energy released from the fuel but not converted to electrical energy is converted to heat in the environment. Power plants are located on large bodies of water so that they can dissipate most of this heat into the water. This *thermal pollution* can have an adverse effect on local conditions, and must be controlled carefully.

In a hydroelectric plant, the process is somewhat different than in a steam plant (Fig. 13.2). There, the gravitational potential energy of the water at the top of the dam converts to kinetic energy as it falls from the dam. This kinetic energy turns water turbine blades, providing the work to turn the generator. The overall conversion is from gravitational potential energy to electrical energy.

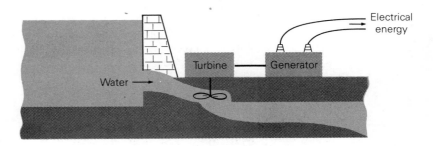

Figure 13.2 The basic principle of a hydroelectric power plant: The conversion of gravitational potential energy to electrical energy.

If you want to compare the fuel you consume when you use electrical energy with that consumed when you use energy in some other form, you have to consider the power plant and transmission efficiencies. For example, electrical baseboard heaters are nearly 100% efficient: They convert all the electrical energy supplied to them to heat. On the other hand, the typical oil-fired home furnace is probably no more than about 50% efficient. If your power plant burns fossil fuels, by which of these two methods of heating your house will you consume *less* of the world's fuel supplies? As we have said, a typical power plant is about 36% efficient, and about 8% of the electrical energy is lost in transmission. Therefore, in operating the 100%-efficient baseboards, you would use about 1½ times as much fuel. Unfortunately, the choice of which of these methods gives the biggest overall benefit is not quite so clear-cut. For example, the power plant might use a much more plentiful fuel than you would use at home, thereby reducing the advantage of less consumption. The answer to which method is most advantageous overall may involve some complicated technical, economic, and social issues that are discussed in some of the end-of-chapter references.

You might question whether or not you really save the power plant any fuel when you turn your lights off, since the generator is turning anyway. From an overall energy viewpoint, if less electrical energy is used, less must be generated, and therefore less fuel is burned. You can understand why by looking at the generator diagram of Fig. 12.32. If you use less current from the generator, less charge is pushed around the rotating loop, and therefore less work is needed to push the charge. Since the loop rotates at a constant speed, it takes less force to rotate it; that means the turbine needs less steam, and therefore the boiler less fuel. This line of reasoning is the same as you would use in answering Question 32 of Chapter 1—that it takes less power to operate a stopped-up vacuum cleaner because you're pushing less air through the cleaner.

Question   Even though you probably can't feel the difference, explain why you pedal a little easier on your bicycle if the bulb on your generator-operated light burns out. Why do you use *slightly* less fuel in your car if you don't use the electrically operated fan on your heater? (Again, you won't notice the difference.)

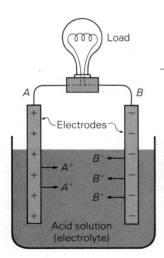

Figure 13.3 The components of a single-cell battery.

Batteries

A battery converts chemical energy directly to electrical energy. It is made up of one or more *cells*, each with the parts shown in Fig. 13.3—two rods called *electrodes* of different metals in an acid solution called an *electrolyte*. (Strictly speaking, a battery means more than one cell; the word is used nowadays to mean one or more.) In 1800 the ·Italian physicist Alessandro Volta used copper and zinc electrodes in a saltwater solution for the first humanmade battery.

The metals dissolve slightly in the solution, each dissolving atom leaving one or more electrons on the electrode and going into solution as a positive ion. Being of *different* metals, one rod dissolves more than the other. If the metal of rod *B* dissolves more, it leaves behind more electrons, becoming more negatively charged than

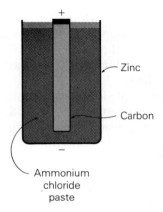

Figure 13.4

+

Zinc

Carbon

−

Ammonium
chloride
paste

Figure 13.4 The main components of a flashlight battery: a "dry" cell. The zinc case is the negative electrode.

rod *A*. This process means that rod *B* is at a lower potential than rod *A*.

The parts of the electrodes extending outside the battery are called *terminals.* If you connect a wire from the positive terminal through a light bulb (or other load) and back to the negative terminal, electrons will flow through this wire and bulb from the negative to the positive terminal. This flow corresponds to a conventional current from the positive to the negative terminal. The continued dissolving of electrode *B* provides a continuous supply of electrons. Electrons arriving at electrode *A* from the wire are absorbed by positive ions in the solution. Eventually, electrode *B* completely dissolves, and the battery quits operating. Some types of batteries can be *recharged* by forcing current in the opposite direction through the battery. This reverses the chemical reaction, restoring the electrodes and electrolyte. Car batteries are, of course, of this type.

The output voltage (emf) of the cell depends on the relative extent to which the electrodes dissolve in the electrolyte. This, naturally, depends upon what substances are used for electrodes and electrolyte. In a conventional car battery, the positive electrode is lead peroxide, the negative terminal spongy lead, and the electrolyte dilute sulfuric acid. The emf of such a cell is about 2 V. Six cells connected in series give a 12-V battery.

As the battery discharges, the chemical reaction uses up sulfuric acid and produces water. Eventually, the solution is too dilute to support the reaction—the battery is "dead." Since sulfuric acid is denser than water, the density of the electrolyte increases with the concentration of sulfuric acid. When service station operators check the "charge" of your battery, they use a *hydrometer* that measures the electrolyte's density. This indirectly measures the extent of discharge of the battery.

A flashlight battery is called a *dry* cell because the electrolyte is in a paste that is thick enough not to run out when turned upside down. As Fig. 13.4 shows, the container itself is zinc, which forms the negative electrode. A carbon rod down the center is the positive electrode. The electrolyte is ammonium chloride. This arrangement gives an emf of 1.5 V.

Many car batteries on the market today have no plugs for adding water, and many that do have them seldom need filling. You may wonder why the old type has to be refilled occasionally, but the new type does not.

When a battery is being recharged, small amounts of hydrogen and oxygen are given off at the electrodes and vented to the outside through the top of the battery. These gasses come from a decomposition of water during the charging process. Water needs to be added periodically to replenish that used up.

Manufacturers use various techniques that basically involve two modifications to minimize the need for adding water. First, the lead plates that form the electrodes have a small amount of "hardener" added to them to make them stronger. That hardener has traditionally been antimony. Many of the newer batteries use calcium as a hardener. This changes the chemical processes so that they produce less gas and therefore use less water. The second technique that some manufacturers use is to redesign the structure of the battery so that the plates rest on, or near the bottom of, the case. The electrolyte level can then be higher above the plates so that the battery can lose more water without affecting its operation.

Inside a battery, charge moves *against* the potential difference: Negative charge moves toward the negative electrode, positive charge toward the positive electrode. The chemical reactions provide the energy to do the work of moving this charge. Outside the battery in the external circuit, the charges flow freely in the direction of decreasing potential energy. The work done per unit of charge moved through the battery equals its emf.

Other Electrical Energy Sources

Even though generators and batteries are the main sources of everyday electrical energy, other sources either are used for special purposes, or are being developed for large-scale power production in the future. We'll briefly comment on a few of these.

Fuel Cells   *Fuel cells* are first cousins to batteries, in that they too produce electrical energy from chemical energy. However, the fuel for the chemical reaction in a fuel cell is continuously supplied from outside the cell rather than being contained within the cell.

A fuel cell combines hydrogen and oxygen to form water, with the excess energy given up in the reaction going to electrical energy. If hydrogen and oxygen are merely combined, they interact violently. In a fuel cell, this reaction is controlled. One type of cell has porous carbon electrodes separated by an electrolyte (Fig. 13.5). Hydrogen is fed to one electrode, oxygen to the other. Chemical reactions at the electrodes liberate electrons at one, and absorb electrons at the other, thereby giving positive and negative terminals, as in a battery.

Because fuel cells are compact and portable, they are well suited for applications that have limited power needs, such as manned space flights. As the techniques for building and operating these cells improve, they may assume a more important role in the production of electrical energy for everyday use.

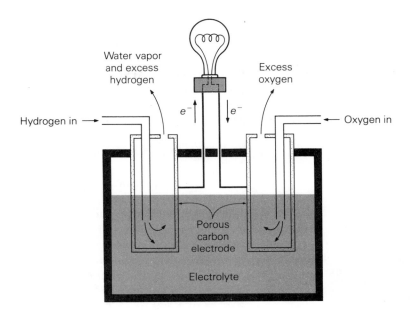

Figure 13.5 A hydrogen-oxygen fuel cell.

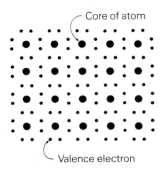

Figure 13.6 The atoms in pure silicon are arranged in a regular crystal lattice so that each atom shares one electron with each of its four nearest neighbors.

Solar Cells Tremendous quantities of energy reach the earth every day in the form of radiation from the sun. People use various schemes to capture some of this radiant energy to product heat. But we can also use this radiation to produce electrical energy.

It is possible to let the heat from the radiation produce steam to run a turbine to turn a generator and generate electrical energy in the conventional way. It is also possible to convert radiant energy directly to electrical energy in a device called a *solar cell*. Solar cells are made of materials in a class we mentioned earlier—**semiconductors.** To understand how solar cells work, we first need to understand semiconductors.

One abundant semiconductor is silicon. Its atoms have 4 valence electrons: that is, 4 electrons in their outermost orbits. A pure piece of silicon is a crystal with atoms arranged in a regular array. (See Fig. 13.6. A real array is three-dimensional rather than two-dimensional as shown.) Each atom is bound tightly to its neighbors by sharing one valence electron with the four nearest atoms. Since this arrangement binds the electrons tightly, pure silicon is a very poor conductor.

Suppose we add to the silicon a small amount of an element such as arsenic whose atoms have 5 valence electrons. As Fig. 13.7 shows, each arsenic atom has an extra electron. These loosely attached electrons are free to move about as negative charge carriers, making the impure silicon a good conductor.

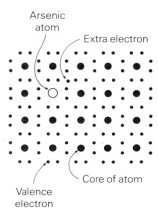

Arsenic atom

Extra electron

Core of atom

Valence electron

Figure 13.7 If an arsenic atom with 5 valence electrons replaces a silicon atom, the extra electron is free to roam about.

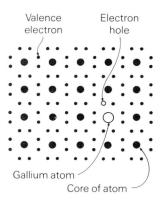

Valence electron

Electron hole

Gallium atom

Core of atom

Figure 13.8 If a gallium atom with 3 valence electrons replaces a silicon atom, it leaves a hole in the electron structure that acts as a positive charge carrier.

We can also make silicon a conductor with *positive* charge carriers. Suppose that, instead of a valence-5 impurity, we add one whose atoms have 3 valence electrons, such as gallium. At the position of the gallium atom one electron is missing, leaving a "hole" in the electron structure (Fig. 13.8). Another electron can move in to fill this hole, leaving a hole where *it* was. Some other electron can fill this hole, leaving a hole in its old slot (Fig. 13.9). Thus the hole acts as a charge carrier as it moves around freely in the silicon. Since the hole represents an *absent* electron, it acts as a *positive* charge carrier.

Impurity semiconductors with excess electrons are called *n-type* (for negative) semiconductors; those with excess holes are called *p-type* (for positive) semiconductors. Transistors and other "solid state" electronic components are made from combinations of n- and p-type semiconductors.

We can now see what semiconductors have to do with solar cells. The heart of a solar cell is a *p-n junction*—a layer of p-type and a layer of n-type semiconductor in contact (see Fig. 13.10)—with a wire connected to each side. When the two layers first make contact, some electrons flow from the n side to the p side, filling some holes on the p side and creating some holes on the n side. This causes a net negative charge on the p side, and a net positive charge on the n side. The charge separation produces an electric field across the boundary from the n to the p side, and this field eventually stops the flow of charge.

When light strikes the vicinity of the junction, it can knock electrons loose from the lattice, leaving *electron-hole pairs*. If this happens on the p side, the electron is accelerated across the boundary by the electric field there. If it happens on the n side, the hole moves across the boundary because of the field. Consequently, an excess of holes builds up on the p side; an excess of electrons builds up on the n side. Thus, when the junction is exposed to light, the p side acts as the positive terminal of a battery, the n side as the negative terminal. The emf of such a cell is about ½ V when in sunlight.

As the technique for making solar cells improves, and the cost of their production comes down in the future, solar cells may become a significant source of electrical energy. Keep in mind, however, that *the output power from a solar cell is limited by the rate of energy hitting it from the sun.* The sun beams an average of about 200 watts per square meter over the United States. If we assume a 20% conversion efficiency in solar cells (about the best that can be hoped for in the near future) this method would require an area of about 25 square kilometers covered with solar cells to equal the power output of a 1000-MW power plant.

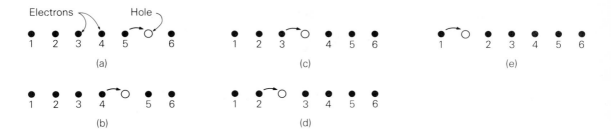

Figure 13.9   Conduction by a hole in a line of electrons. (a) The hole is between electrons 5 and 6. (b) As electron 5 fills the hole, the hole moves to the left. In each successive step—(c), (d), and (e)—the hole is conducted one step to the left.

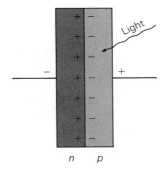

Figure 13.10   The heart of a solar cell is a p-n junction made from layers of p-type and n-type semiconductors.

MHD—Magnetohydrodynamics   A conventional power plant converts chemical energy to thermal energy to mechanical energy to electrical energy. Since some energy is wasted in each conversion, the elimination of any of the steps would seem to be desirable. *Magnetohydrodynamic* (MHD) generators eliminate the mechanical step—they convert thermal energy directly to electrical energy.

Figure 13.11 shows the process schematically. Pulverized coal is burned in preheated air, reaching a temperature of several thousand degrees Celsius. At this temperature, the combustion gases ionize from collisions between atoms, forming a *plasma*—a mixture of electrons and positive ions. This expanding plasma leaves the nozzle at supersonic speeds, and passes through a strong magnetic field. The field bends the positive ions one way and electrons the other so that they hit opposite sides of the channel. The sides become positive and negative electrodes that act as a source of emf.

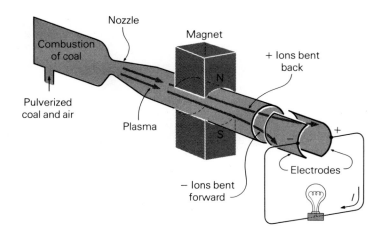

Figure 13.11 In a magnetohydrodynamic (MHD) generator, high-speed positive and negative ions from the high-temperature combustion are bent in opposite directions by the magnetic field. They strike opposite sides of the channel, charging opposite electrodes positively and negatively.

Considerable research effort is underway in several countries to learn the details of making MHD generators practical and reliable for large scale power generation. Researchers expect an overall efficiency around 60%, meaning half again as much electrical energy per kilogram of coal or oil as in conventional power plants.

****13.2   Using Electrical Energy**

Since electrical energy does so many kinds of jobs, it helps to know just how much energy different kinds of devices use in doing their particular tasks.

You've learned several laws about the conversion of energy from one form to another. Now let's use some of these laws to see how ordinary quantities of energy in different forms compare in size. In this section, I'd like you to gain some understanding in at least two areas.

First I'd like you to get an intuition—a feel—for the sizes of the numbers describing ordinary quantities of different forms of energy, and how they compare in the different forms. Here are some analogies. If you have to buy a new suit, you probably have some feeling for what it will cost. If you have to mow the lawn, you probably have some intuition about how much effort is involved. If you have to make a 500-mile trip by car, you have some idea how tired you're going to be when you get there. This section is intended to give you the same kind of intuition about various forms of energy and their interdependence—how much energy it takes to operate a light bulb, sleep under an electric blanket, heat water for a shower, or air condition your room. We'll develop some typical relative numbers that will assist you in comparing the energy needed for various jobs.

The second thing I'd like you to get from this section is the know-how to figure out the energy needed for a specific job—maybe a job you do regularly, or a job you're thinking about doing, and are counting the cost in advance.

Remember, power is the rate of doing work: power = work/time. Thus: work done = energy used = power × time. Since we usually give electrical power in watts (W) or kilowatts (kW), it's convenient to express energy or work in kilowatt-hours (kWh), gotten by multiplying power in kilowatts by the time in hours. For comparison, 1 kWh = 3.60×10^6 J = 860 kcal. For "in-your-head" estimates, a handy conversion factor is that 1 Wh is roughly the same as 1 kcal.

Now, let's look at some general ways of using electrical energy. Some of these jobs we'll consider could be done with other forms of energy, but the same ideas apply anyway. Our calculations will be rough approximations intended to give a feel for the amounts of energy involved.

Light Production   You're certainly familiar with electric light bulbs, and probably have a fair intuition about how much light you get from a 100-W incandescent bulb. The electrical energy used by such a bulb in 1 h of operation is

energy = power × time = 100 W × 1 h = 100 Wh = 0.1 kWh.

If you buy energy at 5¢ per kWh, that 1 h of operation will cost you ½¢. We'll compare the energy needed for other jobs with this familiar one.

Exercise   Since most of us identify better with how hard our pocketbook gets hit than with kilowatt-hours, you will benefit more from the illustrations in this section if you find out the average price of electrical energy in your area. Each time we calculate an energy use, translate that into cost. One way to estimate the price is to look at a monthly electric bill. Divide the charges by the number of kilowatt-hours used as stated on the bill. (This price probably won't be exact, because the rate usually varies with the total energy used.) You can find out the exact rate by contacting your power company.

Notice that 100 W (or whatever the power rating of a light bulb) is the electrical power, *not* the rate of *light output*. Different "sized" bulbs operate at different filament temperatures, and therefore differ in light production efficiency. You can find out the relative efficiencies at the light bulb counter of the grocery store. The package on each bulb tells you not only the power in watts, but the average number of *lumens,* a measure of the rate of visible light output (Fig. 13.12). From the information on the packages, you can see that it takes two 60-W (not 50-W) bulbs to give about the same light output as one 100-W bulb. Worse still, four 40-W bulbs give just about the same amount of light.

Exercise   If you replace two 100-W bulbs with enough 40-W bulbs to give the same light, how much *more* energy will you use in 150 h of operation? How much *more* will it cost at 5¢ per kWh?   **Ans:** 8 bulbs; 18 kWh; 90¢

Figure 13.12   The bulb package tells the light output in lumens.

Mechanical Work   To get a feel for the energy needed to do mechanical jobs, we'll first pick a job that you can do both by hand *and* with a power tool, so that we can compare the energy needed. Suppose you want to drill a 2-cm-diameter hole (3.4 in) in a 4-cm-thick piece of wood (for example, a "two-by-four"), and you have available both a brace and bit (Fig. 13.13) for hand drilling, and a portable electric drill rated at 3 A on 115 V: 345 W. Doing this job with the brace and bit might take 30 turns of the handle, which moves in a circle of radius 13 cm. We'll estimate the average force on the handle to be about 50 N (11 lb). The work done in turning the brace—force × distance per turn × number of turns—is

$$50 \text{ N} \times (2\pi \times 0.13 \text{ m}) \times 30 = 1220 \text{ J.}$$

To convert to kilowatt-hours, divide this quantity by 3.6×10^6 J/kWh to get 0.00034 kWh.

With the electric drill, this job might take about 10 s. Operating the drill at full power (certainly an overestimate) for 10 s gives an energy use of 0.345 kW × 10/3600 h, or 0.00096 kWh.

The answer we get for drilling by hand is about ⅓ that for the electric drill—not a bad comparison considering the estimates made, the difference in the nature of the drill bits, and the somewhat less than full power needed to drill the hole with the power drill. The important point for this discussion is the general amount—"order of magnitude"—of energy used, which is about 1/100th the energy used in our 100-W bulb in an hour.

Consider another mechanical example. How much work is needed by the electrically operated elevator to lift Walter Cronkite to the top of the 380-m-high Empire State Building. We'll assume his mass to be 70 kg: a weight of 700 N (157 lb).

Figure 13.13 A brace and bit for hand drilling holes in wood.

$$\text{Work} = \text{force} \times \text{distance} = \text{weight} \times \text{distance}$$

$$= 700 \text{ N} \times 380 \text{ m} = 266 \ 000 \text{ J}.$$

At 3.6×10^6 J/kWh, this is 0.074 kWh, roughly the energy used by our 100-W bulb in ¾ hour. (We haven't included the energy needed to lift the elevator.)

Estimate the work done in other mechanical jobs that you're familiar with, and see how it compares with the ones done here.

Water Heating Most of us would have a hard time getting along without hot water. In electric water heaters, current in resistive heaters produces heat that raises the water temperature. Typically there are two heating elements, each rated between 1 and 5 kW.

Recall from our chapter on heat energy that the heat needed to raise the temperature of a substance is

$$\text{heat} = \text{mass} \times (\text{specific heat}) \times (\text{temperature change}).$$

For a specific quantity—say, 1 liter—of water, let's find the energy (heat) needed to raise its temperature an amount typical of a household water heater. Since that "typical amount" will vary with climate and temperature setting, we'll estimate some reasonable values. I measured the water temperature entering my house on a winter day and found 7°C (45°F). In a moderate summer this temperature might average 20°C (68°F). Let's choose the *average* incoming water temperature to be 13°C and assume it is heated to 60°C (140°F). One liter of water has a mass of 1 kg, and the specific heat is 1 kcal/kg per C deg, which is 0.00116 kWh/kg per C deg. Raising the temperature of 1.0 liter of water by 47 C deg requires:

$$\text{heat} = \text{mass} \times \text{specific heat} \times \text{temperature change}$$

$$= 1.0 \text{ kg} \times 0.00116 \text{ kWh/kg·C deg} \times 47 \text{ C deg}$$

$$= 0.054 \text{ kWh for each liter.}$$

In other words, it takes *roughly* 1/20 kWh for each liter of water heated. If you think better in gallons, that's about 1/5 kWh per gal, since 1 gallon is about 4 liters.

That may not seem like much, but try estimating how much hot water a family uses in a month. Suppose, for example, a family takes four baths a day at 60 liters per bath, washes dishes in a dishwasher once a day at 50 liters per wash, and washes six loads of clothes a week at 30 liters of hot water per wash. Aside from other uses, that comes to about 9400 liters in a month, needing about 470 kWh for heating. (Calculate the cost of this at your local rates.)

Cooking   Most Americans cook with electrical energy. Surface cooking units on electric ranges are typically rated between 1400 W and 2600 W for full-heat settings. Conventional (not microwave) oven heating elements typically use around 3000 W. Both surface and oven units are resistive heaters that maintain their lower than full-heat settings by cycling on and off for various times according to the setting of the controls.

Space Heating   The energy needed for space heating—the heating of living, working, and playing space—depends not so much on the energy needed to change the temperature of the air and other contents of the space, but on the energy needed to replace the heat that leaks out through the walls, roof, and floor of the building. We considered the factors influencing that heat loss in our chapter on heat transfer. The needs vary widely with climate, the size house, and its insulation. For example, in Philadelphia with about 2500 C degree-days per year, a typical well-insulated house needs about 15 000 kWh of heat. One possible source of that heat is the conversion of electrical energy to thermal energy in resistance heaters. Essentially all the electrical energy delivered to such heaters is converted to thermal energy. On the other hand, with an electrically operated heat pump under moderate temperatures, you can get more heat delivered than electrical energy used (Chapter 9). That house in Philadelphia could get its 15 000 kWh of heat by supplying only about 8000 kWh to a heat pump.

Cooling   Most of us take refrigerators run on electrical energy for granted as a necessity of life. In the past several years, air conditioners have become almost as commonplace in many areas of the United States. The energy needed to pump heat from the inside of a refrigerator depends greatly on the refrigerator and how it is used. A typical refrigerator uses about 500 W of power when the compressor is running, which is about one-third of the time.

An operating air conditioner typically removes about 2½ times as much heat from the room as the electrical energy supplied to it. The amount that needs to be removed depends, of course, on the climate and on the room or house. A central air conditioner for a house might use about 5 kW, a one-room air conditioner about 1 kW. One way that air conditioning engineers describe the extent of air conditioning needed in a given climate is by the number of "cooling hours" per year needed to maintain comfort in a properly air conditioned home. In Houston that's about 1900 h; in Seattle it's about 100 h.

Summary   The illustrations we have included should give you a feeling for the relative energy needs of different kinds of everyday jobs. You might also look again at the comparisons in our boysenberry pie illustration on page 12 and see if they now seem reasonable. As you see, mechanical jobs of the type we do regularly need comparatively small amounts of energy. When we think of conserving electrical energy, many of us think first in terms of turning off unused lights. Of course, every little bit helps, but as our examples have shown, lights are not the worst culprits. Aside from space heating and cooling, water heating is the biggest domestic consumer. This is not a plea for going around dirty, but conservation in the areas of hot water use, as well as in space heating and cooling, will help where it counts the most.

Table 13.1 gives some average power ratings of various devices, and their average total electrical energy consumption per year in the United States. The average time of use and percentage "on" time when in use are built into the values given for energy consumption. Compare the entries in Table 13.1 with the illustrations discussed here and for other devices that are of interest to you.

Table 13.1   Average Power Ratings and Annual Energy Consumption of Various Appliances

Appliance	Average Power (W)	Estimated Annual Consumption (kWh)
air conditioner (room)	860	860[a]
blanket	150	150
blender	300	0.9
broiler	1140	85
carving knife	95	0.8
clock	2.5	22
clothes drier	4856	993
clothes washer	512	103

Table 13.1   *(continued)*

Appliance	Average Power (W)	Estimated Annual Consumption (kWh)
coffee maker	600	138
dehumidifier	257	377
dishwasher	1201	363
fan		
attic	370	291
window	200	170
food mixer	127	2
frying pan	1200	100
furnace (electric)	—	19 000[b]
hair drier	600	25
heat pump	—	11 000[c]
heater (portable)	1322	176
hot plate	1200	90
humidifier	177	163
iron	1100	60
oven (microwave only)	1450	190
radio	71	86
range (with oven)	12 200	730
refrigerators/freezers		
standard (12.5 ft^3)	—	1500
frost-free (17.5 ft^3)	—	2250
sewing machine	75	11
shaver	15	0.5
television (solid state)		
black-and-white	45	100
color	145	320
toaster	1100	39
toothbrush	1.1	10
vacuum cleaner	630	46
water heater	4474	4811

[a] Based on 100 hours of operation.

[b] Assuming 3100 C degree-days.

[c] Heating only, assuming 3100 C degree-days.

Exercise   Compare the electrical energy consumption of small appliances intended for specific jobs with that of general purpose appliances intended for many jobs. Examples: a 1000-W toaster rather than a 3200-W broiler; a 1150-W electric skillet rather than a 2000-W range surface unit.

**11.3   Do We Pay for Power or for Energy?

Almost everything we have said in this book related to consumption of expendable energy has implied that the important consideration is the total energy used to do a job, not the power or rate of using this energy. Some examples: when you pay your electricity bill, you pay for the number of kilowatt-hours—the energy—used during the month; when you are concerned about the operating cost of some appliance, it is the total energy it uses in some time period that counts and not its operating power; the value of fuel is based on its energy content, not the rate at which it can be used. The implication of all this is that power is not important, and that the answer to our section title should be "energy."

However, that isn't the whole story. Suppose, for example, that in the middle of a hot summer afternoon, most people in Metropolis turn on their air conditioners, later their ranges for cooking dinner, and maybe their color TV sets. The *power* for operating these, and everything else electrical in town, must be supplied by the power plant. The mechanical power from turbine to generator must *exceed* the combined electrical power used by the consumers. (Why exceed?) In the wee hours of the morning when most of these devices are off, that high-power generator is used at a small fraction of its capacity. The power company needs large enough generators to meet the "peak" demands, but these generators are either unused or underused at off-peak hours.

You and I pay indirectly for this peak-power capacity. We pay financially because the power company charges enough per kilowatt-hour to let them buy the needed generating equipment. We pay in world fuel supplies because energy is needed to manufacture this equipment. This is why power companies encourage, either financially or morally, the scheduling of electrical energy consumption for off-peak hours.

The answer to the question in this section title is therefore "yes"!

Key Concepts

The main sources of electrical energy in present use are electromagnetic **generators** and **batteries.** In a steam power plant, generators carry out one step in a process that converts *chemical* or *nuclear* energy to *thermal* energy, then to *mechanical* energy, and then to *electrical* energy. Batteries convert *chemical* energy directly to *electrical* energy. Other sources of electrical energy include *fuel cells, solar cells,* and *magnetohydrodynamic (MHD)* generators.

Solar cells use **p-n junctions,** formed by a layer of **n-type** and **p-type semiconductors.** The n-type semiconductors have valence-5 impurity atoms and conduct by means of electrons; p-type semiconductors have valence-3 impurity atoms, and conduct by means of *holes* in the electron structure that act as *positive* charge carriers.

Electrical energy consumption by ordinary devices generally ranks in the following order from lowest- to highest-energy users: appliances that do mechanical jobs, light sources, cooking and other heat-producing appliances, refrigerators/freezers, water heaters, space cooling and heating systems.

Questions

Electrical Energy Sources

1. Draw a block diagram like Fig. 13.1 for a hydroelectric plant.

2. Given the efficiencies for each step in the power plant diagram, (Fig. 13.1), how do you get the overall efficiency to be 36%?

3. Suppose Ann Landers and Dear Abby retire in identical side-by-side houses in Concord, Mass. and that each house needs 25 000 kWh of heat in a particular winter. If Ann heats her house with a 70% efficient natural gas furnace and Abby heats hers with a 92% efficient electric furnace, who uses the most fuel if the power plant also uses natural gas?

4. A fuel cell, like a battery, converts chemical energy directly to electrical energy. In what way is it more like a power plant than a battery?

5. Why do battery electrodes have to be made of two different metals?

6. The emf of a flashlight battery is 1.5 V. Does that mean flashlight bulbs normally operate on 1.5 V? Why not?

7. When a dry cell operates, hydrogen and other gases build up around the carbon electrode. What would you expect this to do to the emf of the cell?

8. John Solarwatt claims to have invented a solar cell so efficient that an array of his cells 10 meters square can provide an

average of 1 MW output. Is there some reason you should be skeptical about this claim?

9. Explain how the movement of a hole in a p-type semiconductor is like the movement of an available auditorium seat when everybody on a row with one seat on the left side moves, one at a time, one seat to the left.

10. What is the direction of the electric field across a p-n junction?

Electrical Energy Uses

11. Charlie Warmears keeps his bedroom at such a temperature that he will be warm under a regular blanket and not "waste" electrical energy on an electric blanket. Jane Coldfoot uses an electric blanket and sleeps in a cold bedroom. Discuss the economics—both energy and financial—of the two approaches.

12. When you turn on a hot water faucet after it's been off awhile, you get cold water at first and then the water gets hot. The farther the water heater is from the faucet, the more cold water you get. At bill paying time, does that cold water *count* as cold water or hot water?

13. Mrs. Savenergy insists that the house be kept at a temperature of 15°C, but Mr. Savenergy refuses to take a bath unless the bathroom temperature is at least 20°C. If the Savenergys have a gas-fired central heating system so that they must heat the whole house in order to heat the bathroom, might they come out ahead if they installed an electric heater in the bathroom?

Problems

1. If a power plant overall efficiency is 38%, and 7% of the electrical energy is lost to heat in the transmission line between plant and user, what is the efficiency for the entire process of generating and delivering electrical energy?   **Ans:** 35%

2. How much chemical energy in the fuel do Ann Landers and Dear Abby use in the situation of Question 3? Assume the local power plant is 38% efficient, including transmission to the house.   **Ans:** Ann—36 000 kWh; Abby—71 000 kWh

3. If the heat of combustion of natural gas is 0.40 kWh/ft³ and that of coal is 8.2 kWh/kg, how much fuel do Ann and Abby

consume in the situation of Problem 2 if the power plant burns coal rather than natural gas?

Ans: Ann—90 000 ft³; Abby—8600 kg (9.5 tons)

4. The Peachtree Power Company normally operates at an efficiency of 39.0% and supplies an average of 1000 MW to its customers at a price of 4¢/kWh. What would be the financial loss to Peachtree's stockholders in a year if the plant's efficiency dropped by 0.1%? **Ans:** $900 000

5. The Jones family uses an average of 1200 kWh of electrical energy per month. They decide to switch to solar cells to get this energy, and use batteries to store the sunny-day energy for nights and rainy days. Assuming they have 20% efficient solar cells, and the incident solar energy density averages 200 W/m² for an average of 8 h per day, how much area would their solar cells have to cover? **Ans:** 125 m²

6. Compare the energy used in toasting two pieces of bread in a 1050-W pop-up toaster that takes 2.0 min with that used to toast them in a 3200-W broiler that takes 3.5 min, including warmup time. **Ans:** 35 Wh versus 190 Wh

7. If your water heater is 10 m from your bathroom, and the hot water pipe is 1.27 cm (½ in) in diameter, roughly how much energy do you waste in running the cold water out of the pipe?

Ans: 0.07 kWh

8. Assume the Savenergys, of Question 13 fame, have a house with hot-water radiators: a total of 10 radiators averaging 20 liters of water per radiator. How does the energy needed to heat the water in all 10 radiators from 15 to 60°C compare with that needed to operate a 1.5-kW electric heater for ½ h?

Ans: 10 kWh versus 0.75 kWh

9. Refer to Problem 8. If the Savenergys' heating system is 50% efficient, and their power plant is 35% efficient, compare the fuel energy used in heating their bathroom by heating their whole house with that used by the bathroom electric heater.

Ans: 20 kWh versus 2.1 kWh

10. Mrs. Typinglady proposed saving electricity costs by switching to a small gasoline-powered generator at night when there are only a few appliances to keep running, rather than using energy from the power company. If the gasoline-engine generator set is 15% efficient, how much electrical energy do you get per liter of gasoline? The heat of combustion of

gasoline is 7820 kcal/liter. At your local gasoline prices, how much would this cost per kilowatt-hour?

Ans: 1.36 kWh/liter; 19¢/kWh @ $1/gal

11. Show that 1 kWh is equivalent to 3.6 MJ.

Home Experiments

1. Stick a piece of copper and a piece of zinc into a lemon or tomato. A copper wire and a galvanized (zinc coated) nail or a paper clip will do. Bring the other ends of each piece close together and touch your tongue. You feel a tingle and a distinct taste similar to that from a battery (see Home Experiment 2, Chapter 11). If you use well-cleaned strips of copper and zinc about 1½ cm wide, and connect several such batteries in series, you might be able to light a small bulb.

2. Run your cold water until the temperature stabilizes and then measure its temperature with a thermometer. Then measure the hot water temperature. (Be sure you have a thermometer with high enough temperature range.) If you have access to the water heater that heats your water, turn off the power to it, remove the cover plates, and check the thermostat settings. Compare with your measured temperature. Try to estimate whether any difference is from heat loss between heater and faucet or from faulty calibration on either the thermometer or thermostat. Calculate the energy needed per liter to heat *your* water and compare with the estimates in the chapter.

References for Further Reading

Derven, R., "Heat Pumps," *Popular Science,* Sept. 1976, p. 92. Includes information about the heating and cooling needs in various parts of the United States.

Edison Electric Institute (90 Park Ave., N.Y. 10016), "Annual Energy Requirements of Electric Household Appliances," EEI-Pub #75-61. Information on average energy consumption of appliances.

Fowler, J. M., *Energy and the Environment* (McGraw-Hill Book Co., New York, 1975). Excellent discussion of all aspects of energy.

Johnston, W. D., Jr., "The Prospects for Photovoltaic Conversion," *American Scientist,* Nov.-Dec. 1977, p. 729. Can solar cells contribute significantly to a solar energy economy?

Pansini, A. J., *Basic Electrical Power Transmission* (Hayden Book Co., Inc., Rochelle Park, N.J., 1975). The problems and techniques of transporting electrical energy.

Powell, E., "How to Keep Your Appliances on an Energy Budget," *Popular Science,* Nov. 1975, p. 106. Gives suggestions for efficient use of appliances.

Saperstein, A.M., *Physics: Energy in the Environment* (Little, Brown & Co., Boston, 1975). Considerable amount of practical information on electrical energy sources and uses.

14

Electrostatics— a Pest, a Tool, a Natural Spectacle

A well-known authority on electrostatics begins the preface to one of his books as follows:

> What is electrostatics? To the mathematician it is an absorbing line of theory. To the physicist it is a science of many and varied manifestations. To the engineer it is a nuisance or hazard he must eliminate, or again, it represents the perfecting of an application the world needs. To the ecologist it is a major means of preventing pollution of the atmosphere. To the secretary, it is liberation from coping with batches of carbon copies. To the housewife it is a nuisance when it makes dust stick to her walls. To me it is an endless fascinating mixture of science and art . . .*

In this chapter we'll try to get a glimpse of what that author had in mind.

So far in our study of electricity, we have looked mostly at things related to electric currents: charges in motion. Current is the basis for most practical uses of electricity. In this chapter we will look at **electrostatics.** Even though the "-statics" in the name implies that the subject deals with charges at rest, most phenomena that come under the heading electrostatics include moving charges. A better definition is that electrostatics deals with those effects resulting purely from the presence of electric charges and their relative positions, *independently of whether or not the charges are moving.*

---

*A. D. Moore, ed., *Electrostatics and Its Applications* (John Wiley & Sons, New York, 1973), p. xi.

You have no doubt experienced electrostatic effects. If you have walked across a carpet and then gotten a small shock when you touched the doorknob, you have felt the effects of electrostatics. Perhaps on a dry day you've noticed that your hair tries to stand up, or follow the comb, as you comb it. You may have been frustrated by the extreme difficulty of getting all the dust off a phonograph record, or wondered why your clothes stick together when you take them out of the drier. On a dry winter night, you might have seen sparks fly between your pajamas and the sheets as you turned over in bed. All these nuisances result from electrostatic effects.

On the other hand, you've also experienced the benefits of electrostatics. You've certainly made copies or used copies made on a Xerox or other electrostatic copying machine. You've breathed more freely because an electrostatic precipitator captured crud that otherwise would have been released into the air by industrial smoke stacks, or because an electrostatic air filter in your home heating or air conditioning system filtered out pollen and dust. And you very likely have been either delighted or petrified by nature's spectacular display of electrostatics—lightning.

To understand why electrostatics can be both a nuisance, as well as a boon, we need to know about the ways of giving a net charge to an object, and then to understand why a charged object is attracted to one that is uncharged.

14.1   Methods of Charging

As you read this section, keep in mind that two like charges—two positive or two negative charges—repel each other and two unlike charges—a positive and a negative—attract each other.

Contact Electrification

Contact electrification is responsible for most of the nuisances associated with electrostatics. For example, as you walk across a carpet, the soles of your shoes come in close contact with the carpet fibers, which are of course made of a different material than the shoe soles. It is usually true that the atoms of one substance have a stronger attraction for electrons than do those of another substance. If the two come into close contact, the substance with the stronger attraction steals some electrons from the other, giving the thief a net negative charge, and leaving the victim with a net positive charge. The close contact between shoe soles and carpet allows such a charge transfer. This net charge on the soles is conducted upward and spreads out over your body. As your hand, which is then charged, approaches the uncharged doorknob (a conductor), the charge jumps across the small gap between hand and knob just before touching, and you feel an annoying but harmless shock.

We often demonstrate this contact electrification in the classroom by rubbing together two objects with different electron affinities. For example, if you rub a rubber rod with a piece of fur, the rubber attracts electrons from the fur, becoming negatively charged and leaving the fur positively charged (see Fig. 14.1-a). If you rub a glass rod with silk, electrons transfer to the silk from the glass, leaving the glass positive, and the silk negative (Fig. 14.1-b).

Figure 14.1 (a) When fur and rubber come into close contact by rubbing, the rubber atoms attract some electrons from the fur atoms, leaving the rubber charged negatively, and the fur charged positively. (b) For glass and silk, the silk is the electron "thief."

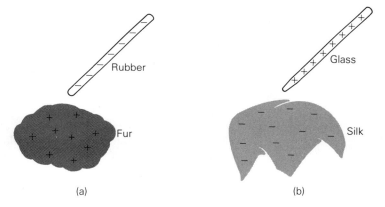

We can show the presence of these charges by what is called an electroscope (Fig. 14.2-a), which consists of a metal rod with two very thin "leaves" of metal—usually gold—attached to the bottom. The end of the rod containing the leaves is housed inside a metal box with windows for viewing. The rod is insulated from the box.

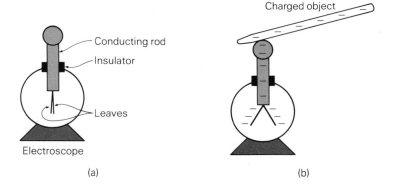

Figure 14.2 When you charge an electroscope, the leaves spread out because of mutual repulsion of the charges.

If we touch a negatively charged object to the conducting rod, some of the charge from the object distributes itself along the electroscope rod and leaves. The repulsion between the charges on the leaves causes them to spread apart (Fig. 14.2-b). When the charged object is removed from contact with the electroscope terminal, excess charge stays on the electroscope leaves, and they remain separated.

We get the same effect by touching the electroscope with a positively charged object. Then electrons escape the electroscope leaves, causing them to diverge due to the repulsion of the net positive charge.

Question   What would you expect to happen to the electroscope leaves if a charged object is brought near, but does not touch, the electroscope? What would happen if the object were then removed?

If the electroscope is charged negatively and a positively charged object brought near, the leaves will come together. You see the same effect if you bring a negative charge near a positively charged electroscope.

Your instructor may ask you to volunteer to become a human electroscope by standing on an insulating stool and putting your hands on some kind of charging device (Fig. 14.3). Your hair becomes the electroscope leaves as each charged strand repels the others. If charged and later discharged *slowly*, your body can be safely raised to thousands of volts above the potential of your surroundings. Remember, it's current you need to worry about; voltage doesn't matter if it has no way of producing a current.

Charging an object by rubbing is sometimes called charging by *friction*. Even the experts in the field do not understand everything that goes on in this process, but most of them think the main effect of the rubbing is to bring more of the atoms of the different materials into contact. It is the *contact* that counts, *not* the friction.

Contact electrification not only bugs each of us in little ways from time to time—shocks from walking across a rug or sliding across a car seat, hair sticking to the comb, clothes clinging together in the drier, and dust sticking to the phonograph record—but is quite a problem in certain industrial processes. Powdered materials such as flour and sugar pick up a charge moving through chutes. Conveyer belts pick up charge as they turn over their pulleys. Oil or fuel sloshing around in tankers can pick up a charge and cause sparks that ignite violent explosions. Airplanes have static dischargers on their wings to dissipate charge picked up in flying through dust or some form of precipitation.

Figure 14.3 A person with a net charge becomes a human electroscope.

Things that would ordinarily be only an annoyance can be a potential hazard in places such as hospital operating rooms. There, gases used as anesthetics or excess oxygen content in the air can easily turn a normally harmless spark into a disastrous explosion. Operating room personnel usually wear special shoe coverings that minimize charge pickup.

You may have noticed that the electrostatic effects we have discussed show up more in winter than in summer. Do you think charges prefer cold weather, or is there possibly some indirect effect? You can get a clue if you also notice that these effects are more pronounced when the air is very dry—that is, when the humidity is low.

When the relative humidity is more than about 50%, a thin layer of moisture forms on the surface of most solids. This moisture greatly increases the electrical conductivity of the surface of insulators. Even though contact electrification still charges the bodies in high humidity, the highly conducting surface layers quickly drain off the charge. If you recall that *indoor* air is often quite dry in winter, you can see why electrostatic effects are noticeable mostly in winter.

Charging by Induction

Suppose you have two metal balls in contact with each other but insulated from ground as shown in Fig. 14.4(a). Bringing a positively charged rod nearby causes electrons to be attracted from both balls to the near side of the ball closest to the rod. This leaves a net positive charge on the ball farther away. If the balls are then separated as in Fig. 14.4(b), you end up with a net positive charge on one ball, and a net negative charge on the other *without* either ball making contact with the charged rod (Fig. 14.4-c). We call this process **charging by induction.**

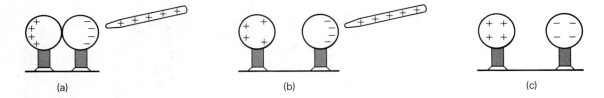

(a)   (b)   (c)

Figure 14.4   Charging two metal balls by induction.

You could charge a single ball by induction as follows. Bring the positively charged rod near the insulated ball (Fig. 14.5-a). This will induce a negative charge on the side near the rod, leaving a positive charge on the other side. By touching the ball with your hand or with a wire connected to the earth, ground it at the point where the positive charge is concentrated (Fig. 14.5-b). Negative charge flows from the ground connection to the ball, neutralizing the positive charge. Remove the ground (Fig. 14.5-c) and *then* the charged rod (Fig. 14.5-d). You have a net negative charge on the ball.

Question   Explain how you would charge two balls and one ball by induction using a negatively charged rod. What is the sign of the charge on each ball?

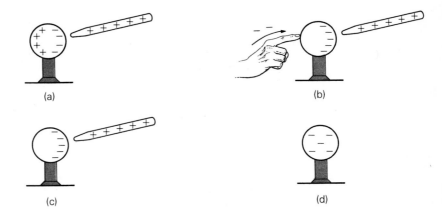

(a)   (b)

(c)   (d)

Figure 14.5 Charging one metal ball by induction.

When we were discussing the shock you get from walking across a carpet and then touching a doorknob, you may have wondered how you could pick up enough charge that it would actually jump across an air gap? Why doesn't it just jump back to the floor as you walk? Would you get the same shock if you touched the wooden door rather than the metal knob?

To get much of a shock, you need to touch metal (or some conductor). Induction plays an important role in the process. The charge you pick up spreads over your body (which is a conductor) including some to your hand. Assume your body to be negatively charged. As your charged hand comes near the knob, it induces a positive charge on the knob nearest your hand (see Fig. 14.6). This charge attracts more negative charge to your hand from the rest of your body, which in turn attracts more positive charge into the nearest knob. This redistribution is extremely fast, giving a larger concentration of charge, the nearer the hand is to contact. Just before actual contact is made, the concentration is large enough to cause charge to jump across the gap.

Figure 14.6 If your body is charged negatively it induces positive charge on the knob nearest your hand, which in turn induces more negative charge onto your finger tips. The closer your hand gets to the knob, the greater this effect.

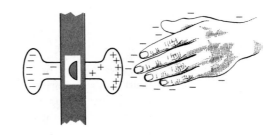

These charging processes work equally well, whether it is the positive or the negative charge that moves. Negative charge moving out of a region charges the region positively just as well as positive charge moving in. We know that in metal conductors, it is the negative charge that moves. Thus, when we talked about attracting positive charge into one of the doorknobs, in reality negative charge was being repelled from it.

Corona

If we have a positively charged conductor in the shape of a sphere, the charges spread out uniformly over the sphere. The resulting electric field is along radial lines just as it is for point charges (Fig. 14.7-a). But if the conductor has a sharp point, as in Fig. 14.7(b), the charges are much more concentrated at that point. Since the point is farther away from most of the conductor than any other part, mutual repulsion pushes more charge onto the point. Because of this charge concentration, the electric field near the point is much stronger than elsewhere, as indicated by the dense field lines there.

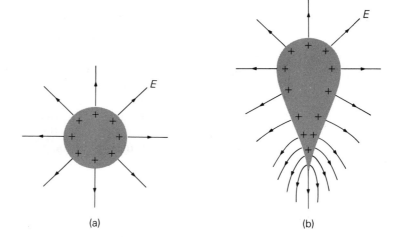

Figure 14.7 In attempting to get as far as possible from each other, the charges concentrate at the point, giving a much stronger electric field near it. The denser the spacing of electric field lines, the stronger the field.

(a)   (b)

Any charges in the presence of the field are accelerated along the field lines, positive charges in the field direction, negative charges in the opposite direction. Since there are always *some* charged particles in air, sharp points on a charged conductor rapidly leak the charges from the conductor. For example, if the conductor is positively charged, electrons and negative ions in the air will be attracted along the field lines to the conductor, neutralizing its charge.

If the conductor is negatively charged, electrons stream from the point. If this field is strong enough, these electrons gain enough kinetic energy to knock electrons from atoms with which they collide. The freed electrons are accelerated and collide with other atoms, similarly ionizing them in an avalanche fashion. This breakdown of the air, called a **corona** discharge, makes it highly conducting. Farther out from the point of the conductor, where the field is weaker, most of the electrons don't have enough energy to knock out other electrons from atoms, but rather attach themselves to the atoms forming negatively charged ions. As we will see, this is an important part of the removal of polluting particles from gases released into the air by certain industrial processes. In the next section, where we discuss the process of *polarization,* we will see why it's necessary to be able to give a charge to such particles.

14.2   Attraction Between a Charged and an Uncharged Body—Polarization

Let's now find out why a charged body is attracted to an uncharged one: why your comb will pick up little bits of paper after you comb your hair, why a balloon will stick to the wall after you rub it on your sweater or hair, why the dust sticks to the phonograph record, and why dust and pollutants can be filtered out with an electrostatic filter or precipitator.

Consider again the charges induced on different parts of a conductor by a nearby charge. Suppose you wad a small piece of aluminum foil into a ball and hang it from a string. If you then run a comb through your hair a few times and bring it near the foil ball, the ball will be attracted to it (Fig. 14.8). The amount and the sign of the charge picked up by the comb will depend on the material of the comb. Assuming the comb has a negative charge, it induces positive charge on the side of the ball nearest the comb, leaving the other side negative. That is, the ball becomes **polarized:** There is a separation between the positive and negative charge. Since the foil's positive charge is closer to the comb than its negative, there is a net attraction toward the negative charge of the comb, causing it to swing to the right.

If you then remove the foil ball from the string and bring the charged comb near the string, you will find the string itself—an *insulator*—attracted to the comb. The reason is that the charge in each molecule of the string behaves as the charge on the conduct-

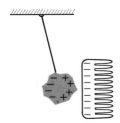

Figure 14.8 The negatively charged comb attracts the uncharged conductor because the conductor's charge gets polarized: positive charge nearer the negatively charged comb.

ing ball. That is, the charge within a molecule shifts so that the *molecule is polarized.* As Fig. 14.9 shows, all the molecules near the comb get polarized and line up with their positive ends toward the negatively charged comb. The interior charges cancel each other, leaving a net positive charge on the right, a net negative charge on the left. Thus, the insulating string *as a whole* is polarized, with the positive side closer to the negative charge of the comb, giving a net attraction.

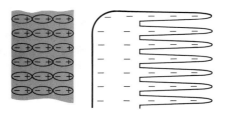

Figure 14.9   If the charged comb approaches an insulator, the comb's negative charge attracts the positive charge of each molecule and repels the negative. The individual molecules polarize, giving a net positive charge closer to the comb, a net negative charge on the far side. If the comb were positively charged, the insulator would polarize the other way, still giving a net attraction.

The *polarization* of insulating materials in an electric field accounts for the attraction of uncharged bodies—insulators or conductors—to charged bodies. For some materials the molecules are permanently polarized: The center of the positive charge is separated from the center of the negative charge. When this type material is put in an electric field, the molecules merely rotate to align with the field as in Fig. 14.9. The overall result is the same whether the polarization of the molecules is permanent or induced.

Question   What are the similarities and what are the differences between the way a charged body attracts a piece of insulator, and the way a magnet attracts a piece of iron?

14.3   Applications of Electrostatics

In describing some of the basic properties of static charges, we have used several of the nuisance aspects of electrostatics as examples.

Now we'll look at some ways electrostatic effects are useful. We won't discuss any new principles in this section, but will illustrate the principles we've covered.

Xerography   In the mid-1930s Chester Carlson invented a process that many of us now take to be one of the necessities of life. This process, called *xerography*, is the dry-copying method used in the Xerox machine and in similar electrostatic copying machines.

In addition to electrostatic principles, such a machine uses one other process we have not yet talked about: the *photoelectric effect*. We will consider this effect in more detail later. For now, we'll just point out that when light strikes certain materials—for example, selenium—electrons are given off. Thus, those parts of a piece of negatively charged selenium exposed to light would lose their charge.

In a typical electrostatic copying machine, a corona uniformly charges a selenium-coated drum. The image of the writing to be copied is focused onto the drum, lighting the entire drum *except* for the position of the image. The light removes the charge, leaving only the image on the drum charged. This image then passes near a black dust called toner, whose particles get polarized and attracted to the image. A precharged sheet of paper then contacts the drum and attracts the toner to the paper, where it is later heated briefly and fused to the paper. The mechanism for carrying out these few steps is quite complicated, as you can see by looking inside one of these machines.

Electrostatic Precipitators   An effective means of reducing certain pollutants in industrial stacks is to use an electrostatic precipitator. In a typical precipitator a negatively charged wire extends along the center of a cylindrical metal duct through which flue gas passes. The wire has enough negative charge to produce a corona discharge, thereby charging the waste particles as they move along. The duct is grounded so that the charged particles are attracted to it and collect on the walls. An occasional rap on the duct shakes the particles loose to fall into a hopper. Figure 14.10 illustrates the effectiveness of such a precipitator.

An electrostatic air filter for the home uses basically the same approach, but usually applies a charge to the dust and pollen particles in the air as it passes through a positively charged grid. The particles are then trapped as the air passes near negatively charged plates.

Other Electrostatic Applications   The list of electrostatic applications is almost endless. It includes: electrostatic paint sprayers that

Figure 14.10 The electrostatic precipitators on this large alumina-processing plant were (a) momentarily turned off and then (b) on again. (Photograph courtesy of Research-Cottrell, Inc.)

give a charge to paint particles so that they will be attracted to all sides and into the crevices of an object being painted, thereby saving paint and improving the paint job; mineral separators that separate conducting from nonconducting substances; electrostatic motors; nonimpact printing that charges a jet of ink by induction and then deflects it up and down, back and forth electrostatically to write on a page at high speeds; carpet making by the electrostatic attraction of pile to adhesive coated backing; and electrostatic loudspeakers. Some writers have suggested that the Ark of the Covenant that held the Ten Commandments written on stone tablets was designed so that it picked up a large electrostatic charge. Priests would be protected by their vestments adorned with gold and other metals, but trespassers would be zapped by discharges if they came near.

Many useful and interesting applications are discussed in the references at the end of this chapter.

14.4 Atmospheric Electricity

Atmospheric electricity is an extremely complex and, in some ways, mysterious field. It is by no means thoroughly understood by anyone. We certainly cannot explore all the details here, but will look at a few overall features. We will consider first some atmospheric effects that are independent of weather conditions, and then consider nature's own spectacular fireworks—lightning. We'll see that atmospheric electricity is not all electrostatic in nature, but a combination of electrostatics and slowly varying currents.

Charges and Electric Fields in the Atmosphere

Ordinary air is a very poor conductor (good insulator) of charge, but it does conduct some. The reason is that about one of every few million billion air molecules is an ion—a molecule that has either lost or gained an electron and therefore has a net charge. These charges conduct a current if there is an electric field in the air.

Where do these ions come from, and why don't they merely exchange electrons and become neutral? They do this, but new ion pairs are continually being formed. Small amounts of naturally radioactive materials near the earth's surface give off electrons and gamma rays that collide with air molecules, knocking off some electrons and thereby forming ions. In addition, cosmic rays—high energy particles from space—enter the atmosphere and knock electrons from molecules. Near the earth's surface about 10 million ion pairs are formed this way per second per cubic meter of air. The rate of formation increases with increasing altitude, since the cosmic ray density increases.

Experiments have shown that the surface of the earth carries a net negative charge that averages about $-\frac{1}{2}$ million coulombs over the entire earth's surface. The upper atmosphere carries an average net positive charge of the same amount, making the earth and atmosphere together electrically neutral (Fig. 14.11). Because of this charge separation, there is a downward electric field in the atmosphere. In clear weather, this field is about 130 volts/meter near the earth's surface. The overall potential difference between the earth and the upper atmosphere averages about 300 to 400 kV.

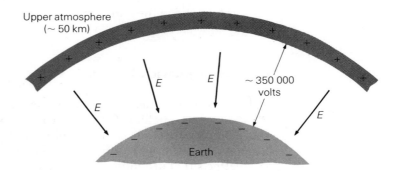

Figure 14.11 The earth's surface has a negative charge of about $-\frac{1}{2}$ million coulombs, the upper atmosphere an equal positive charge. The potential difference between them is 300-400 kV.

Since you are probably close to 2 m tall, does this mean there is a 250-V potential difference between your head and your feet when you stand up? Why don't you get zapped? Your body, being a conductor, stays at the same potential as the earth with which you are in contact. This potential "gradient," or electric field, starts above your body or any other conductor in contact with the earth.

Since there is a voltage between the earth and upper atmosphere and the air in between is a conductor (admittedly poor), there is a continual current from the upper atmosphere to ground. That current averages about 1800 A through the entire earth's atmosphere. From the equation for electrical power ($P = VI$) the power dissipated in this current is around 600 megawatts, roughly half the output of a large modern power plant.

The variation of this current with time of day is interesting. When the current over the oceans is measured, and this value is averaged over many days, we find that it has its maximum value all over the world at about 2:00 P.M. eastern standard time—11:00 A.M. Pacific standard time. No matter *where* you make the measurement, its largest value usually comes when it's 2:00 P.M. in Washington, D.C.

This rather surprising occurrence is related to something else that has its peak at that time. That something else, we'll discuss next.

Charging the Earth

The current of 1800 A reaching the earth would discharge the ½ million-coulomb negative charge on its surface in about five minutes. Yet the current continues, and the earth keeps its rather constant negative charge. How is this possible? Where is the "battery" that keeps this charge flowing?

Figure 14.12 is a picture of this "battery."* Lightning flashes all

Figure 14.12   The "battery" that keeps the earth charged negatively.

---

*The idea of lightning as a "battery" is borrowed from R. P. Feynman, R. B. Leighton and M. Sands, *The Feynman Lectures on Physics,* Vol. II (Addison-Wesley Publishing Co., Reading, Mass., 1964). See Chapter 9.

Cloud

Ground

Figure 14.13 The stepped leader of a lightning flash follows a zigzag path.

over the earth carry a negative charge to earth at an average rate of about 1800 A. This keeps the charge separation between earth and atmosphere relatively constant. As you might guess from the last section, the average rate of thunderstorm activity all over the earth has its maximum value when it's about 2:00 P.M. in Washington, D.C. Even if you're tempted to attribute this effect to the return from lunch of the United States Congress, this is also the time of day when the sun is over the tropical regions of main thunderstorm activity. Such storms are mainly generated by convection currents in the air that is heated near the surface of the earth.

Just how thunderstorms come about is a complicated process discussed in the references at the end of the chapter. However, we will discuss the events in a lightning flash.

Lightning

Thunderclouds are extremely complex systems that are usually polarized so that there is mostly positive charge near the top, mostly negative charge near the bottom. This large negative charge at the cloud bottom, usually at a height of about 5 km, induces a positive charge on the part of the earth just below the cloud in the same way your charged hand induces an opposite charge on the nearby doorknob as your hand approaches. A potential difference between 25 million and 300 million V exists between the cloud base and ground.

It takes about 300 kV per meter for a spark to develop through air. Therefore, continuous sparks from cloud to ground cannot form. Instead, electrons in the base of the cloud are accelerated by the tremendous electric field. They collide with air molecules, knocking out electrons and producing a conducting path. This path, filled with electrons, progresses about 50 to 100 m in less than a microsecond (10^{-6} s). After a pause of about 50 μs, this *stepped leader,* as it's called, advances another step of 50 to 100 m. This process continues in a zigzag pattern, often with several branches, until it reaches ground (Fig. 14.13).

When the leader touches ground, the negative charge at its bottom immediately leaks off. Then the charges from successively higher parts of the leader flow to ground, until all the negative charge in the leader has been deposited in the ground. This process, called the *return stroke,* can have currents that exceed 300 000 A, but typically are about 25 000 A and deposit an average of 10 to 20 coulombs of charge. Most of the light is produced during the return stroke.

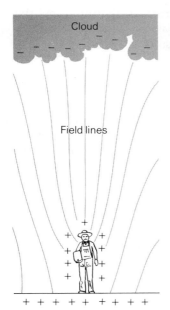

Figure 14.14   The effect of a person, or other conductor, on the electric field under a thundercloud.

The return stroke is usually followed in about 0.04 s by another leader, called a *dart leader,* that darts smoothly along the original channel to ground. It's also followed by a return stroke that usually carries much less charge than the original one. A typical flash has three or four leaders and return strokes, but occasional flashes have twenty or more.

The surge of current in the flash can heat the air in the channel to a temperature of 30 000°C. The gases expand quickly into the surrounding air, creating the sound waves we call thunder.

What Does Lightning Strike?   Suppose you're standing in an open field with a thundercloud above, as shown in Fig. 14.14. (Your height relative to the distance to the cloud is greatly exaggerated.) Positive charge will flow onto your body—actually, electrons will leave—concentrating the electric field at the highest part of your body. If a lightning strike is developing in the area, it will probably follow the path of the strongest electric field; namely, the path to your head. For the same reason, lightning is more likely to strike any object sticking up that has a conductivity greater than that of air—a lone tree, a house, a mountaintop, or particularly a metal tower.

What about *lightning rods?* They were invented by Benjamin Franklin over 200 years ago, and still are used in essentially the same form. But do they *really* do any good aside from giving their owners a *feeling* of security?

The rod is a well-grounded conductor with a pointed end that sticks up higher than any part of the building it protects. It has two intended functions. The first is related to the fact that, as we discussed earlier, charge concentrates at the tip of a pointed object, increasing the surrounding electric field. On the lightning rod, the other end of these pointed conductors is connected to ground. The strong electric field around the points leaks some of the earth's charge into the surrounding atmosphere. The charge leaked during a storm may amount to several coulombs. The idea is to neutralize some of the atmospheric charge, reduce the electric field, and thereby reduce the probability of a lightning strike.

The second function of the rod is to provide a good conducting path between a point higher than the building and ground. Should lightning then strike in spite of the charge leakage, it will strike the rod and be conducted to ground.

There is no doubt about the beneficial effect of the second function. It has been proven many times. There is also no doubt that substantial amounts of charge are leaked into the atmosphere by the pointed rods. Specialists in the study of lightning feel that this leakage may help in some circumstances; in others, however, it may intensify the electrification of the storm.

Electrical Shielding   We're sometimes advised that a safe place during a lightning storm is inside a metal automobile, airplane, or train. The metal body conducts the charge around the person, where it jumps the small distance to ground. You may get the same benefit if a power line falls on your car. You are safe from electrical shock if you stay inside because the car body conducts the current around you. (However, you may not have this option if sparking ignites a flammable part of the car.)

Key Concepts

Electrostatics deals with those effects resulting purely from the presence of electric charges and their relative positions, independently of whether or not the charges are moving.

Contact electrification occurs during contact between two different substances, and results from the different electron affinities of the two substances. Bodies can be charged by **induction** because of the attraction between unlike charges, and repulsion between like charges. **Corona** discharge occurs in the regions of high electric field near the sharp points of a conductor.

An atom, molecule, or body is **polarized** when the centers of the positive and negative charges are separated from each other. Polarization causes the attraction between a charged and an uncharged body.

The earth's surface has a charge of about -0.5×10^6 C and the upper atmosphere an equal positive charge. This charge separation causes an electric field of about 130 V/m in the atmosphere near the earth on a clear day. There is an average downward current of about 1800 A through the entire earth's atmosphere. The charge separation between earth and upper atmosphere is maintained by lightning flashes that carry an average of about 1800 A of negative charge to earth.

A lightning flash to earth starts with a *stepped leader* that is followed by a *return stroke*. There usually follows one or more *dart leaders*, each followed by a return stroke. Lightning flashes average

25 000 A, and deposit an average of 10 to 20 C of charge. Any conducting object projecting above the earth increases the electric field directly above it, and is more likely to be struck by lightning.

Questions

Charging Methods

1. Dust particles in the air usually have an electric charge they have picked up by rubbing on other objects or from ions in the air. What does this have to do with the fact that dust will collect on vertical walls? When dust settles on a table, why do you have to wipe it off to get rid of it; why can't you just blow it off since it's so light?

2. Why is dust so hard to remove from phonograph records?

3. Blowing sand or snow on the plains sometimes charges wire fences enough to knock down a man or animal who touches such a fence. How does the fence get this charge?

4. Smooth well-waxed floors are preferred in hospital operating rooms to help keep down the spread of infection. Smooth well-waxed floors are the kind you can easily pick up an electrostatic charge from. Do you see a problem here for hospitals and their patients?

5. Thunderstorms happen mostly in summer. Does this mean that other electrostatic effects that we see—for example, picking up of static charge by walking on a carpet—happen mostly in summer?

6. After you've picked up a static charge by walking on a carpet, why do you get a shock when you touch a conductor but not an insulator? If the earth were an insulator, how would that influence the occurrence of a lightning flash, assuming the charged cloud still developed?

Polarization and Attraction between Charged and Uncharged Bodies

7. Would you guess that electrostatic effects have anything to do with the "stickiness" of clear plastic food wrapping?

8. If you rub a balloon on your hair or sweater and touch the rubbed spot to a wall, the balloon stays there. Why?

9. If you run a comb through your dry hair on a dry day and then hold the comb near (not in) a thin stream of water from a faucet, the stream will bend toward the comb. Why? Why "dry hair" and "dry day"? Why will the comb also pick up bits of paper?

10. When you make several copies on a Xerox machine, or other electrostatic copier, why are the pages sometimes hard to stack on top of each other immediately afterwards?

Atmospheric Electricity

11. What keeps the earth charged negatively relative to the upper atmosphere?

12. Discuss the relative probability of lightning striking the following objects during a thunderstorm: a mountaintop; a valley between mountains; a flat open field; a single tree in a flat open field; a person standing in a flat open field; a person squatting in a flat open field; a man and a boy walking across a field; a particular tree in a forest of many similar trees; a building on a flat terrain; a metal tower; a lightning rod.

13. Why is there a stronger electric field near the point of a lightning rod than anywhere else nearby?

14. Why is an automobile or airplane a safe place to be during a lightning storm?

Problems

1. If a lightning bolt flashes across a potential difference of 200 MV between cloud and ground, and carries down a charge of 20 C, how much energy is dissipated? How high could this much energy lift a 20 000-kg bulldozer if converted to mechanical work?   **Ans:** 4.0GJ; 20 km

2. It takes about 7800 kWh of energy to heat the average house in Atlanta for a year. How many lightning flashes of the intensity described in Problem 1 does it take to dissipate this amount of energy?   **Ans:** 7

3. From the average potential difference between the earth and upper atmosphere (about 350 kV), and the average current through the atmosphere (1800 A), what is the resistance of the atmosphere as a whole?   **Ans:** 194 Ω

4. What is the rate of heat generation in the atmosphere from the current and resistance described in Problem 3? **Ans:** 630 MW

5. How close to the ground would cloud bases have to be in order for continuous sparks from cloud to ground to form? (Assume the potential difference between cloud base and ground to be 150 MV.)   **Ans:** 500 m

6. Based on the answer to Problem 1, how many such lightning flashes would need to occur over a 1-km² area to dissipate the amount of energy released in the atmosphere as 2 cm of rainfall condenses? (Heat of vaporization of water = 540 kcal/kg.)

Ans: 11 000

Home Experiments

1. Try the three experiments mentioned in Questions 8 and 9. Try combs of different materials, and test them under different conditions of humidity.

2. Salt and pepper unmixer. Sprinkle some salt and ground pepper on the table together. How can you get them unmixed? Rub a plastic spoon on your sweater or your hair, and hold it just above the mixture. Both salt and pepper will be attracted, but the pepper, being lighter, will jump up to the spoon.

3. Charge a plastic spoon as in experiment 2, and hold it above a bowl of puffed rice. The rice will first be attracted to the spoon, and then will fly off in all directions. Why?

References for Further Reading

Few, A. A., "Thunder," *Scientific American,* July 1975, p. 80. Describes how the sound of thunder is used to give information on the lightning flash.

Feynman, R. P., R. B. Leighton, and M. Sands, *The Feynman Lectures on Physics*, Vol. II (Addison-Wesley Publishing Co., Reading, Mass., 1964). See Chapter 9. An interesting and readable account of atmospheric electricity.

Garr, D., "Electronic Air Cleaners: What They Can Do For You," *Popular Science,* Sept. 1972, p. 58. The workings and advantages of electrostatic air filters.

Malan, D. J., *Physics of Lightning* (The English Universities Press Ltd., London, 1963). A detailed description of lightning and atmospheric electricity. Many of the numerical data cited in this chapter were taken from Malan's book.

Moore, A. D., *Electrostatics* (Doubleday & Co., Inc., Garden City, N.Y., 1968). Easily readable discussion of the full range of electrostatic phenomena and their applications.

Moore, A. D., Electrostatics," *Scientific American*, March 1972, p. 47. Some modern applications of electrostatics.

Moore, A. D., ed. *Electrostatics and Its Applications* (John Wiley & Sons, New York, 1973). Various aspects of electrostatics and its applications. Each chapter written by a specialist in that particular topic. Some sections are quite mathematical, but most require minimal mathematical skills for understanding.

Orville, R. E., "The Lightning Discharge," *The Physics Teacher*, Jan. 1976, p. 7. Franklin's study of, and the current understanding of, the lightning flash, including some interesting lightning photographs.

Part IV

Waves, Sound, and Light

. . . The principle of relativity requires, in particular, that mass be a direct measure of a body's energy. Light carries mass. A fascinating and intriguing idea. I only wonder if God might not just be jesting and making a fool of me.

A. Einstein

Albert Einstein was born in 1879 in the former German state of Württemberg. Even though his prep school teachers considered him uneducable, at age 17 he entered the Swiss Federal Polytechnic School after considerable difficulty getting in because of his inadequacy in mathematics. After graduation, Einstein worked for eight years in a patent office in Bern, Switzerland, where his duties left him ample time to develop ideas that have revolutionized our understanding of our universe. Even though he is best known for his theory of relativity, Einstein won the 1921 Nobel Prize for his explanation, given in 1905, of the quantum nature of light. After holding several European professorships, he accepted a position at the Princeton Institute for Advanced Study in 1933, which he held until his death in 1955.

15

Sound

When you think of sound, what comes to your mind? Is it the chirping of birds in spring? Or maybe the rumble of city traffic? The rock band that's hitting campus next week, or an album of Beethoven's Fifth Symphony? The drone of your physics teacher's voice, or the bell that will end his or her monologue for another day? Or maybe that one voice more special than all the others in the world?

In this chapter we'll look at some of the physical properties that determine the nature and behavior of sound. I hope that reading this chapter will not only help you to understand sound better, but will help you to notice and appreciate some aspects of sound that you've heard all your life, but may not have been aware of.

Since sound exists as a wave, our first job is to understand waves. The general wave characteristics we'll discuss will also apply in the next chapter when we consider light. Another general phenomenon called *resonance*, that we need to understand as a background for sound, also applies to other areas. We can then discuss the nature of sound waves and how they are produced. After briefly describing how the ear functions, we'll contrast the properties of sound that exist as measurable physical quantities with those that a listener perceives. We'll see how these ideas apply to sound that most of us like—the production of music—and then to sound that most of us do not like—noise. Finally, we'll consider some special effects that result from the wave nature of sound.

15.1   **Waves**

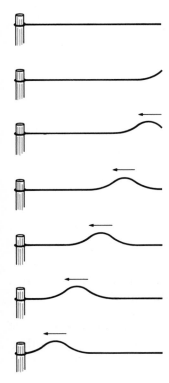

Even though most of us learned in elementary school that sound and light are waves, about the only kinds of waves that we often see—in the sense that you *see* this page—are water waves. You probably have some feeling for *their* physical properties from dropping pebbles into a stream, watching people in a pool, or playing in ocean waves at the beach. This chapter and the next should help to extend your intuition to sound and light waves.

Waves on a String

Preferably in reality, but otherwise in your mind, take a long rope, tie one end to a post or doorknob, and hold the other end so that there is a slight tension in the rope. Give your end an upward jerk and quickly return it to where it was. You will see a *pulse*—a lump—in the rope (as in Fig. 15.1). This pulse travels from your hand to the other end.

Now suppose that, instead of just one wiggle, you move the end of the rope quickly but smoothly up and down, continuously. You then get a **traveling wave** (Fig. 15.2) that moves along the rope. Each successive line of the figure shows the waveform of the rope at a later point in time. When the wave gets to the end of the rope, it reflects back in the opposite direction. The general shape of the rope then gets messed up, and no longer has the smooth form of Fig. 15.2. We'll consider reflected waves a little later.

Figure 15.1   A traveling pulse in a rope. Successively lower lines show the pulse position as time goes on.

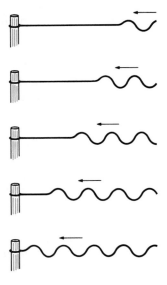

Figure 15.2   A traveling wave in a rope. Each line shows the wave at a later time as it travels toward the left.

When you pluck a guitar string, waves like these travel back and forth in the string. The effect is not exaggerated as in our diagrams, but the overall behavior is the same. In describing this kind of wave, we usually refer to it as a wave on a *string*, even though that "string" may be a rope, a spring, or some other carrier of the wave. Compare the shape of the string with the shape of the water surface when you jiggle your finger in an otherwise-quiet pool.

The Language of Waves   Discussing waves calls for a new language. Even though the vocabulary is fairly brief, we need to understand it before we go further. Figure 15.3 shows three of the important terms. The **wavelength,** usually identified by λ (lambda), is the distance between any two points **in phase** with each other—the distance between two crests, two troughs, or between two cross-over points where the string is moving down, for example. The **displacement** of a point on the string, shown as y for one point, is the distance of the point from the equilibrium (central) position for that string particle. It varies from point to point along the string, and varies with time at any one point. The **amplitude,** represented by A in the figure, is the maximum value of the displacement.

Figure 15.3   The wavelength, displacement, and amplitude of a wave. Wavelength and amplitude characterize the wave as a whole, but each particle has its own displacement, that varies with time between zero and the amplitude.

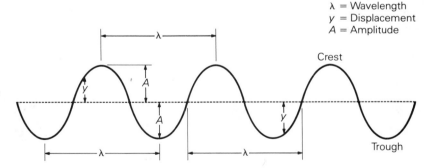

λ = Wavelength
y = Displacement
A = Amplitude

Crest

Trough

Another important characteristic of a wave is its **frequency.** The *frequency* of a *sound* wave mainly determines its *pitch*. The *frequency* of a *light* wave determines its *color*. We define frequency in two equivalent ways: as the number of wave pulses passing some point in the string per unit of time, or as the number of complete oscillation cycles per unit of time of a point on the string. The usual unit of frequency is the hertz (Hz), defined as one cycle/second and named after Heinrich Hertz who, in 1886, first demonstrated the transmission of radio frequency electromagnetic waves:

$$1 \text{ Hz} = 1 \text{ cycle/s}.$$

Even though it gives the same information as the frequency, the **period** T sometimes fills the bill better. The period is the time for one complete cycle. Since frequency is the number of cycles per second,

$$\text{period} = \frac{1}{\text{frequency}},$$

or

$$T = \frac{1}{f}.$$

An analogy will show the relationship of these quantities to the speed with which the wave travels, the **wave velocity.** If you're waiting for a train to pass at a railroad crossing, you can find the train's speed by multiplying the number of train cars passing per minute by the length of one car (Fig. 15.4-a). For example, if 25 15-m-long cars pass per ½ min—a frequency of 50 cars/min—the train speed is 50 cars/min $\times$ 15 m/car = 750 m/min or 45 km/h. Similarly, Fig. 15.4(b) shows that the wave velocity v is the number of wave pulses passing a point per second, f, multiplied by the length of each pulse, λ. That is,

$$\text{wave velocity} = \text{frequency} \times \text{wavelength};$$

$$v = f\lambda.$$

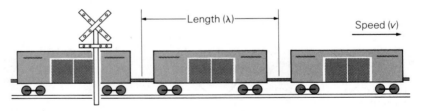

f Cars pass per minute

(a)

Figure 15.4 Whether you're talking about (a) a freight train, or (b) a wave train, the speed v is the number of cars (pulses) passing in each unit of time multiplied by the length of each car (pulse).

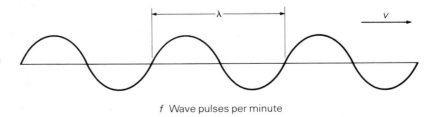

f Wave pulses per minute

(b)

Example 15.1   The strings for middle C on a piano vibrate with a frequency of 262 Hz. If their wavelength is 2.00 m what is the speed with which the waves move in the string? From the above equation,

$$\text{velocity} = \text{frequency} \times \text{wavelength}$$

$$= 262 \text{ Hz} \times 2.00 \text{ m} = 524 \text{ m/s} \ (1890 \text{ km/h}).$$

Exercise   CB radio waves are electromagnetic waves of frequency near 27 MHz (M = 10^6). If the speed of electromagnetic waves in air is 3.0×10^8 m/s, what is the CB wavelength?   **Ans:** 11 m

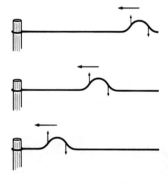

Figure 15.5 The pulse moves to the left as the string particles move up and down. The vertical arrows show the direction of motion of the string particles.

Which Way Does the String Move?   If you think again about the pulse you gave the string (Fig. 15.5), you realize that the string itself didn't move to the left; only the pulse did. As the pulse passed any particular little piece of the string, that piece merely moved up, and then back to its original position. For the continuous traveling wave of Fig. 15.2, each piece of the string continually moves up and down with a periodic motion. The waveforms in Figs. 15.2 and 15.3 describe the distance from equilibrium of any point along the string at some instant of time. But we could just as well think of them as graphs giving the distance of some point on the string from its equilibrium as time increases. Any point oscillates smoothly up and down with time, as shown by the graphs.

Wave Speed along a String   On a guitar, the bass note strings are more massive than those that produce a higher pitch. Tightening a particular string gives it a higher pitch. Both of these properties, mass and tension, influence the pitch by influencing the wave speed. As we've just seen, the frequency, which determines the pitch, is directly related to wave speed. Let's see how mass and tension determine the speed.

Figure 15.6 focuses on a short section of the string that, at the instant shown, is at the lowest point in its motion. This segment, shown darkly in the figure, needs to move upward as the wave moves to the right. The tension T in the string pulls on both ends of the segment. Since the tension slopes upward, it provides the string segment the force needed to accelerate it back toward equilibrium (recall Newton's second law, $F = ma$). The stronger the tension, the more quickly the segment gets moved back to equilibrium, and hence the faster the wave moves along the string. On the other hand, the more mass the segment has, the less its acceleration toward equilibrium and the slower the wave moves.

In summary, the stronger the tension the higher the wave speed, and the greater the mass of a given length of string the

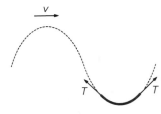

Figure 15.6 The tension, *T*, acting on a segment of string at its lowest point. The more quickly the tension can move the segment back to equilibrium, the higher the wave velocity. This speed increases with increasing tension and decreases with increasing mass of the segment.

lower the wave speed. Does that fit with what we said about guitar strings?

Transverse Waves The kind of waves we've been talking about are **transverse waves:** *Waves for which the particles of the medium vibrate in a direction perpendicular to the direction of wave propagation.* Electromagnetic waves are also transverse waves. An important difference in their case is that they do *not* involve a transverse oscillation of a *medium,* but rather transversely varying electric and magnetic fields.

Longitudinal Waves

A **longitudinal wave** is one in which *the particles of the medium oscillate in a direction parallel to the direction of wave propagation.* You may have played with a "slinky," a loosely wound springlike toy enjoyed by nine-year-olds and physics professors (Fig. 15.7). If you can get your professor to share his with you, ask him to hold one end, and

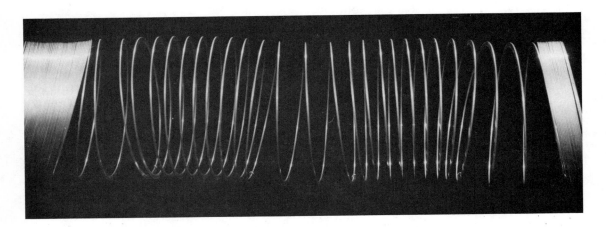

Figure 15.7 A Slinky.

stretch it out across the floor (Fig. 15.8-a). Then give your end a quick push toward the other end and then back (Fig. 15.8-b). A pulse made up of a compression followed by an expansion will travel down the slinky as shown in Fig. 15.8(c), (d), and (e).

If you then shake the end of the slinky back and forth continuously, you will set up a traveling *compressional,* or *longitudinal,* wave (Fig. 15.8-f).

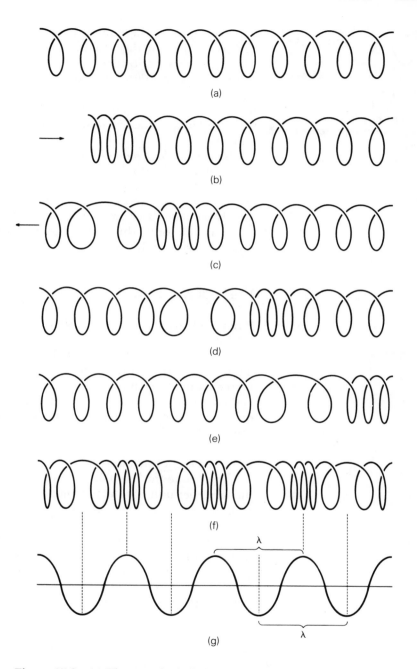

Figure 15.8 (a) The stretched Slinky. (b) through (e) By wiggling the left end one time, you can set up a longitudinal pulse that travels from left to right, as in the successive lines. (f) A traveling longitudinal wave. (g) The waveform representation for this wave.

Figure 15.9 A longitudinal wave in a "line of molecules," and its waveform representation.

Even though the motion is parallel to the wave direction, we can still represent the situation by a wave diagram (Fig. 15.8-g). The crests then represent regions of compression, the troughs regions of expansion. The wavelength λ is the distance between either. The other properties—frequency, period, velocity—have the same definition as for a transverse wave. A particle's displacement and amplitude are now parallel to the direction of wave motion.

Sound is a *longitudinal* wave. You can visualize how such a wave can exist in air by thinking of a long line of air molecules, as shown in Fig. 15.9. As the wave passes along the line, there are regions of compression followed by regions of expansion. Any one molecule merely oscillates back and forth. In a real sound wave in the air, the effect is not only three dimensional, but the wave motion is superimposed on the random thermal motion of the molecules.

The Energy of a Wave

When you shake the end of a string to set up a wave, you do work on the particles in the end of the string. They in turn do work on the neighboring particles, that do work on those down the line. Thus, the work you do generates a form of energy that is transferred along the string.

We know that a wave carries energy because it can do work on—give energy to—an object with which it interacts. A water wave causes you to bob up and down as it passes you. A sound wave causes your eardrum to vibrate back and forth. Radiation from the sun heats bodies that it strikes.

We define the **intensity** of a wave as the energy transmitted by the wave per unit of time through a unit of area. Since power is the energy per unit of time, *intensity is the power transmitted through a unit of area.* You would probably guess that the intensity increases with increasing amplitude of the wave. It turns out that the intensity is directly proportional to the amplitude *squared.* Doubling the amplitude of a wave means it carries four times as much power.

No matter what the nature of the wave, it carries energy. Some of that energy is absorbed by whatever the wave interacts with.

15.2   **Resonance**

When you push somebody in a swing, you don't just push at random times, or at some arbitrary frequency; you push at the natural frequency of the swing. A small periodic push at that frequency makes the swing "go." To force it to swing at another frequency would take an extreme effort. Similarly, the pendulum of a grandfather clock and the balance wheel of a mechanical watch oscillate continually under small impulses from the spring mechanism at rates near their natural frequencies. We depend on this constant natural frequency to keep the watch or clock running at the right speed. When you jump up and down on a diving board, there is only one frequency at which you can jump and build up an appreciable height. If the tires of your car are out of balance, the vibration is strongest at certain distinct speeds. Even a graduate student on a pogo stick has no control over the frequency of his jumps; it depends entirely on his weight and the spring in the stick.

These examples illustrate that it is easy to excite large vibrations or oscillations in a body if the driving force has a frequency near a natural frequency of vibration or oscillation of the body. We call this phenomenon **resonance,** and the frequency or frequencies at which the vibration of the body is easily excited we call **resonant frequencies.** When the body and driving force are in resonance, we can maintain a constant amplitude of vibration if the amount of work done each time we apply the driving force exactly equals the energy lost to friction in each cycle. Even the winds can excite resonant vibrations in large structures, as Fig. 15.10 illustrates. We'll see that resonance is essential in most sound sources.

15.3 Sound

Now that we know about waves and about resonance, we can discuss the characteristics of sound waves, and how they are produced.

What Is Sound?

Sound is a longitudinal wave that travels through a material medium. We usually limit the word sound to mean those waves that have the right frequency to produce the sensation of hearing in the human ear. This **audible region** of sound includes frequencies between about 20 and 20 000 Hz, although it takes an extremely good ear to hear the lower and higher frequency sounds within this range. Sounds with frequency below the audible region are called *infrasonic* and those above it *ultrasonic*.

Figure 15.10 Only four months after it opened in 1940, a mild gale set up resonant vibrations in the ½-mile-long center span of the Tacoma Narrows Bridge. Because of the resonant vibrations, this bridge collapsed within a few hours (opposite page).

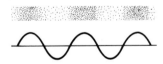

Figure 15.11 The effect of a sound wave on air molecules. We have greatly exaggerated the difference in density.

Figure 15-11 illustrates a sound wave passing through air molecules. The waveform at the bottom shows the pressure variation in the air. The peak-to-valley pressure difference in the faintest audible sound is only about 2×10^{-5} N/m² (3×10^{-9} lb/in²). For such a sound, a given molecule oscillates over a distance of only about 10^{-11} m: about 1/10 the diameter of a typical atom. These exceedingly small numbers show the incredible sensitivity of the ear!

Sound can travel through any material medium, but it usually gets to our ears by traveling through air. The speed of a sound wave depends on the density of the medium, and on how easily the medium can be compressed. Its speed in air is 332 m/s at 0°C, and increases by about 0.6 m/s for each Celsius degree increase in air temperature. Thus, its speed in 20°C air is 344 m/s.

We can calculate the wavelength range *in air* of the audible frequencies from

$$\text{velocity} = \text{frequency} \times \text{wavelength}.$$

Figure 15.10   *(continued)*

At 20 Hz,

$$\text{wavelength} = \frac{\text{velocity}}{\text{frequency}} = \frac{344 \text{ m/s}}{20 \text{ Hz}} = 17 \text{ m}.$$

At 20 000 Hz, the wavelength is 0.017 m, or 1.7 cm.

Exercise   The speed of sound in water is about 1500 m/s. What is the range of audible wavelengths in water?   **Ans:** 0.075 to 75 m

---

As sound travels from one medium to another, its frequency cannot change. (Reason out why that wouldn't be possible.) Since its speed does change, the wavelength of sound of a particular frequency depends on the medium.

Sources of Sound

Anything that vibrates with a frequency in the audible range can be a source of sound. That includes not only objects intended as sound sources, but things like electric shavers, automobile engines, boiling water, and wind "whistling" through the trees.

A tuning fork (Fig. 15.12-a) is a handy sound source for classroom demonstration because it vibrates with a single fixed audible frequency. As the sides vibrate, they set the surrounding air molecules in vibration at the same frequency (Fig. 15.12-b).

In a loudspeaker, sound is produced by a cone-shaped diaphragm that is set in vibration electrically (Fig. 15.13). We often represent sound, as well as other kinds of waves, emanating from a source by **wavefronts** as shown in the figure. We usually draw them one wavelength apart so that all points on all the wavefronts are in phase with each other: We can think of them as representing wave crests.

(b)

Figure 15.12   (a) A tuning fork. (b) The vibrating tuning fork is a source of single-frequency sound established by the vibrating arms.

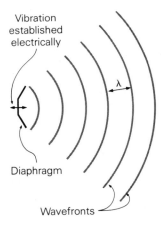

Figure 15.13   In a loudspeaker, the vibrating cone-shaped diaphragm produces the sound. We show the spreading of sound from the source by wavefronts drawn one wavelength apart.

In most musical instruments, the sources of sound are either vibrating strings (in guitars, pianos, violins) or vibrating air columns (in pipe organs, trumpets, trombones, tubas). In either strings or air columns, the vibration occurs at the resonant frequency that is characteristic of the vibrating body. We'll now look at the way strings, and then air columns, vibrate.

Vibrating Strings   When you pluck a guitar string, its vibrations set up sound waves that travel through the air to your ears. *The frequency of the sound is the same as the frequency of the vibrating string.* A string can vibrate easily only at certain distinct resonant frequencies. Notice the plural—a string has *many* resonant frequencies.

If your telephone has a stretchy coiled cord from the receiver to the body, hold the receiver with a slight tension in the cord and wiggle it back and forth. With a little experimenting, you can find the right frequency (probably around a cycle per second, depending on the nature of cord) that will make it vibrate between two extreme positions as in Fig. 15.14(a). Notice that once you find the resonant frequency, you need to move the end only slightly to have large vibrations, but you get only a jumble with a slightly different frequency or with an irregular frequency.

If you then wiggle it about twice as fast, you can get the pattern of Fig. 15.14(b). At about three times the original frequency, you get pattern (c). If your cord is long and heavy enough, you might

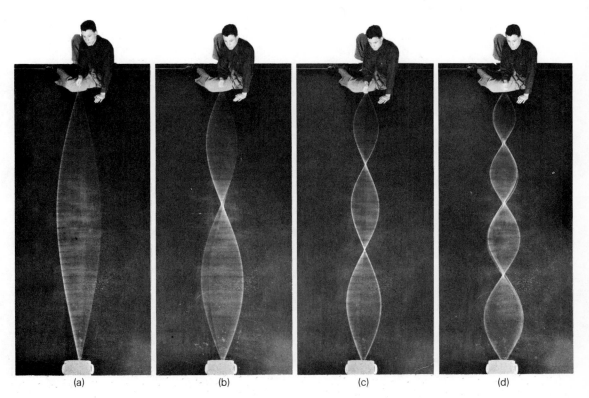

(a)   (b)   (c)   (d)

Figure 15.14   Photographs showing the vibration of a rubber hose at its four lowest frequencies.

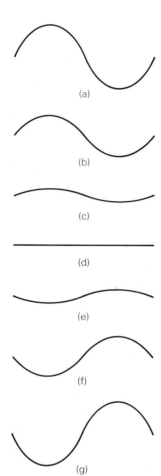

Figure 15.15 If you took snapshots of the cord of Fig. 15.14 (b), successive shots would look as shown from (a) through (g).

get 4, 5, and so on, segments for frequencies of 4, 5, and so on, times that needed for shape (a).

Notice that when one segment of the cord is up, the adjacent one is down. Figures 15.15(a) through (g) simulate "snapshots" of the way two neighboring segments would look at short separated time intervals.

A string of an instrument vibrates the same way between its two fixed ends, when you pluck, strike, or bow it. It vibrates only with those frequencies that are some whole number, 1, 2, 3, 4, . . . , times the frequency that gives it a shape like Fig. 15.14(a). In fact, it usually vibrates with several such frequencies at once, as Fig. 15.16 shows. We call the lowest vibrational frequency the **fundamental** frequency. The whole series of frequencies we call the **harmonics**—the fundamental being the *first harmonic*, the next highest frequency the *second harmonic*, and so on. (The strings of an instrument usually don't sag as much as the phone cord does.)

Whatever harmonics are present in the vibrating string, are also present in the sound that results. For example, if a guitar string vibrates with a 260-Hz fundamental frequency plus the third, sixth, and seventh harmonics, the sound will have frequency components of 260, 780, 1560, and 1820 Hz.

Figure 15.16 A string vibrating with several frequencies at the same time.

Exercise   On a particular grand piano, the note A above middle C vibrates with a fundamental frequency of 440 Hz, and the second, third, fourth, fifth, and seventh harmonics are also present. What frequencies exist in the sound heard from the note?   **Ans:** 440; 880; 1320; 1760; 2200; 3080 Hz

Standing Waves   *At any instant* the shape of our vibrating string looks suspiciously like its shape when a wave travels through it, and for good reason. The shape results from *two* waves in the string moving in opposite directions. There are certain points in the string, called *nodes,* that never move. Between these nodes the string moves up and down to form the pattern of Fig. 15.14.

What do two waves moving in opposite directions have to do with vibrating strings? As you shake one end of the string, the wave reflected from the fixed end forms the second traveling wave. When the wavelength is right for the crests and troughs of the waves going one way to always overlap at the same point the crests and troughs of the waves going the other way, this pattern that we call a *standing wave* exists. The various segments of the string move back and forth between the extreme positions shown by the solid and dashed lines of Fig. 15.17. If you concentrate on either the solid or the dashed line—the actual shape of the string at any instant—you see that the separation between nodes is ½ wavelength. For a string fixed at both ends and set in vibration, the waves reflect back and forth from each end to give the standing wave pattern. The points of maximum motion we call *antinodes.*

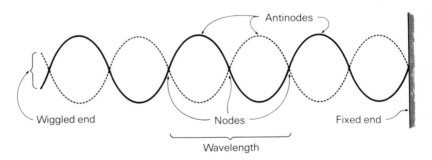

Figure 15.17   A standing wave in a string fixed at one end and wiggled at the other. The string oscillates back and forth between the two "maximum" positions shown by the solid and dashed lines.

We can now see why the harmonic frequencies are whole-number multiples of the fundamental. Figure 15.14 shows that the wavelength for the fundamental frequency is twice the length of the string—twice the distance between its nodes. The wavelength of the second harmonic is ½ that of the fundamental, that of the third harmonic is ⅓ that of the fundamental, and so on. From the relationship between wavelength and frequency,

$$\text{frequency} = \frac{\text{wave velocity}}{\text{wavelength}},$$

we see that reducing the wavelength to ⅓ its fundamental value triples the frequency. Thus, the frequency of the third harmonic is three times the fundamental. A similar relationship holds for other harmonics.

Vibrating Air Columns   Suppose you put a vibrating tuning fork in front of a pipe that is closed at one end by a movable piston as in Fig. 15.18. A sound wave will travel through the air in the pipe and be reflected from the closed end. Even though this wave is a longitudinal one, if the crests (high-pressure regions) and troughs (low-pressure regions) of the reflected wave always overlap the crests and troughs of the original wave at the same point along the air column, you get a standing wave in the air. Therefore, if you start with the piston near the tuning fork and move it to the right in the pipe, there will be certain lengths for the column of air at which it will be in resonance with the tuning fork. For these lengths, the vibrating tuning fork sets up standing waves in the air column, that greatly increase the intensity of the sound emitted.

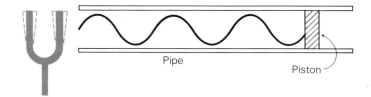

Figure 15.18   Sound waves travel through the air in the pipe and reflect from the moveable piston.

Pipe

Piston

Question   When the air column is the right length for standing waves, much more sound is given off by the system consisting of tuning fork and air column than at other lengths. From the energy conservation principle, how would you expect the length of time the tuning fork vibrates to be influenced by whether or not the column length is right for resonance?

If our air column were a fixed length, and we could vary the frequency of the sound source, we would find increased sound intensity from standing waves only at certain distinct frequencies. The longer the pipe, the longer the wavelength for the lowest, or fundamental, resonant frequency. For a given pipe length, all other possible frequencies are harmonics of the fundamental: They are whole numbers multiplied by the fundamental frequency.

Exercise   The sound from a particular 25-cm-long organ pipe has frequencies of 344 Hz, 1032 Hz, 1720 Hz, and 2408 Hz. What harmonics are present in the standing waves in the pipe, assuming the lowest frequency given to be the fundamental.

Ans: 1, 3, 5, and 7

Exercise   Blow into an empty Coke bottle so that you get a reasonably loud sound, indicating resonant vibrations (Fig. 15.19). Then add varying amounts of water and repeat each time, noticing how the pitch at resonance changes with height of *air* in the bottle. (We'll see later that higher pitch means higher frequency.)

Figure 15.19   With a little experimentation you can set up resonant vibrations in the air in the bottle by extending your upper lip over the bottle as shown to direct air into it.

Vibrating air columns are the sound sources in pipe organs and in all other wind instruments.

The Human Voice   Even though there surely are *some* voices you'd rather hear less of, you will probably agree that the human voice is the most important of all sound sources. We'll briefly discuss how it works.

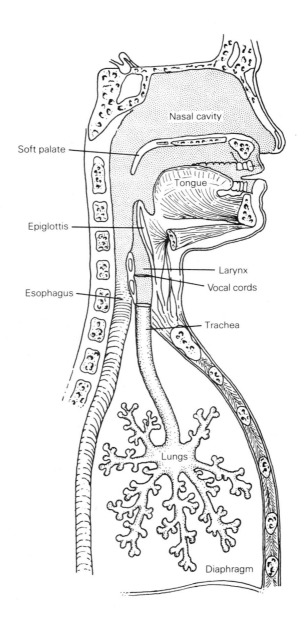

Figure 15.20 An outline of the human vocal tract.

Figure 15.20 is an outline of the vocal tract. The vocal apparatus is essentially a wind instrument in which standing waves are set up in the air in the *larynx* and various cavities of the mouth and head. Air forced up from the lungs has to pass through the opening between the *vocal cords*—two band-shaped membranes stretched across the larynx.

As air passes between the vocal cords, they vibrate back and forth, opening and closing the gap between them. We control the frequency of this vibration by controlling the tension in the cords. The frequencies in the sound emitted depend not only on the frequencies of the vibrating vocal cords, but upon which of these frequencies resonate in the air within the throat, mouth, and nose. As we talk or sing, we simultaneously control the vocal cord tension and the shapes of the various air cavities, in order to produce the resonant frequencies needed for the sound we want at any instant.

15.4   **The Human Ear**

Now that we've discussed sources of sound, let's look at the other end of the picture—the ultimate *receiver* of sound, the human ear.

Figure 15.21 is a diagram of the human ear. Sound waves entering the auditory canal set up vibrations in the eardrum that are transmitted to, and through, the bones of the middle ear. A membrane called the oval window separates the middle and inner ear regions, and is in contact with the fluid in the cochlea. Within the cochlea are hairlike cells that sense the vibrations and convert

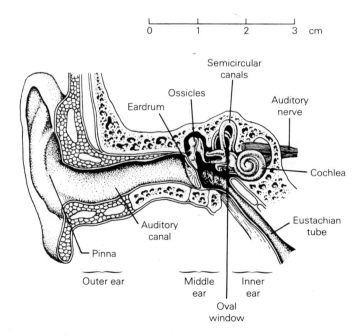

Figure 15.21   A diagram of the human ear.

mechanical vibrations to electrical signals for transmission to the brain. The mechanism of this mechanical-to-electrical energy conversion process is not well understood, but some of the theories concerning the process are discussed in the references at the end of this chapter.

15.5   The Objective and Subjective Properties of Sound

You might have heard someone claim: "If a tree falls in a forest and there is no person or animal nearby to hear it, there is no sound." The heated arguments this issue has been known to arouse center around the two different ways of defining sound: (1) the wave disturbance set up in a sound-transmitting medium; or (2) the sensation interpreted as sound by the brain. Whichever way you go, the question focuses on the two complementary aspects of sound that we'd like to compare and contrast in this section. One side of sound deals with its **objective** properties—those properties such as frequency and intensity, which are definite physical characteristics of the wave itself that can be measured with instruments and agreed upon by all observers. The other aspect of sound deals with its **subjective** properties that are partly psychological and *depend upon the responses of the particular person who hears the sound.* Properties in the second category include pitch, loudness, and quality.

For each *subjective* property, there is one *objective* property that mainly determines it, although the other objective properties have an influence. The situation is complicated because the relationship between subjective and objective properties is different for each listener. In this section we will look at each of the three subjective properties—loudness, pitch, and quality—and see what objective properties influence them.

Loudness and Intensity Level

You probably would guess that the **loudness** of sound would increase with an increase in the energy carried by the wave. That guess is correct, but loudness also depends on the sound's frequency. To see how this all fits together, we need to define a new term, **intensity level.**

The *intensity* of a sound, like that of any traveling wave, is defined as the power transmitted by the sound through a unit of area. For example, if a rock band is playing in a room, and 0.30 J of sound energy emerge per second from an open 2.0 m by 1.0 m door, the sound intensity is:

$$\text{intensity} = \frac{\text{power}}{\text{area}} = \frac{0.30 \text{ W}}{2.0 \text{ m}^2} = 0.15 \text{ W/m}^2.$$

At a frequency of 1000 Hz, the faintest sound a normal ear can hear has an intensity of about 10^{-12} W/m². The loudest sound most of us can tolerate without pain has an intensity of about 1 W/m². In other words, our ears, if normal, are sensitive to sounds that vary in intensity by a factor of 1 000 000 000 000!

To avoid dealing with this broad range of numbers for intensity, we use the term *intensity level*, which is expressed in the unit *decibel* (1/10 bel), abbreviated dB. On this scale we assign the intensity of 10^{-12} W/m² an *intensity level* of 0 dB. Then, each time the intensity is *multiplied* by a factor of 10, we *add* 10 dB to the intensity level. If the sound intensity is 10^{-11} W/m², that's 10 times that for 0 dB, or the intensity level is 10 dB. If the intensity is 10^{-9} W/m², that's 10×10 or 100 times that for 10 dB. The intensity level is then 10 dB + 20 dB, or 30 dB. If the intensity of a sound goes down from 0.37 W/m² to 0.037 W/m², the intensity level drops by 10 dB. Sounds with intensity level 120 dB have an intensity 10^{12} times those of intensity level 0 dB. The smallest *change* in intensity level that a person with normal hearing can detect is about 1 dB. Table 15.1 gives the average intensity level of a few well-known sounds.

Table 15.1   Intensity Levels of Certain Sounds

Sound Source	Approximate Intensity Level (dB)
barely audible sounds	0
rustling leaves	10
whisper at about 1 m	20
city noise at night	40
ordinary conversation	60
noisy factory	85
Niagara Falls	90
riveter about 10 m away	100
rock band	110
sound becomes painful	120
jet plane take off, 30 m away	140
mechanical damage to ear	160
large rocket (nearby)	180

<table>
<tr><td>Exercise</td><td>Because of complaints from the old folks over 30 in the audience, a rock band turns the amplifiers down so that the sound energy output rate is only 1% of what it was. If the original intensity level were 110 dB, what is the new intensity level?   **Ans:** 90 dB</td></tr>
</table>

For a given frequency, the loudness is roughly proportional to the intensity level. A sound of a particular intensity level is loudest at frequencies around 3000 to 4000 Hz. If you decrease or increase the frequency away from this region, the intensity level must increase somewhat to maintain the same loudness.

Pitch and Frequency

The **pitch** of a sound is its position on a musical scale, *as judged by a listener.* For example, a bass voice has a low pitch, a soprano voice a high pitch.

Pitch depends *mostly,* but not entirely, on the frequency of the sound. A low-frequency sound has a low pitch, a high-frequency sound, a high pitch. Frequency is an objective, measurable property of the sound; it is the number of actual vibrations in the sound wave per second. But pitch is a subjective property that depends upon the listener's response.

Pitch is influenced also by the loudness of the sound, particularly for low frequencies. In the frequency range of 200 to 300 Hz, an increase in intensity level of 40 dB can drop the pitch of a single-frequency sound by 10% for a sensitive listener.

Since most sounds are not *pure tones* of a single frequency, but are made up of a fundamental frequency and various overtones, you might wonder which frequency or frequencies determine the pitch. Experiments have shown that, even if one of the overtones is more intense than the fundamental, *the pitch depends mainly on the fundamental frequency.*

Our ears play a trick on us that sometimes turns out to be very handy. Suppose we hear a sound that has frequency components 400, 600, 800, 1000, and so on Hz. These frequencies are the overtones for a sound of frequency 200 Hz. For this kind of situation, our ears often add the missing fundamental frequency, and we hear a pitch that corresonds to 200 Hz.

One result of the production of these *subjective tones* by the ear is that the ear can partly compensate when we listen to inexpensive audio equipment. Small loudspeakers cannot produce low-frequency (below about middle C) sound. Yet we hear some bass from even the cheapest radios. Our ears use the high-frequency harmonics from the bass instrument to produce the sensation of bass even though it's not physically present in the sound hitting our ears.

Quality and Harmonic Content

If your mother, your rich aunt, and your closest female friend all hummed a note at exactly the pitch of middle C on the piano, you would have no trouble distinguishing one voice from the other, even though they sang with the same pitch and loudness. Also, you would have no trouble picking out the sounds of a piano, a trombone, and a factory whistle, all sounding the same pitch at the same loudness. The property that distinguishes two sounds of the same loudness and pitch, we call **quality,** or sometimes **timbre.**

We can get a hint at the cause for difference in quality from Fig. 15.22—the waveforms of the sound from three different sources. Each of these sounds has the same fundamental frequency, but the relative amounts of each of the higher harmonics are different. In general, *the relative amounts of the different harmonics present in the sound mainly determine its quality.* The difference in waveforms for different sound sources, as in Fig. 15.22, arises from the combined effect of the harmonics.

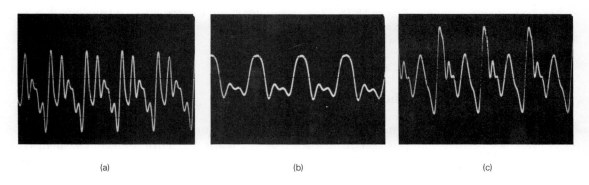

(a)   (b)   (c)

Figure 15.22   The waveforms of the sound of (a) a male voice singing the vowel ō, (b) a flute, and (c) a saxophone (from Hall, *Musical Acoustics*).

Figure 15.23 shows bar graphs of the harmonic content of the sound from the two instruments of Fig. 15.22. Considering the differences in harmonics present, it isn't surprising that the instruments have a different sound.

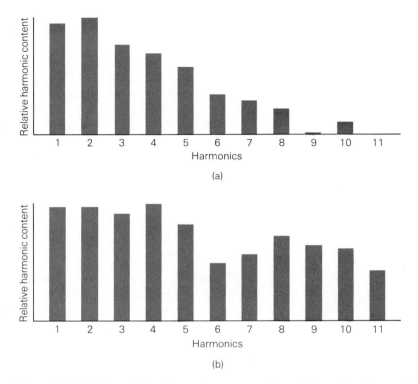

Figure 15.23   The relative harmonic content of sound from (a) a flute and (b) a saxophone as they produce the sound waves of Fig. 15.22 (based on Hall, *Musical Acoustics*).

**15.6   The Sound of Music*

Music: The sound that really counts. We've already discussed the physical principles that form the basis for most musical instruments. We'll now point out in a general way how these principles apply to instruments.

Almost all the instruments that we ordinarily play or listen to are either stringed instruments or wind instruments. Some exceptions are those that use vibrating rods (the xylophone), those that use vibrating membranes (drums), those that use vibrating plates (cymbal and bell), and those in which the sound is generated electronically. We'll take a look at stringed and wind instruments separately.

—————————

*Sections with the ** symbol are for illustration only, and contain no new physics.

Stringed Instruments The guitar, the violin, and the piano, for example, all use vibrating strings as their sound source. These three instruments illustrate the three ways that standing waves are set up in the strings: by plucking, by bowing, and by striking, respectively. (Have you ever looked inside a piano and noticed the little hammers that strike the strings when a key is pushed?) The method of exciting the vibrations, and where on the string the excitation is applied, influence the quality of the sound: the relative harmonic content.

But if all you had were the strings, you wouldn't get much sound. In portable instruments, the vibrating strings set up standing waves in the body of the instrument and in the enclosed air volume. The air and the body together greatly amplify the sound and determine its harmonic content. In the piano, the strings are attached to a large, heavy, wooden *sounding board* that resonates to a broad range of frequencies. (That's why the piano is so heavy!) The strings excite vibrations in the sounding board that supply the main part of the sound energy.

In principle, drums have a lot in common with strings. Rather than waves in a one-dimensional string, they have standing waves in a two-dimensional membrane that can have various harmonics present. Again, waves established in the enclosed air volume amplify the sound and influence its harmonics.

Figure 15.24 A saxophone has a tapered, cone-shaped air column that allows the existence of many harmonics.

Wind Instruments Wind instruments—for example, the pipe organ, the saxophone, and the tuba—all use standing waves in vibrating air "columns" as their main source of sound energy. When we considered standing waves in uniform air columns, we found that the length of the column determines the pitch of the sound. Wind instruments differ in the way we change the length of the air column. Pipe organs have different length pipes for each note. But we change the *effective* length of portable instruments by opening or closing holes along the pipe—the clarinet—or by valves that add or remove sections of the column—the french horn. The particular shape of the instrument determines which harmonics exist and thus the quality of the sound: Notice, for example, the gradual increase in the size of the "pipe" that makes up a saxophone (Fig. 15.24).

Most wind instruments use one of two methods of setting up vibrations in air columns: either *edge tones* or *reeds*. When air blows against a sharp edge, vortices are formed first on one side and then the other. Their intermittent formation provides the vibration to set up standing waves in the air column.

In a reed-type instrument, air blowing across a thin flexible strip—a reed—causes it to vibrate and start standing waves in the air. For brass instruments, the players' lips form a pair of reeds that open and close quickly to set up the vibrations. The way the player blows can influence which overtones are set up.

Most of us would probably guess, just from playing a wind instrument or watching it played, that the blowing of air through the body of the instrument and out the end is intimately connected with the sound given off—you sort of blow the sound out. This idea is probably reinforced in the type of instrument on which you open and close holes along the side, seeming to let the air out in different places. But *now you know* that the blowing merely starts the vibration by using reeds or edge tones, and that the main sound source is the standing waves in the air. They have nothing to do with whether or not air is moving through the instrument. Opening and closing holes along the side changes the standing wave pattern by changing the effective length of the air column, but letting the air out there has little, if any, effect.

The references at the end of the chapter will give you more details on the effects we have just mentioned, and will show you how the various principles apply in specific instruments.

15.7   Noise

There is little doubt that what is music to one person is mere *noise* to another. You've probably experienced this difference of opinion between yourself and your parents. The discrepancy fits with one definition of noise: *any* unwanted sound.

Technically, however, we usually define **noise** *as sound having random frequencies mixed in random amounts.* The various frequencies are *not* harmonics of some fundamental frequency; rather, they have no particular relationship to one another. Noise produces no particular pitch or quality sensations.

Noise is not necessarily bad. It is important in the formation of consonants in speech. The sound from certain instruments—snare drums and cymbals, for example—is mostly noise. And experiments have shown that the noise burst during the first fraction of a second following the sound of a piano or harpsichord note are very important in our recognition of the sound.

You might be surprised to learn that noise has *color. White noise* consists of a fairly uniform mixture of frequencies throughout the audible range. The term is borrowed from its use with light: *White light* is light such as sunlight, which is a mixture of all colors. Just as pink light is light of mostly lower frequency and longer wavelength, *pink noise* is low-frequency noise; *blue noise* has the higher frequencies emphasized.

The concept of noise carries over into other fields. For example, if you are observing an electric current or voltage that varies smoothly with time, there may be random wiggles in the signal—called *noise*—superimposed on the smooth variation. For instance, in amplifying the small voltages measured across nerve cell membranes, random electrical activity in the amplifier might produce noise with larger variations than the signal being measured. For reliable measurements, it's important to have a large *signal-to-noise ratio*. The same is true if you're listening to a recording. You want the signal-to-noise ratio (music-to-scratches ratio) to be as large as possible.

In our increasingly technical society, *noise pollution* is becoming an ever more important consideration. The more gadgets we have, the more noises there are to pollute and mask the sounds we enjoy hearing. The psychological effects of noise pollution are yet to be thoroughly understood, but prolonged exposure to noise levels above about 85 dB often leads to permanent hearing damage. Rock musicians have been known to develop "boilermaker's ear," a name for a hearing loss whose origin you can easily imagine.

15.8 Special Effects in Sound Waves

There are certain things that only waves can do. The names of some of these phenomena are *reflection, refraction, diffraction,* and *the Doppler effect.* We'll define or describe these effects and show how they play a role in the behavior of sound.

Reflection

You might argue with the statement that only *waves* can reflect; and I wouldn't protest too strongly. Objects do bounce off others: basketballs off backboards, pool balls off the cushion. But we usually reserve the term *reflect* for *waves,* because they rebound in a particular way. Depending on how your opponent hits it, a racquetball may come off the wall at a weird angle. But waves from a smooth surface always reflect so that *the angle the reflected wave makes with the surface exactly equals the angle the incident wave makes with it.* This **law of reflection** applies to any kind of wave.

We're all familiar with the reflection of light from mirrors, from water surfaces, and various other objects. Even though the effect is not as obvious, the reflection of sound waves is just as commonplace.

The most familiar example of sound wave reflection is the echo you hear when sound bounces off a distant large object. Reflection is important for proper acoustics in an auditorium. You want some reflection from the walls to make the sound "live"; too much reflection makes the auditorium like an echo chamber.

Sonar uses the reflection of sound waves (usually at ultrasonic frequencies) from underwater objects to determine their depth. Knowing the speed of sound in water and measuring the time between the emission and return of a sound pulse, you can calculate how far away the object is.

Exercise	A ship sends out a pulse of sound of frequency 50 000 Hz that is reflected from a submarine and returns 1.00 second later. If sound travels at a speed of 1450 m/s in water, how far away is the submarine? **Ans:** 725 m

You may have been accused of being "blind as a bat." You also may not be as skillful as the bat in making up for visual deficiencies. Bats detect prey and obstacles in their paths by sending out short bursts of ultrasonic sound, and detecting the echo reflected from the object. They can tell how far away their target is by timing the interval between the emission and return of the ultrasonic pulse.

Medical applications of ultrasonic sound, or *ultrasound,* already abound and are continuing to increase. One example is shown in Fig. 15.25. Ultrasonic waves reflected from boundaries between two internal media can show the exact location of different organs, of abnormalities, or of "foreign" objects such as those in the picture. Ultrasonic techniques not only are sometimes more sensitive than x rays to variations in density of the tissue, but do not produce the harmful effects associated with x rays. Many interesting applications are discussed in the end-of-chapter references.

Refraction

Refraction is the bending of a wave's direction because it crosses a boundary between two media in which the wave travels at different speeds, or because the wave speed is different in different parts of the same medium. The second condition would apply to sound if, for example, the temperature were different in different parts of the medium. It would also apply if parts of the medium itself were moving faster than others, since the overall sound speed is the speed of sound through the medium plus (or minus) the speed of the medium.

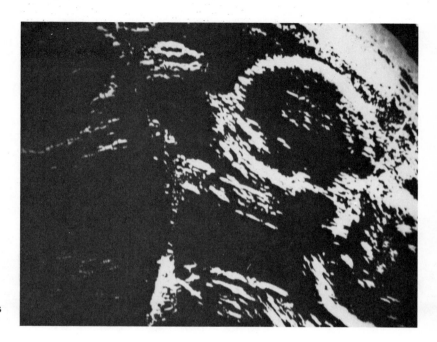

Figure 15.25 One use of ultrasound in medicine. Reflected ultrasonic waves show the presence of triplets inside the mother's uterus.

To see why a speed difference causes a change in direction, imagine that you and a good friend are riding on your bikes side-by-side and holding hands. If your friend is on your right and moving just a little faster, the only way you can continue to hold hands indefinitely is for you both to curve gradually to the left.

Now suppose a sound wave, shown by the wave fronts in Fig. 15.26, is moving to the right. If, as in Fig. 15.26(a), the sound speed is higher closer to the ground, the wave will gradually bend upward, and a listener a moderate distance away won't hear it. If the speed is slower near the ground (Fig. 15.26-b), the sound wave will tend to bend downward, reflect from the ground and travel a long way near the surface.

But what would cause the sound to travel faster or slower above ground level? One factor is a temperature difference. Sound speed increases with temperature increase. On a hot day when the ground is warm, the air near the ground is likely to be warmer than it is up a short distance; the sound curves up. On a cold day, or even better a cold night, the opposite is true. Do these conclusions agree with your observations of how far away you can hear particular sounds on cold nights and warm days?

The wind has a similar effect. The speed of sound relative to the ground is its speed in air plus that of the wind (minus that of the wind if it and the sound are moving in opposite directions).

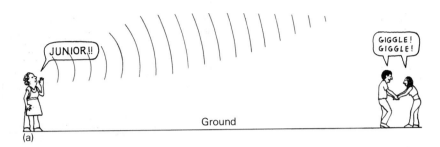

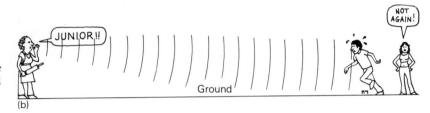

Figure 15.26 (a) If the wave speed is higher near the ground, the wave bends upward. (b) If it's lower near the ground, the wave bends downward and travels farther along the ground.

Suppose in Fig. 15.26 the wind is blowing from Junior to Mother. When he doesn't hear, she assumes the wind has held back the sound. In reality, the air is moving slower near the ground because of the frictional resistance of the ground. Thus, the net sound speed is a little higher near the ground. The sound curves up as in Fig. 15.26(a). With the wind blowing from Mother to Junior, the net sound speed is slower near the ground and the sound curves down as in Fig. 15.26(b).

Diffraction

A tree or post casts a sharp shadow in sunlight. This is another way of saying that light waves do not bend around the object. But that same post in the path of water waves has almost no effect: The waves simply bend around the post and go on their way almost undisturbed (Fig. 15.27). The difference in behavior is caused *not* by the difference in the nature of light and water waves, but by the *difference in wavelength relative to the size of the post.* When an object in the path of waves is much larger than the wavelength—as for the post in light—the waves do not bend around it but cast a sharp shadow. When the object is comparable in size or smaller than the wavelength—as for the post in water waves—the waves merely bend around it. This bending effect has the name **diffraction.**

Figure 15.27 The posts cast sharp shadows in light waves, but have little effect on the much-longer-wavelength water waves.

The wavelength of most audible sound is between a few centimeters and a few meters. Sound, therefore, diffracts around ordinary-sized objects—people, trees, walls, and the like—and through ordinary-sized openings—doors, windows, and people's mouths.

The Doppler Effect

If you've ever heard the "VAAROOOM!!" of a passing racing car, you've maybe noticed that the "ROOM" part has a lower pitch than the "VAA" part. Of if you've been passed by a train blowing its whistle, you may have heard the drop in pitch of the whistle as it passed. Turning the last scene around, if you're on a train passing a clanging crossing bell, the bell's pitch drops as you go by. All these examples illustrate the **Doppler effect.**

Sound source

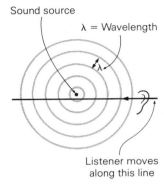

λ = Wavelength

Listener moves
along this line

Figure 15.28   A traveler
moving toward a
stationary source
intercepts waves at a faster
rate than the source
frequency. Going away he
intercepts them at a slower
rate.

Figure 15.28 shows a situation like that of the clanging bell: A
listener moving past a stationary sound source. The circles are
wavefronts of the sound emanating from the source at their center.
A traveler moving along the straight line meets the waves coming
toward him as he *approaches* the source. He's going away from the
waves that pass him as he *leaves* the source. Therefore, wavefronts
pass him more often—he hears a higher pitch—as he approaches,
and they pass him less often—he hears a lower pitch—as he leaves.

For a moving sound source—such as the passing racer—the
effect is similar but the cause is different. Figure 15.29 shows what
happens. The wave source at the center moves to the right with a
speed lower than the wave speed. This movement causes the
wavefronts to be crowded together with a shorter wavelength on
the right of the source, and to be spread out with a longer
wavelength on the left. Because the wave speed is the same on both
sides, a listener hears a higher frequency (higher pitch) on the right
than on the left. As the source passes, the pitch drops.

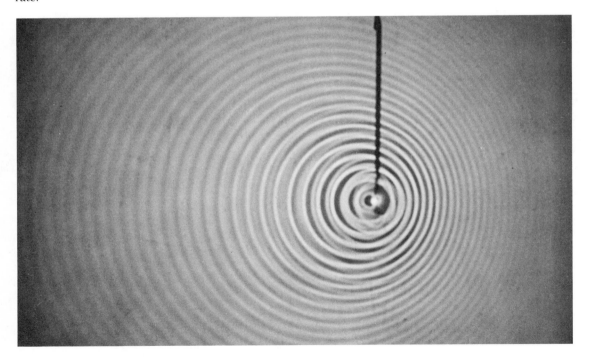

Figure 15.29   When the source is moving, the waves bunch up in front of
the source and spread out behind it. This gives a shorter wavelength
(higher frequency) in front, and longer wavelength (lower frequency) in
back. The photograph is of water waves.

Sonar (that we discussed in connection with reflection) can do more than just tell the distance away of an underwater object. By measuring the *Doppler shift* in frequency between the emitted and returning pulse, you can tell the speed of the underwater object that reflects the sound. Police radar uses the Doppler shift of 16-GHz electromagnetic waves reflected from your car to monitor your speed.

Shock Waves and Sonic Booms

The age of supersonic transport (SST) has brought livened interest in sonic booms, which were previously a concern only with military aircraft.

Figure 15.30(a) shows the wavefronts for a sound source moving to the right with speed lower than the speed of sound—the situation we described in the last section. Circles F_1, F_2, F_3, etc. are the wavefronts of the sound emitted when the source was at points 1, 2, 3, etc., respectively. Point 6 is the "present" source position. The waves bunch together on the right. The faster the source speed, the closer the bunching. If the source speed exactly equals the wave speed, the wavefronts all crowd on top of each other at the source, and can never spread out in front of it (Fig. 15.30-b). When the source speed exceeds the wave speed, the individual circular waves spread out behind the source in a cone-shaped region, as in Fig. 15.30(c). This cone-shaped wavefront on which the individual waves pile up on each other is called a *shock wave*.

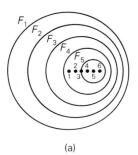

(a)

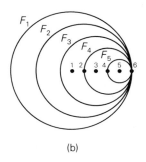

(b)

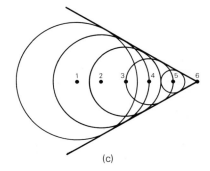

(c)

Figure 15.30 Wavefronts when the source is moving with speed (a) less than the wave speed; (b) equal to the wave speed; and (c) higher than the wave speed. The circles F_1, F_2, etc. are the wavefronts of the sound emitted when the source was at points 1, 2, etc. In each case point 6 is the *present* position of the source.

When a supersonic airplane passes by, this conical shock wave spreads out from the plane. There are actually two shock waves, one from the front of the plane—a bow wave—and one from the back—a tail wave. As the shock waves pass, there is a rapid rise and then fall in the air pressure—typically, a variation of the order of 100 N/m² above and below atmospheric for the *Concorde* SST. We hear a loud double "crack" that we call a **sonic boom.** For the *Concorde,* the perceived noise level for the boom is usually equivalent to about 120 dB.

Question   A lot of people erroneously think a sonic boom occurs only as an aircraft "breaks the sound barrier," not later. Explain why this belief is wrong.

There are other shock waves that we sometimes observe. As a high-speed bullet passes near you, the "crack" you hear is a shock wave. The V-shaped bow wave following a boat that is moving faster than the waves *in water* is an easily visible shock wave of the type we've described. You can even make them in your bathtub by running your finger along the surface of the water.

Key Concepts

A **traveling wave** is a disturbance that propagates from one region to another in a medium. The distance of any point in the medium from its equilibrium position is its **displacement.** Wave properties include: (1) their **wavelength,** the distance between two points **in phase** with each other; (2) their **amplitude,** the maximum displacement; (3) their **frequency,** the number of wavelengths passing some point per unit of time; (4) their **period,** the time for one wavelength to pass; and (5) their **wave velocity,** the speed with which the wave disturbance travels. In **transverse** waves the displacement is perpendicular to the direction of wave propagation; in **longitudinal** waves it is parallel to the propagation direction. For any type wave,

$$\text{wave velocity} = \text{frequency} \times \text{wavelength}.$$

The **intensity** of a wave is the power transmitted by the wave through a unit of area, and is proportional to the square of the amplitude.

Resonance occurs when the periodic impulses that excite vibration in a body are at a frequency near one of the natural, or **resonant,** vibrational frequencies of the body.

Sound is a longitudinal wave disturbance in a material medium. The **audible** range of frequencies—the frequencies to which the normal human ear responds—is between about 20 and 20 000 Hz.

Standing waves occur when two identical waves move in opposite directions in the same region. There are *nodes*—points of no motion—and *antinodes*—points of maximum motion—on a standing wave. Nodes are one-half wavelength apart.

The possible standing-wave frequencies of a string or air column are called **harmonics,** the lowest harmonic being called the **fundamental.** All harmonic frequencies are whole numbers multiplied by the fundamental frequency. The sound given off by a vibrating string or air column has the same frequency components as does the vibrating body.

The **subjective** properties (those that depend on the response of the listener) of sound—**loudness, pitch, quality**—depend mostly on the **objective** properties (those that are physically measurable): **intensity level,** *frequency,* and *harmonic content,* respectively. The intensity level of the faintest audible 1000-Hz sound is assigned a value of 0 decibel (dB), and the loudest 1000-Hz sound that does not cause pain has an intensity level of 120 dB. When the intensity is *multiplied* by 10, 10 dB is *added* to the intensity level.

Most musical instruments use either vibrating strings or vibrating air columns as their sound source.

Noise is sound having random frequencies mixed in random amounts.

Sound waves, as well as other kinds of waves, can **reflect, refract,** and **diffract.** The **Doppler effect** is the change in frequency resulting from the relative motion of the wave source and observer. *Shock waves* occur when the wave source travels at a faster speed than the wave speed in the medium. Shock waves from a sound source give rise to **sonic booms.**

Questions

Waves

1. How does the wavelength of waves in the ocean compare with that of the waves you cause when you drop a pebble into a pond?

2. The low-pitch strings on some instruments consist of metal wire wrapped around catgut strings. What is the purpose of the wrapping?

3. Tightening the strings on an instrument makes them vibrate faster and give a higher pitch. If you hold the ends of a rubber band in each hand and stretch it while someone plucks it, there is almost no change in pitch as you increase the tension by stretching the band longer. Why?

4. How does a bottle bobbing up and down on a water wave demonstrate that the wave carries energy? How do you know electromagnetic waves from the sun carry energy?

Resonance

5. Why do you have to know about resonance to ride a pogo stick, even though you may not call it by that name?

6. What does resonance have to do with the following things you can observe riding in an automobile? (a) When your tires are out of balance, the vibration is much worse at certain speeds. (b) A bump in the road may cause the car to bounce wildly at one speed, but hardly be felt at a *higher* speed. (c) If your car is stuck in mud or snow, you can sometimes get it out by alternating between forward and reverse. The chance of success is better if you alternate quickly and rhythmically at a particular frequency. (d) Young drivers sometimes find it fun to rock the car when stopping, by applying and releasing the brakes at regular intervals. However, the frequency of brake application must be just right to get good bounce.

7. If you are trying to carry a pan of water without spilling it, why do some frequencies of shaking make the water spill so much more readily than others?

8. Certain singers have reportedly shattered crystal glasses by sounding a particular note, although this is probably possible only if the voice is amplified (Fig. 15.31). Is it only the intensity of the sound that counts?

Sound Waves

9. I was taught in my high school biology class that if a tree falls in a forest and there is no person or animal nearby to hear it, there is no sound. Was I taught correctly?

10. Suppose Neil Armstrong had decided to take a nap on the moon's surface, and had laid his alarm clock on the ground

Figure 15.31 An amplified voice can shatter a glass.

beside him. Do you think he would have made his next appointment?

11. Sound will travel not only through air but through any material medium—for example, bone. What does that have to do with the reason *your* voice sounds so horrible on a tape recorder, while everybody else's sounds fine?

12. The speed of sound in water is about four times what it is in air. A waterproof bell ringing in air has a sound frequency of 400 Hz. What frequency will you hear if both you and the bell are underwater?

13. Bowing a violin string produces sound as the bow alternately grabs and then slips a little. Describe how this same grab-and-slip process can produce some other types of sounds: chalk on the blackboard, squealing tires, creaking doors.

Sound Sources

14. Why does pushing a guitar string down against a fret increase the pitch?

15. On a given violin string, you can play a note of any pitch within a certain broad range. On a guitar string, you play only discrete pitches within a certain broad range. Why?

16. The highest piano note has a pitch corresponding to a frequency of 4186 Hz. The upper limit of other ordinary instruments is lower than this. Does this mean that the only value in having a sound system that can reproduce frequencies higher than 4186 Hz is to give the owner or manufacturer something to brag about—not to give better sound?

17. What accounts for the gradual increase in pitch of the sound as you fill a bottle with water? (Hint: Think about what's happening in the air left in the bottle.)

18. Give some reasons why a boy's voice deepens as his mouth and nasal cavities get larger, and as his vocal cords get heavier.

19. Imagine you live on Mars and have never seen an earthling. If you hear a recording identified as an earth man's voice and one identified as an earth woman's voice, what could you predict about the relative length and/or massiveness of the male and female vocal cords, and about the relative size of their nasal and mouth cavities?

The Ear

20. Your ear canal leading the sound to your ear drum is probably about 2.5 cm long. An air column this long at 37°C, closed at one end and open at the other, has a fundamental resonant frequency of about 3500 Hz. Do you think these statements have anything to do with your ears being most sensitive in the 3 to 4-kHz frequency range?

21. The basilar membrane in the cochlea of your inner ear is made like a harp. It has many tiny fibers that are longer and thicker

as the distance from the oval window increases. The vibration of these fibers sets up impulses in nearby nerve endings, and these impulses are sent to the brain. Why do different frequencies excite vibrations in different parts of the basilar membrane? From which end of the membrane does your brain get impulses if the sound is of high frequency?

Objective and Subjective Sound Properties

22. What physical property of sound mostly determines its loudness? Its pitch? Its quality?

23. Suppose you call your dog with an average intensity level of 80 dB. How much more sound energy would you have to put into each call to increase that to 90 dB?

24. Table model radios seldom have speakers larger than about 15 cm in diameter. Even though such speakers probably cannot reproduce sound of pitch lower than about middle C (262 Hz), we certainly hear lower pitches from such radios. Why?

25. How is the sound of a saxophone physically different from that of a trombone playing the same note?

Music and Noise

26. Why does the pitch of wind instruments increase, and that of stringed instruments decrease, as an orchestra warms up?

27. Explain how edge tones can cause the sound of wind whistling through the trees.

28. Which would you take to be pinker and which bluer, the noise of a jet airplane overhead, or that of a bulldozer leveling ground for a new runway?

Special Sound Effects

29. Why is it so hard to understand what your opponent says to you on a racquetball court, unless you're very close to her?

30. A Polaroid SX-70 camera sends out "chirps" of inaudible ultrasonic sound that reflect from the subject back to the camera. The time between emission and return of the sound pulse determines the distance for which the camera automatically focuses. Compare this with the way bats locate their prey.

31. Moths, a favorite epicurean delicacy of bats, have a furry covering on their bodies that absorbs sound waves. Their wings, on the other hand, are good reflectors. Why do you suppose moths instantly fold their wings and drop when they detect ultrasonic sound?

32. Even though the speed of sound is much higher than the speed of air in wind currents, why does sound not travel nearly as far along the ground "against" the wind as "with" the wind?

33. When you're in your room and your friends are making lots of noise in the hall, does *partially* closing the door decrease the noise in the room an appreciable amount? Why not? Does it matter where you are in the room? (Assume soundproof walls and door.)

34. The pulse of ultrasound emitted by some bats varies in frequency during the pulse. How does this variability give the bat more information about the objects from which the waves are reflected?

35. If you're on the side of the road and a police car passes with its siren sounding, why does the siren pitch drop as it passes?

36. A *Concorde* SST leaves Washington, D.C., and passes over Philadelphia and New York before heading out over the ocean. It accelerates through the speed of sound just as it passes over Philadelphia. True or false: The people in Philadelphia hear a sonic boom but the people in New York do not.

Problems

1. Waves in a particular banjo string travel at a speed of 600 m/s. If the vibration frequency is 400 Hz, what is the wavelength of the wave in the string? What is the wavelength of the sound in air produced by this string? (The speed of sound in room temperature air is 344 m/s.)   **Ans:** 1.5 m; 0.86 m

2. If you hear a clap of thunder 0.50 s after you see the lightning flash, how far away did the lightning strike? (Speed of light = 3.0×10^8 m/s.)   **Ans:** 172 m

3. A car with snow tires is traveling at a speed of 50 km/h. If the tread pattern on the tires repeats every 5.5 cm, what is the frequency of the hum from the tires?   **Ans:** 250 Hz

4. The sound from a 1.0-m-long piano string has a frequency of 440 Hz. What is the wavelength of the waves *in the string*? If the tension in the string is doubled, what is the wavelength of the waves *in the string*? Why does it not change, even though the pitch of the sound certainly does? **Ans:** 2.0 m

5. The volunteer firefighters in Smithville are called to a fire by a siren mounted on a tall pole in the center of town. If the intensity level of the sound 1.00 m from the siren is 120 dB, what is the intensity level 3.16 m away? Assuming no wind or other disturbances in the air, within what distance of the siren would the sound have an intensity level of at least 50 dB? (Comment: The intensity of the sound from a small nondirectional source *decreases* in proportion to the *square* of the distance from the source, provided there is nothing around for the sound to reflect from. That is, the intensity follows an inverse square law. Recall, $\sqrt{10} = 3.16$.)

   Ans: 110 dB; 3.16^7 m, or 3.1 km

6. A hoarse mockingbird sings loud enough that the intensity level is 20 dB 1.00 m away. What is the intensity level 3.16 m away? Notice the comment in Problem 5, and explain why the relative effect of moving by $\sqrt{10}$ is so much greater here than in that problem. **Ans:** 10 dB

7. The Apostle Paul reported that when he was about to be shipwrecked in the Adriatic Sea, the sailors "sounded and found it twenty fathoms." (The Acts of Apostles 27:28). At that depth, how long would it be between the emission and return of a sound reflected from the bottom? A fathom is 1.8 m (6 ft) and the speed of sound in water is 1450 m/s. From your answer, do you think it likely that their sounding was by way of sound reflection? **Ans:** 0.05 s

8. A bat emits an 80-kHz pulse of sound that reflects from a tree and returns in 0.030 s. What is the wavelength in air of the sound? How far away is the tree? **Ans:** 0.43 cm; 5.2 m

9. An airplane passes 1 km over your head at twice the speed of sound. How far from the point directly overhead will the plane be when you hear the sonic boom? **Ans:** 2 km

Home Experiments

1. Borrow a tuning fork from your instructor. Set it in vibration and notice how long it takes for the sound to die out. Set it in

vibration again with about the same intensity, and hold the base in contact with a tabletop. Notice the difference in the total sound output, and in the time required for the vibration to stop. Explain the time difference in terms of the energy conservation principle.

2. Measure the speed of sound by timing the return of an echo. Find a flat field bounded on one side by a building. One timing method that is fairly accurate is to clap rhythmically, starting slowly and increasing the frequency until the echo returns from the wall on the "off beat." With your watch lying where you can observe it, count the number of claps per 10 s. Step off the distance to the wall. From your data, calculate the speed of sound and compare with the accepted value.

3. The next time you're bored with a telephone conversation, try some wave experiments. If the receiver is connected to the telephone with a coiled wire, notice that by varying the frequency with which you shake the receiver, you can set up various harmonics of standing waves in the cord. Estimate the frequency change for each different harmonic.

4. Try jumping on a diving board, pushing someone in a swing, rocking a boat, or riding a pogo stick at some frequency other than the resonant frequency.

5. Take a cardboard mailing tube, or the tube on which wrapping paper is wound, and hold it near your mouth. Sing a steady "aah" into the tube and vary the pitch slowly from low to high until you pass through resonance. You should not only hear a slight increase in intensity, but you can feel the difference in your throat as the standing waves are set up. To help convince yourself that you are passing through resonance, cover the other end and notice the absence of a change at the old resonant frequency. Find a new resonance with the end covered. It should be about an octave higher (that is, at twice the frequency). Try a tube of another length, and compare the pitch at resonance.

6. Hold down the damper pedal of a piano and yell various kinds of sounds into the region containing the strings and sounding board. Yell out the various vowels, and notice that the particular frequencies in the sound are picked up and maintained for several seconds. Discuss the relationship of this to the frequency response mechanism of the ear (Question 21). Caution: Be very careful not to be seen yelling into the piano, lest someone haul you off to the funny farm.

References for Further Reading

Crawford, F. S., Jr., *Waves* (McGraw-Hill Book Co., New York, 1968). Gives many interesting problems and home experiments that show the interaction of sound and other forms of waves with our daily lives. Several of the home experiments in this chapter are variations of Crawford's experiments.

Culver, C. A., *Musical Acoustics,* Fourth Ed. (McGraw-Hill Book Co., New York, 1956). Discusses the basics of sound and how it applies to music and musical instruments.

Devey, G. B., and P. N. T. Wells, "Ultrasound in Medical Diagnosis," *Scientific American,* May 1978, p. 98.

Gelatt, R., *The Fabulous Phonograph* (J. B. Lippincott Co., Philadelphia, 1955). A history of the reproduction of sound.

Hall, D. E. *Musical Acoustics: An Introduction* (Wadsworth Publishing Company, Belmont, California, 1980).

Kock, W. E., *Radar, Sonar, and Holography, an Introduction* (Academic Press, New York, 1973). In addition to the topics in the title, the author discusses wave properties in general, including some interesting photographs of diffraction and other effects in sound.

Kryter, K.D., "Sonic Booms from Supersonic Transport," *Science,* Jan. 1969, p. 359. A study of the effects of sonic booms on people.

Roederer, J. C., *Introduction to the Physics and Psychophysics of Music* (Springer-Verlag, New York, 1973). Gives not only the physics of sound, but considers specifically the sensation of hearing and how we react to various sounds.

Stanford, A. L., Jr. *Foundations of Biophysics* (Academic Press, New York, 1975). Includes details of the hearing mechanism.

Villchur, E., *The Reproduction of Sound* (Dover, New York, 1965).

Walker, J. "Some Whispering Galleries are Simply Sound Reflectors, but Others are More Mysterious," The Amateur Scientist Column, *Scientific American,* Oct. 1978, p. 179.

16

Light

And God said, "Let there be light"; and there was light. And God saw that the light was good; and God separated the light from the darkness. God called the light Day, and the darkness he called Night. (Genesis 1:3–5)

What is *light*? Is it merely the absence of darkness, as the sign on one dentist's darkroom implies (Fig. 16.1)? Where does it come from, and how does it behave? We will answer these questions in this chapter and the next.

Since light has a wave nature, we first need to understand waves. If you have not read Chapter 15, you should read Sections 15.1 and 15.2 of that chapter, which deal with waves in general, and with resonance. We can then start this chapter by discussing the particular wave properties of light. We'll next consider light sources, and then how we see other objects, as well as what determines their color. The phenomena of polarization and interference result from the wave nature of light, and produce some interesting effects, which we will consider. Finally, we'll see that, even though the wave nature of light gets the most publicity, light also has a particle nature. We can describe light completely only by considering its dual nature, both wave and particle.

The next chapter treats light from the viewpoint of rays that travel in straight lines in a given medium. We can use that viewpoint to discuss mirrors and lenses, and how they are used in optical instruments.

Figure 16.1 The sign on this dentist's darkroom tells the story of one theory of the nature of light.

16.1   The Wave Nature of Light

Light is a **transverse electromagnetic wave.** Saying that the wave is *transverse* means that whatever oscillates to form the wave, oscillates in a direction perpendicular to the direction of wave travel. In contrast, a *sound* wave is *longitudinal;* that is, the oscillating molecules move parallel to the direction of wave travel.

When we say that the wave is *electromagnetic,* we mean that it consists of oscillating electric and magnetic fields. Figure 16.2 illustrates this point. Rather than the movement of particles, as on a string, the "displacement" consists of two parts. First, an electric field varies smoothly from zero to a maximum in one direction, then smoothly to zero again, and then to a maximum in the opposite direction. Second, in a plane perpendicular to that of the electric field, a magnetic field varies exactly *in phase* with the electric field: It has its crests at the same points as the electric field does, but in a perpendicular direction. This electromagnetic field variation propagates along just as do waves on the ocean or in a string.*

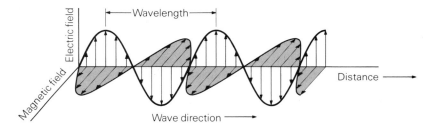

Figure 16.2   Light is an electromagnetic wave: It consists of electric and magnetic fields that oscillate at right angles to each other.

An electromagnetic wave, unlike all other known kinds of waves, needs no medium to travel through. As a matter of fact, it travels at its fastest through vacuum—*nothing.* Expecting that light, like all other waves, needed a medium for propagation, nineteenth-century physicists went on many "wild goose chases" in search for the medium. You might find these searches interesting to read about in a science history book.

---

*In Chapter 12 we described how to produce an electromagnetic wave in the radio frequency range.

The Characteristics of Light Waves

The *speed of light in vacuum* is 3.00×10^8 m/s or 186 000 mi/s. This is roughly 30 000 times the highest speed reached by a rocket on its way to the moon. The speed of light in air is almost the same, but in glass is about ⅔ this value.

The frequency range of *visible light*—light to which the human eye is sensitive—is between about 4×10^{14} Hz and about 7×10^{14} Hz. We call this the **visible region** of the *electromagnetic spectrum*. As we discussed in Chapter 12, other forms of electromagnetic radiation range in wavelength from above 10^{22} Hz (x rays and gamma rays) to below 100 Hz (long radio waves). Even though we will be concerned mostly with visible light in this chapter, much of what we say will apply also to electromagnetic radiation of other frequencies.

From the speed and frequency range, we can calculate the wavelength range of visible light in vacuum or air to be between about 7×10^{-7} m and 4×10^{-7} m. As with any type of wave, the frequency stays the same as light goes from one medium to another. If the speed changes in the new medium, the wavelength must change also. When we talk about the wavelength of some particular light, we will mean its wavelength in vacuum, unless we specifically state otherwise.

The Speed of Light—a Fundamental Constant The speed of light *in vacuum* is a very special quantity. It has much more significance than just the speed at which light happens to travel. We'll mention three of these special areas in which this quantity—to which physicists assign the symbol c—has special significance:

1. The speed of light in vacuum is the same for all observers, regardless of their state of motion (see Fig. 16.3).

2. No material object can be accelerated to a speed equal to, or exceeding, the speed of light in vacuum.

3. Strange things happen when one object moves relative to another at speeds near the speed of light. For example, the length of the "moving" object gets shorter, and time *slows down*.

These three statements are all part of Albert Einstein's theory of relativity. The first is a basic postulate of the theory; the other two are predictions. Many experiments have shown either directly or indirectly that as far as we can tell, all three are true. They are mentioned here in connection with light to point out the special significance of its speed. Maybe a better way to think of this quantity we call **c** is as a **fundamental universal constant.** *One* role of this constant in nature is the speed at which light travels in vacuum.

Spaceship speed = 200,000 km/s

MY LIGHT SPEED METER ON THE DASH REGISTERS 300000 km/s FOR THAT FLASHLIGHT...

MY LIGHT SPEED METER CLOCKS THIS LIGHT SPEED AT 300000 km/s.

Light

Figure 16.3 Both people, the one standing on the ground and the one in the spaceship moving at speed ⅔ *c*, measure a light speed of *c* relative to themselves.

EINSTEIN RIGHT: TRAVELING TWIN NOW YOUNGER

WASHINGTON, D.C., (October 7, 1971)—One of two identical twins returned from a trip today to find its twin who stayed home considerably older. To the scientists accompanying the traveler or waiting with the twin that stayed home, this was a happy confirmation of exactly what Einstein's theory of relativity predicted early in this century.

Relativity theory tells us that if one of two identical twins travels around at high speeds while the other stays home, the returning traveler will be younger than the twin who stayed home. This effect is more pronounced, the closer the traveler's speed is to the speed of light.

Scientists J. C. Hafele and R. E. Keating tested this theory by an exper-iment. They sent one of two identical twins on commercial jet flights around the world while the other stayed home. After the flight, the home-bound twin was found to be older than the other by 0.000 000 059 second, just the amount Einstein's theory predicted. Hafele and Keating had to use a rather special breed of twins in order to detect such a small age difference: They were twin atomic clocks capable of measuring time accurately to within a few billionths of a second.

Scientists consider this result to be strong evidence that, *for example*, if: (1) the twins were named Ralph and Maxine; (2) Maxine went to Alpha Centauri, four light-years away, and returned; (3) she traveled at a speed of 8/10 the speed of light; and (4) the twins were 20 years old when she left, then Ralph would be age 30 upon her return, but Maxine would be only 26.

Wavelength and Color

Light from the midday sun appears to have *no* particular color. We say that it is *white light*. But if we examine sunlight with an instrument capable of distinguishing colors, we find this white light is made up of light of several different colors.

For example, if we pass a beam of sunlight through a prism, we find different colors of light coming out in different directions, as shown in Fig. 16.4. (A prism is a triangular-shaped piece of glass or other transparent material.) If we let the outcoming light fall on a screen we will see a *spectrum* made up of bands of light corresponding to the various colors of the rainbow. Each color of light corresponds to a particular band of wavelengths, the red having the longest wavelength, the violet having the shortest. The bands are not the same width, and they gradually merge from one color to another rather than having distinct boundaries.

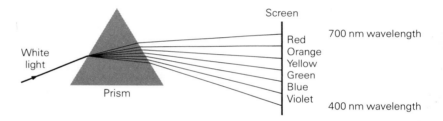

Figure 16.4   White light such as sunlight breaks up into its component colors—wavelengths—as it passes through a transparent prism.

When you look at a rose petal and admire the deep red color, that means physically that light in the wavelength range from about 650 to 700 nm (1 nm = 10^{-9} m)* has entered your eye. The sensation of redness is your brain's interpretation of your eyes' response to that particular wavelength of light. The sensation of green, from looking at the leaves of the rosebush, comes from your eye-brain response to light in the 500 to 550 nm range. When light containing all the colors of the visible spectrum—wavelengths between 400 and 700 nm—falls on your eye, the combined effect is the sensation of white.

---

*Many books use the unit angstrom (abbreviated Å) in describing light wavelengths. Since 1 Å = 10^{-10} m = 0.1 nm, moving the decimal point one place to the left converts angstroms to nanometers.

Question   From the fact that the visible region of the electromagnetic spectrum runs from about 400 to 700 nm, and assuming the same wavelength interval for the six colors of the rainbow, what is the approximate wavelength range for each of the colors?

16.2   Light Sources

Aside from the sun, there are many types of artificial light sources that we use or see each day. Probably the most common type is the *incandescent* lamp, or what we usually call a "light bulb." The light source here is an extremely hot metal filament. However, before considering this very ordinary source, let's first examine a type of source in which we can more easily understand the origin of the light—gas light sources, such as neon signs.

Light Emission from Gas Atoms

Imagine a marble sitting on a step of a staircase (Fig. 16.5-a). At that point the marble has a certain gravitational potential energy. As long as it stays on that step, its energy doesn't change. If it drops to a lower step, it loses some of its potential energy. Since *total energy* is conserved, we know that the loss in gravitational potential energy means an increase in another form (or other forms) of energy by the same amount. In this case, the potential energy goes to kinetic energy of the marble before hitting the next step, and then to heat as it hits. Each step the marble goes down results in the production of an amount of heat equal to the change in the marble's potential energy. The heat production takes place in discrete, fixed-sized chunks of energy. When the marble gets to the bottom of the stairs, it's in the lowest energy state and can't go any lower as far as that staircase is concerned. It then cannot give up any more potential energy to heat.

An electron in an atom has several characteristics in common with the marble on the stairs. First, it can exist stably only in certain *distinct energy levels* separated by *discrete energy intervals*. These energy levels are shown in Fig. 16-5(b) as circles symbolizing electron orbits in the atom. The electron energy is made up of kinetic energy and electrical potential energy.

Second, the electron can *change* its energy only in *discrete steps* given by the difference in energy between the two levels. Third, when the electron goes to a lower energy level, energy is given off. This energy is in the form of *light. The energy carried away by the light exactly equals the energy change of the electron.* When the electron is in

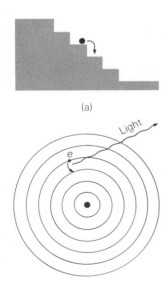

Figure 16.5   (a) Energy levels of a marble on a staircase. (b) Energy levels of an electron in an atom.

the lowest energy level for the atom—called the **ground state**—the atom can give off no light. The levels of higher energy than the ground state are called **excited states.**

When you see that bright orange-red neon sign advertising your favorite soft drink, you are seeing light given off as the electrons in the atoms of neon gas fall to lower energy states. The same process is happening in the mercury-vapor lamps that are used so widely for city street lights nowadays.

The light from the neon sign is reddish, but that from the mercury-vapor lamp is bluish white. From another gas, we get another color of light: Those green "neon" signs aren't really filled with neon, but with a gas that emits light that appears green.

To see why different gases give light of different colors, let's look at the energy level diagram of a particular gas. We'll pick hydrogen, since that's the simplest gas. We'll show its energy level diagram (Fig. 16.6) the way physicists usually draw them: A series of horizontal lines representing the levels, and separated by distances proportional to the energy intervals between levels. The figure shows the lowest five levels, with the energy of each level above the ground state—the **excitation energy**—written on the right. The energy unit used, the electron volt ($1 \text{ eV} = 1.60 \times 10^{-19} \text{ J}$) is convenient for atomic energy levels because it's about the right size.*

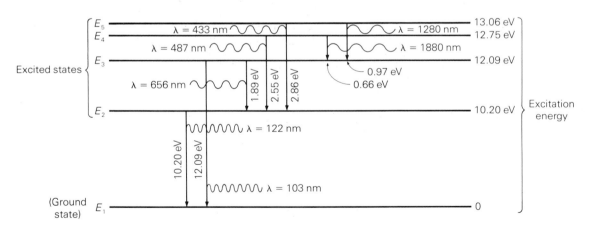

Figure 16.6   An energy level diagram for the five lowest energy states of the hydrogen atom. Vertical lines show possible transitions between these levels. The wiggly lines symbolize photons given off during a transition that have wavelengths λ as given.

---

*One electron volt is defined as the *energy* an electron acquires if it falls through an electrical potential difference of 1 volt.

The vertical arrows show possible downward changes in electron energy, with the numbers on each arrow giving the energy change. The chunk of energy given off as an electron makes a jump between any two levels we call a **photon,** or a **quantum** of energy. We say the light is **quantized,** meaning that the energy is emitted in discrete quanta. (We talked in an earlier chapter about the fact that electric charge is quantized in units of electron charge.)

Experiments show that the frequency of a particular photon's light is proportional to the energy it carries:

$$\text{frequency} \propto \text{energy,}$$

or

$$f \propto E,$$

where $\propto$ means "proportional to."

The proportionality constant is a fundamental constant called **Planck's constant,** for which we use the symbol h. It has the value 6.62×10^{-34} joule-second (J·s). We can then write the equation as

$$E = hf.$$

Since for any wave, frequency equals velocity divided by wavelength, we can calculate the wavelength of the photon's light if we know its energy. This energy is just the *difference* in energy between the two atomic levels involved in the energy change.

For example, let's find the wavelength of the light given off as a hydrogen atom goes from level E_4 at 12.75 eV to level E_2 at 10.20 eV. The photon energy is the energy change, 2.55 eV. First convert this energy to units of J so that all quantities will be in the SI system:

$$E = 2.55 \, \text{eV} \times 1.60 \times 10^{-19} \, \text{J/eV}$$

$$= 4.08 \times 10^{-19} \text{J.}$$

The frequency is

$$f = \frac{E}{h}$$

$$= \frac{4.08 \times 10^{-19} \text{J}}{6.62 \times 10^{-34} \, \text{J·s}} = 6.16 \times 10^{14} \, \text{Hz.}$$

Then the wavelength is

$$\lambda = \frac{c}{f}$$

$$= \frac{3.00 \times 10^8 \, \text{m/s}}{6.16 \times 10^{14} \, \text{Hz}} = 4.87 \times 10^{-7} \, \text{m} = 487 \, \text{nm.}$$

The diagram shows the wavelength for this transition as well as for the others.

Only the transitions *to* the second energy level E_2 give wavelengths in the visible region. The others are in the ultraviolet (beyond-violet in frequency) or infrared (below-red in frequency) regions of the electromagnetic spectrum.

The light from hydrogen has these particular wavelengths. If a beam of it were passed through a prism, the light would break up into several distinct beams of different colors. The colors shown in Fig. 16.7 would emerge, rather than the continuous range of colors we get if the incoming beam is white light.

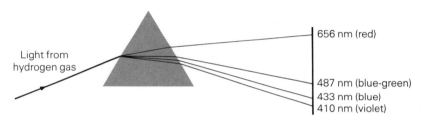

Light from hydrogen gas

656 nm (red)

487 nm (blue-green)
433 nm (blue)
410 nm (violet)

Figure 16.7 Light from hydrogen has only certain distinct wavelengths, as you can see by passing it through a prism.

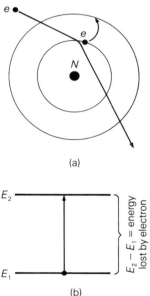

(a)

E_2 —————

E_1 —————

$E_2 - E_1$ = energy lost by electron

(b)

Figure 16.8 The excitation of an atom by a collision from another electron: (a) orbital representation; (b) energy level representation.

For other gases, a different set of distinct colors would be emitted. The light spectrum of an element can serve as its fingerprint for identifying the element—no two elements have the exact same spectra. The overall color appearance of the light from an element results from the combined effect of the wavelengths emitted.

We've left out one necessary ingredient, if the atoms of a gas are to emit light. Some of the atoms have to be in excited states in order to "fall" to lower states and give up energy. And for continuous light emission, it's necessary to continually excite atoms to higher states. Otherwise, there would be no more light once all the atoms were de-excited—a fairly rapid process, since most excited states live only about 10^{-8} s.

One way to excite the atoms is by collisions from other atoms or electrons. If the gas is placed between two metal conductors across which there is a voltage, electrons or ions traveling between these conductors can collide with and excite the atoms of the gas (Fig. 16.8). This process is usually used in lighting applications such as neon signs.

Atoms can also be excited by *absorption* of a photon, *if the photon has exactly the right amount of energy to raise the atom to one of its possible*

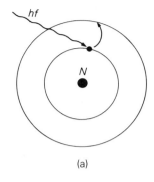

(a)

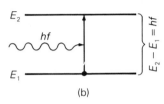

(b)

Figure 16.9 The excitation of an atom by a photon having energy hf exactly equal to the difference in atomic energy levels.

levels (Fig. 16.9). If the photon energy is *either* higher than, or less than, the energy difference between levels, no excitation will occur. This process, which is just the reverse of photon emission, is a form of **resonance** similar in principle to the mechanical resonance we discussed in Section 15.2. We sometimes call this process *resonance absorption*.

If white light passes through a gas, the atoms of the gas absorb light of those wavelengths corresponding to upward energy transitions in the atoms. If, for example, you look at sunlight through a *spectroscope* (a device that measures wavelengths of light), you will find certain wavelengths missing from the continuous spectrum. The missing wavelengths are those absorbed by the gases of the sun's atmosphere as the light passes through. This is the way helium was first discovered.

Fluorescence

There's a fair chance that you are reading this book under the light of a fluorescent lamp. A diagram of a fluorescent tube is shown in Fig. 16.10. This tube contains mercury vapor mixed with an inert gas and heated electrodes at each end. Some electrons in the electrodes gain enough thermal energy to leave the metal—a process something like evaporation. These electrons get accelerated back and forth by the alternating voltage between electrodes. Some collide with mercury atoms, exciting them enough to emit ultraviolet radiation. This radiation is absorbed by atoms and molecules of the *phosphor* coating inside the tube. The phosphor atoms then decay in

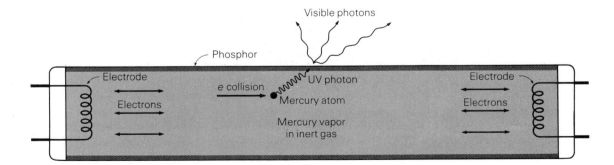

Figure 16.10 In a fluorescent bulb, electrons carrying current between cathodes collide with mercury atoms, thereby raising them to excited states. They de-excite by emitting ultraviolet photons that excite atoms and molecules in the phosphor coating of the tube. The phosphor atoms each emit several visible photons in going back to their ground states.

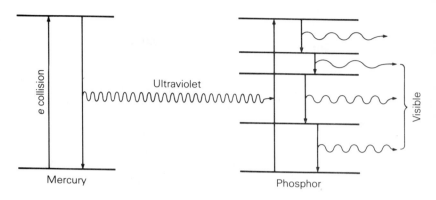

Figure 16.11   An energy level diagram for the processes in a fluorescent bulb (see Fig. 16.10).

smaller steps, emitting photons of visible radiation. Figure 16.11 shows the energy level diagrams for these processes.

Fluorescence is the process by which a material absorbs ultraviolet radiation, and de-excites in smaller steps by emitting visible radiation. The phosphor coating on the inside of the tube does this in a fluorescent lamp. Sometimes one or more excited states live for extended periods of time. They then may glow for hours after exposure to ultraviolet radiation. This process is called *phosphorescence,* but there is not a clear dividing line between it and fluorescence.

I promised in Chapter 12 to discuss why a TV screen glows when the electron beam hits it. There, the phosphors are excited by the electron bombardment, and emit visible light as they de-excite.

A few years back, the person doing the family laundry usually added a little "blueing" to the rinse water. That gave the white clothes a slight bluish tint and made them look whiter. Modern detergents with "whiteners" do the same job by leaving small amounts of a phosphor in the clothes that fluoresces in the blue region under sunlight.

Question   So-called "black lights" emit ultraviolet radiation. Explain how they can cause certain substances to glow without any apparent illumination when the "black light" is shone on them.

Incandescent Light Sources

In discussing light emission from individual atoms, we mentioned several times that the substance needs to be a gas. But we didn't say why. If the substance is a solid, liquid, or even a gas under high pressure, the atoms are close enough together that they interact strongly with each other, and a given electron is not confined to a single atom. As a result, the energy values that an electron can have are not quantized, but are spread over a wide continuous range.

An electron of such a substance can give up any fraction of its energy. This means that photon energies, and therefore wavelengths, do not take just discrete values, but are spread over a continuous range. Figure 16.12 shows how the number of photons given off varies with wavelength. At moderate temperatures, most of the radiation emitted is in the infrared region. As the temperature of the body is raised by adding heat, more of the electrons can gain, and then radiate, higher energy. Therefore, the frequency for the most dominant radiation shifts up, and the wavelength shifts down, with increase in temperature. At 700°C, there is enough radiation in the visible region that the body appears dark red. At 3000°C—a little above the temperature of the filament of an incandescent bulb—a large fraction of the radiation is in the visible region, and the object appears white. At about 5000°C, the peak is near the center of the visible region.

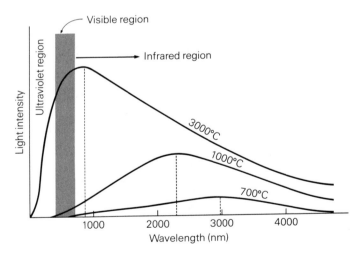

Figure 16.12   The relative amount of light radiated at various wavelengths for objects at three different temperatures. The vertical dashed lines show the wavelength for the most dominant radiation at each temperature.

The relative intensity of the radiation at each wavelength depends only on the temperature and not on what the material is.

Tungsten is used in incandescent light bulbs because of its high melting point. Electrical current passing through the filament heats it, and converts the electrical energy to radiant energy.

**Efficiencies of Light Sources

Figure 16.12 shows that most of the energy radiated by a hot object is *not* in the visible region. This means, for example, that an incandescent bulb converts only a small fraction of the electrical energy supplied to it to light. In fact, a 40-W incandescent bulb converts only 1.7% of the energy to light, a 100-W bulb only 2.5%: Their *absolute efficiencies* are 1.7% and 2.5%, respectively. Most of the rest goes to infrared radiation that serves only to heat the bulb's surroundings. In comparison, a 40-W fluorescent bulb is 9.4% efficient.

The rate of light output of a bulb is usually measured in *lumens*. The lumen is defined in the SI system in terms of the light output of a particular standard source. Rather than giving that definition, we'll point out that at 550 nm, the wavelength at which the eye is most sensitive, 1 lumen is equivalent to 0.00147 W of light energy. You can probably get a better feel for the "size" of the lumen from the fact that a 100-W standard incandescent light bulb has a light output of about 1750 lumens.

To get the most for your money from a light source, you want the ratio of light output to energy input to be highest. This quantity, the *luminous efficiency,* is usually measured in lumens per watt. The watt here is a watt of *electrical* power. A light bulb you buy at the grocery store has the light output in lumens, as well as the power needed, marked on the package. Table 16.1 gives the luminous efficiency of a few ordinary bulbs.

Table 17.1   Efficiencies of Light Sources

Source	Light Output (lumens)	Luminous Efficiency (lumens/watt)
Incandescent bulbs (standard)		
40 W	455	11.4
60 W	870	14.5
100 W	1750	17.5
Fluorescent bulbs		
20 W	1000	50.0
40 W	2560	64.0

Flame

A burning flame has been the source of artificial light during most of human history. Even though the practical importance of this light source has diminished, the intrigue of an open fire or burning candle is not ever likely to escape the person who takes the time to leisurely stare at it.

As you might guess, the processes involved in the production of light in a flame are complicated chemical reactions and are, in some respects, not thoroughly understood. We can, however, describe the general nature of the light emission processes.

There are basically two types of effects that produce the light of a flame. First, the chemical reactions taking place in the flame leave many of the molecules in excited states. These molecules emit visible light in going to lower energy states. This mechanism is responsible for the blue light you see near the base of a candle flame, for example.

The second way that light is produced in a flame is somewhat different. Carbon in the fuel forms solid carbon particles that are heated to incandescence by the hot gases in the flame. These hot particles emit a continuous spectrum of light in the same way the filament of an incandescent bulb radiates. The temperature of the particles is such that the light usually appears yellowish. This mechanism is responsible for most of the yellow light of a burning candle or an open wood fire.

Lasers

The **laser** is a light source that is growing by leaps and bounds in its applications to our technology (Fig. 16.13). Laser light is unique in that its output, which can be extremely intense, is in a narrow beam of *single-wavelength, coherent* light that broadens very little as it leaves the source. Coherent light is light for which the waves are all *in phase* with each other. Figure 16.14 illustrates this point. Part (a) of this figure shows light waves typical of an ordinary source. They have many different wavelengths that have no particular relationship to each other; they are *incoherent*. In part (b) of the figure the waves all have the same wavelength and are exactly in phase with each other: their peaks and valleys occur at the same time and reinforce each other; they are *coherent*.

The name "laser" is an acronym for **l**ight **a**mplification by **s**timulated **e**mission of **r**adiation. Inside the laser, gas atoms are left in a particular excited state by collisions with other atoms. When

Figure 16.13   A laser.

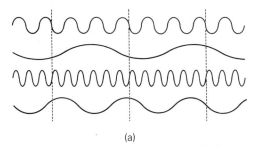
(a)

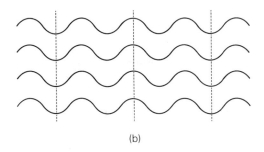
(b)

Figure 16.14   (a) Incoherent waves. (b) Coherent waves.

some of these atoms de-excite, the emitted photons pass other excited atoms, *stimulating* them to de-excite and give off identical photons. These photons *stimulate emission* by other atoms, thereby *amplifying* the light intensity.

Laser Applications   We'll mention a few of the hundreds of important modern applications of laser beams. These include: reading the Universal Product Code (bar code) at the checkout counter of many supermarkets (Fig. 16.15); communicating over long distances by pulsed or intensity-modulated beams; detecting pollutants in the atmosphere by determining what wavelength of laser light is absorbed by the air; measuring the distance to the moon by determining the time for a laser-beam pulse to travel to and return from a special reflector left on the moon by the Apollo astronauts; drilling of tiny holes in hard substances; welding of special materials that are difficult to weld otherwise, including the "welding" of

Bar code

Figure 16.15   The Universal Product Code that appears on most grocery store items can be read by laser beam at the check-out counter. The supermarket's computer then automatically adds the cost to the customer's bill and reduces the store's inventory by one jar of peanut butter.

Figure 16.16
Photographs, taken at two different angles, of a hologram. Different angles show different views of the image, just as they would show different parts of a real object.

detached retinas in the human eye; inducing fusion in small hydrogen pellets as a way of producing and controlling thermonuclear reactions (Chapter 18); routing of airport luggage; and even performing acupuncture with tiny laser beams.

One fascinating application is in the making of *holograms:* recordings on photographic film that, when illuminated, produce a three-dimensional image. An ordinary photograph is two dimensional, in that you see the same thing at whatever angle you look. As Fig. 16.16 shows, by moving your eyes around as you look at a hologram, you see different parts of the image, just as you would in looking at a real object. Holography will probably find many applications in entertainment, science, and industry in years to come.

LEDs and LCDs

In view of the literal explosion in recent years in the availability of digital readout devices such as calculators and watches, we'll briefly discuss the light sources usually used in the displays of these devices. Most early hand-held calculators and digital watches used *light emitting diodes* (**LEDs**) for readout. Because of much lower power requirements and readability in ordinary light, *liquid crystal displays* (**LCDs**) have taken over many of these jobs.

LEDs   In Chapter 13 we discussed semiconductor diodes used in solar cells for converting light energy to electrical energy. Light production in an LED is just the reverse of that process. If you have forgotten or never learned about semiconductors, you might want to read the part of Section 13.1 dealing with the basic properties of p- and n-type semiconductors.

An n-type semiconductor has a small excess of electrons in an otherwise tightly bound atomic lattice. These electrons can act as negative charge carriers. A p-type semiconductor has electron vacancies, or *holes,* that act as positive charge carriers.

A semiconductor diode is a *p-n junction* formed by putting together a layer of n-type and one of p-type semiconductor (Fig. 16.17). When a voltage is applied across the junction in the direction shown, electrons move from the n side to the p side and combine with the holes. (We could equivalently say that holes move in the other direction and combine with the electrons.) This annihilation of the electron-hole pair results in a lower energy state. As a result, energy is released. If the energy released in one electron-hole combination corresponds to that of a light photon, and if the material is transparent to those photons, light is emitted: We have an LED. If these two conditions are not met, the released energy goes to heat in the diode.

LEDs can be shaped to form various segments of a digit or letter. Which segments are lighted determines what letter or digit is displayed. Typical red LEDs are made of gallium phosphide or gallium arsenide phosphide. They operate on about 2 V, and a single digit uses a few milliwatts of power.

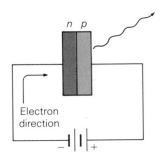

Figure 16.17 A light emitting diode (LED) gives off photons with energy equal to that released when an electron and hole combine near the p-n junction.

LCDs *Liquid crystal displays* are not actually light sources. They merely reflect or transmit light that is incident upon them. From that viewpoint their discussion fits more logically in Section 16.3 than here. However, since they *look like* light sources when used in digital displays, and to contrast them with LEDs, we include them here.

A liquid crystal is *liquid* in the sense that it takes on the shape of its container; it is *crystal* in the sense that there is a definite order to the molecules. Those used in displays are made up of long thread-like organic molecules.

With no voltage across such a crystal, the molecules are regularly arranged and the crystal is transparent. When a voltage is applied, the molecules become randomly oriented and scatter any light incident on them. Figure 16.18 shows how this effect can be used in a display. The part of the liquid crystal with no voltage applied transmits the light to the retroreflective foil—a specially made foil containing tiny spherical reflectors that reflect any incident light directly back along the path of incidence.* This light is not seen, and the crystal appears dark. Where voltage is applied, the randomly oriented molecules scatter light in all directions, making that part of the crystal appear light.

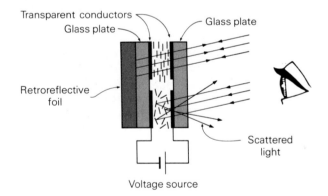

Figure 16.18 In a liquid crystal display (LCD), the molecules become randomly oriented and scatter the light only in those regions where there is an applied voltage across the crystal.

Transparent conductors
Glass plate
Glass plate
Retroreflective foil
Scattered light
Voltage source

A given display digit is usually divided into segments. Any digit between 0 and 9 can be made visible by applying voltage to appropriate sections of the display.

The type of LCD we've discussed here, called the *dynamic scattering display,* typically uses less than a milliwatt of power per square

*See Question 10, Chapter 17.

centimeter of display. We'll describe another LCD technique in our section on polarization that uses less than a microwatt: 1/1000 that needed by an LED.

16.3   How We See Things and What Color They Are

There are two ways that you see things. If you're looking at a light source, you see the light emitted by the source. If you're looking at something that is not a light source—for example, this page—you see it by light reflected from the object into your eyes.

The Color of Light Sources

If the light from a source enters your eyes directly, the color you see is the color of the light emitted. If, as is usual, the source emits a combination of wavelengths, the color you perceive is the combined effect of the various wavelengths.

Suppose you have a white light source, and you put a piece of red glass (or plastic or cellophane) between it and your eyes, as in Fig. 16.19. The light that comes through is red because light of all other wavelengths has been absorbed by the glass. Only the red is transmitted. Any time you look at light coming through a colored "transparent" material, you are seeing the only color for which the material is transparent; the rest of the light is absorbed. The process is called *selective transmission* because the material selects only a specific color to transmit.

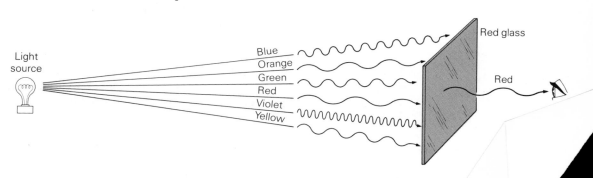

Figure 16.19   If light from a white light source passe
light of all wavelengths except red is absorbed. Onl
(The incoming light is usually not separated into c
as white light.)

The Color of Everything Else

Everything you see that is not a light source, you see because light from a source *somewhere* shines on the object, and then reflects into your eyes. But why are things the color they are? Why is grass green, the sky blue, red rose petals red, and the White House white?

To answer that question, consider the sketch of the daisy in Fig. 16.20. The various parts of the daisy appear the color they are because of *selective reflection.* The leaves and stem absorb light of all

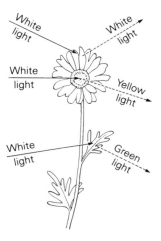

Figure 16.20   Each part of the daisy has its own particular color because all other colors are absorbed. Light of that particular color is reflected.

wavelengths except green, but reflect the green, some of which gets to your eyes. The center of the flower reflects the yellow light and absorbs the rest. The petals of the flower reflect all colors of visible light, and therefore appear white. In general, the colors *not* absorbed by an object give it its particular color.

A black object is black because it absorbs *all* the light hitting it. Actually its impossible to get a truly black object that reflects no light, but that is the definition of black. You see black writing on a white page by the light reflected from the white background and the *absence* of light reflected from the writing.

Question   From the facts that photosynthesis occurs in plant leaves, and leaves are usually green, what colors of light would you expect to be needed for photosynthesis? After you've reached a conclusion, compare it with the graph of Fig. 16.21 that shows the relative amount of radiation of each wavelength used in photosynthesis.

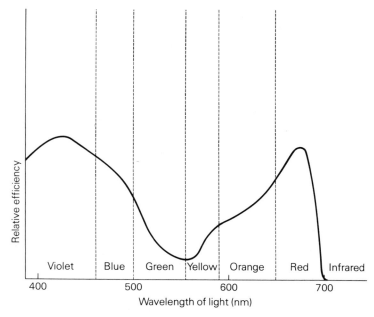

Figure 16.21 A graph of the relative efficiency of each wavelength of light in causing photosynthesis in plants.

Why Is the Sky Blue and the Sunset Red?

Have you ever wondered why the daytime sky is blue and the sunset colors are generally reddish? The reason is actually the same for both. As light passes through air, the electrons of the air molecules absorb and reradiate light. This process *scatters* some of the light in all directions. But the probability of a scattering taking place increases with increasing frequency: It actually increases in proportion to the fourth power of the frequency, f^4. This fact means that blue light is about 16 times more likely to be scattered than red. As Fig. 16.22 illustrates, the light you see from the day-time sky is scattered light, and is therefore dominated by blue. The sunset light comes directly from the sun, and has had much of the blue removed. It, therefore, is redder.

When you see the blue sky, you're actually *seeing* the atmosphere in the same way you *see* a leaf: by the scattering of light from it. Since scattering from an air molecule is a relatively rare event, it takes the entire thickness of the earth's atmosphere to scatter enough light to give the faint illumination you see from the sky.

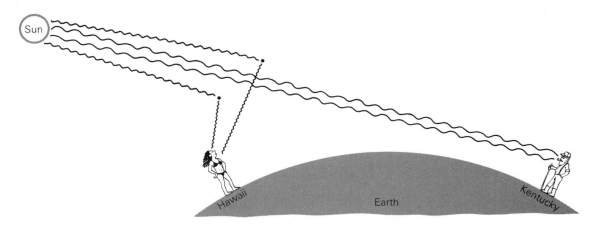

Figure 16.22 Short-wavelength light is more likely to be scattered by air particles than long-wavelength light. The Hawaiian sees the daytime light scattered in the sky, which is predominantly blue. This leaves mostly long-wavelength light (red to yellow) for a Kentucky sunset.

The Mixing of Colors

Color is all in your head! The light emerging from a beautiful autumn forest has no color in it; it's merely electromagnetic radiation of various wavelengths. The leaves themselves do not *contain* the color; they merely reflect light of particular wavelengths. The color comes about when your eyes send different messages to your brain in response to different frequencies of light received. Color perception is your brain's interpretation of the receipt by the eye of certain frequencies. We usually go ahead and talk about the "color" of an object, or the "color" of light. For practical purposes it amounts to that. But keep in mind that, in reality, color comes about only in your head.

A single wavelength of light—monochromatic light—entering your eye gives the sensation of one color. But a combination of wavelengths gives some color sensation intermediate between the responses to the separate wavelengths. White light, as we've said before, is a roughly even mixture of all visible wavelengths.

A color television set produces its colors on this basis. There are three electron beams, rather than one as described in Chapter 12 for black and white TV. The screen is covered with three sets of tiny phosphor dots that glow separately in red, blue, and green. Each beam is aimed to hit only dots of one color. Because the relative intensity of each beam hitting any one spot may be varied, the right color is produced at that point. The three dots of different color are close enough together to appear to the eye to be overlapped and producing only one combination color.

Having an object appear the right color depends on having that color present in the illuminating light. A red object absorbs the short-wavelength (blue-violet) light. If you illuminate it with blue light, it will reflect no light and appear black. If certain wavelengths are accentuated in the illuminating light, they will be accentuated in the light reflected from the object. That's why you may have to take that coat outside the store into daylight to really find out what color it is before you buy it. Fluorescent and other artificial lights are usually not truly white light. Therefore, certain colors are emphasized more in fabrics viewed under them than if viewed under sunlight.

16.4 Polarization

You've probably seen TV commercials attempting to convince you that if you wear "polarized" sun glasses, you can see underwater or see people in front of your car that are otherwise obscured by glare. Let's examine the process called **polarization** and see if these claims have some factual basis.

Consider again, as we did in Chapter 15, waves on a rope. If you tie one end of a rope to a post and shake the other end, you can produce waves. If you shake it back and forth in all directions perpendicular to the rope, you will get wave displacements in all directions perpendicular to the rope. But suppose (as shown in Fig. 16.23), you let the midpoint of the rope pass between two vertical boards with a gap just large enough for the rope. The only waves getting past the boards will be those with vertical oscillations. We

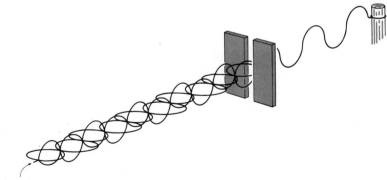

Shake this end.

Figure 16.23 Shaking the near end of the rope in all directions perpendicular to the rope gives wave displacements in all these directions. The boards polarize the waves so that only those with vertical oscillations get through.

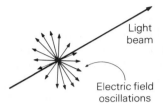

Figure 16.24 As viewed along the line of travel of an ordinary light beam, the electric field oscillates in all directions perpendicular to the beam.

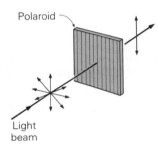

Figure 16.25 Polaroid transmits light with electric field variations in one plane, but absorbs other light. The vertical lines on the Polaroid indicate the transmission direction: The direction along which the electric field variations are transmitted.

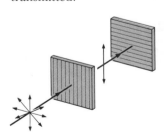

Figure 16.26 Two crossed Polaroids transmit no light.

say that the boards **polarize** the wave: that is, they transmit waves with vibration only in a single plane. We call the boards **polarizers.**

Polarization of Light by Absorption

There is a way to do the same thing to light. Imagine that there's a thin light beam coming directly toward you. From an ordinary light source, the electric field oscillations would be in all directions perpendicular to the beam, just as for the rope vibrations in front of the polarizer of Fig. 16.23. If you could see the electric field as you look toward the beam, it would be oscillating up and down, back and forth in all directions as shown by the arrows of Fig. 16.24.

In 1928 Edwin Land, while a freshman in college, invented a material he called "Polaroid" that does for light what the vertical boards do for the rope wave. (If you've taken pictures with a Polaroid Land camera, you can thank the same inventor.) When light passes through Polaroid, the light with electric field vibrations in one direction is transmitted; other light is absorbed. Figure 16.25 illustrates this point. Light with its electric field vibrations at an oblique angle to the transmission direction is only partially absorbed, leaving a fraction of that light transmitted. That fraction varies from 0 to 100% as the angle with the transmission direction varies from 90° to 0°. This fact means that, since the electric field direction of the incoming light is equally distributed in all directions, on the average half the light is absorbed. The other half is transmitted with its electric field vibrations all in one direction. If the light then falls on another sheet of Polaroid with its transmission direction perpendicular to the first, no light emerges from the second sheet (Fig. 16.26).

"Polarized" sunglasses are made of Polaroid. They therefore cut out half the light from an ordinary unpolarized light source. Try looking through two pairs of polarized sunglasses held perpendicular to each other—you get no light through (Fig. 16.27).

An electromagnetic wave, such as light, consists of both electric *and* magnetic field vibrations that are perpendicular to each other. In discussing polarization, we haven't said anything about the magnetic field variations. These, of course, exist also. A material that transmits electric field vibrations in one direction, transmits magnetic field vibrations in a perpendicular direction. Otherwise, no light would get through.

Figure 16.27 When the transmission directions of polarized sunglasses are parallel (top), you get almost no light *loss* in the second pair. With the directions perpendicular (bottom), you get *no light* through the second Polaroid.

Polarization by Scattering and Reflection

Remember that the scattering of light, for example by atmospheric particles, occurs when electrons are set into vibration by the light and then re-emit light in another direction. The electrons vibrate only perpendicular to the incident light, they reradiate most strongly perpendicular to their vibration, and the electric field oscillations of the scattered light are in the same plane as the electron vibration. Therefore, the scattered light is polarized. You can check this for yourself by looking through your polarized sunglasses at the blue sky. Rotate the glasses and notice that at certain orientations the sky looks almost black.

With polarized sunglasses, check the direction of polarization of blue light from the sky. (The polarizing direction of the glasses is vertical as you normally wear them.)

Figure 16.28 Polarized sunglasses reduce the glare by absorbing the horizontally polarized portion of the light.

The eyes of bees can detect the polarization of the light they receive. They use the direction of polarization of light from the sky to navigate, even when they are not in direct sunlight.

The reflection of light from a surface is basically the same process as scattering, except that it occurs from much more densely packed scattering centers. Therefore, light reflected from a smooth surface is partially polarized. The direction of polarization is *parallel to the surface*. The extent of polarization depends on the material, and on the angle of reflection.

The glare from the surface of roads or lakes, for example, is light from the sky reflected by the surface. It obscures what you may be trying to see *on* or *under* the surface. Since this glare is partially polarized parallel to the surface, you can largely reduce it by viewing through material that polarizes vertically. That's why polarized sunglasses have their polarizing direction vertical, and why they are more effective than nonpolarizing glasses in reducing glare (Fig. 16.28).

✳✳ Polarization in LCDs

One type of liquid crystal display (see Section 16.2) makes effective use of polarization. It's called a *twisted nematic display.* As light passes through crystals used in these displays, the molecular structure of the crystal causes the direction of light polarization to twist through 90° when there is no voltage across the crystal. A display has a sheet of crystal sandwiched between two sheets of polarizing material, oriented with their polarizing directions 90° to each other (Fig. 16.29). A mirror is behind the second polarizer. Incident light is polarized, and then has its polarization twisted so that it goes through the second polarizer, is reflected, and travels back out the front. The crystal would then appear light under external illumination.

In those regions of the crystal where a voltage is applied, the polarization is not twisted, and the light is absorbed by the second polarizer. These regions appear dark in contrast with the lighter surroundings. Look through Polaroid sunglasses at an LCD watch; then rotate it 90°. Any surprises?

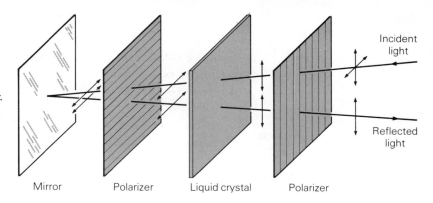

Figure 16.29 The main components of a twisted nematic liquid crystal display are shown separated from each other. Arrows on the light beam show the polarization directions. When *no* voltage is applied across the crystal, it twists the light polarization through 90°.

Mirror   Polarizer   Liquid crystal   Polarizer

Incident light

Reflected light

16.5   Interference in Light

When two waves of the same type in the same region are *in phase,* the overall wave amplitude is the sum of the two individual amplitudes (Fig. 16.30-a). However, if two waves of the same amplitude are exactly out of phase, they cancel each other (Fig. 16.30-b). We call both of these effects **interference;** the first is *constructive* interference, the second *destructive* interference. Intermediate phase differences give effects between these two extremes.

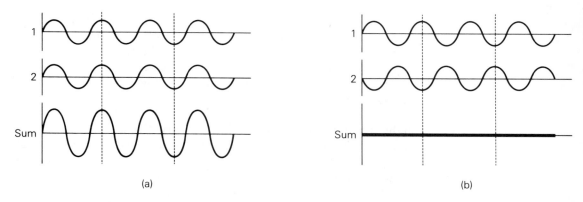

(a)

(b)

Figure 16.30 (a) Two waves undergo constructive interference if they are in phase. (b) They undergo destructive interference if they are exactly out of phase.

Interference from Thin Films

You have probably seen interference effects due to thin transparent films of some material. These effects are responsible for the

479

spectrum of colors you see in soap bubbles or when small amounts of gasoline or oil are spilled on water.

Look at what happens when single-wavelength light hits a thin film of material, such as a soap bubble. Figure 16.31 shows two

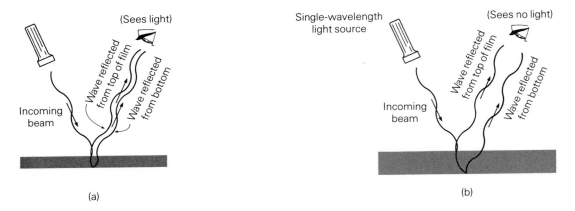

(a)

(b)

Figure 16.31   (a) If the film thickness is such that the light reflected from both surfaces is in phase, an observer sees reflected light. (b) If exactly out of phase, no reflected light is seen.

thicknesses of film, greatly exaggerated in thickness. Even though the film is transparent, some light always reflects from both the top and bottom surfaces. If the thickness of the film is such that the light reflected from both surfaces is *in phase,* constructive interference occurs and you see reflected light (Fig. 16.31-a). When the thickness is right for the reflected waves to be exactly out of phase (as in Fig. 16.31-b), the waves interfere destructively and you see no light. The film thickness for which constructive interference occurs depends also on viewing angle, since the distance traveled through the glass by the wave reflected from the bottom increases as the angle of the light beam relative to the surface decreases.

Now suppose that you have white rather than monochromatic light incident on the film. If the film varies slightly in thickness from point to point, constructive interference will occur at different points for different wavelengths, and you will see light of different colors reflected from different places on the film. Even if the film thickness is the same everywhere, you can see a spectrum of colors because of the slight variation in viewing angle from point to point.

This effect is responsible for the different colors you see in soap bubbles, "blowing" bubbles, oily films on water, and sometimes on dishes that aren't thoroughly rinsed of detergent. For this to happen, the film thickness must be no more than a few wavelengths of light.

Destructive interference gives us the nonreflecting glass that is often used, for example, in picture frames. A picture covered with ordinary glass gives a glare at some angles that inhibits seeing the picture. Nonreflecting glass is covered with a coating that has just the right thickness to give destructive interference for yellow light—the center of the visible spectrum. This interference cuts down on reflection for the part of the visible light to which the eye is most sensitive. Since other wavelengths are still reflected somewhat, the glass has a bluish hue.

16.6   **The Particle Nature of Light**

Remember that when light is emitted from a substance, *one* photon of light is given off as *one* electron goes to a lower energy state in an atom or a group of atoms. Furthermore, the energy carried away by this photon is *hf*—Planck's constant multiplied by the frequency—and is exactly the energy given up by the atom or atoms during the electron transition.

If the energy is *given up* in bundles as photons, you might expect that the energy *carried* by light exists in discrete photons, or *quanta*. This would mean that light, in some ways, would act as if it were made up of particles. In fact, certain experiments bear out this idea. One of the first processes to substantiate the *particle nature* of light was the **photoelectric effect,** which we'll now discuss.

The Photoelectric Effect

When light hits certain solids, electrons are given off, as illustrated in Fig. 16.32. This phenomenon is called the *photoelectric effect.* The number of electrons emitted from a particular substance depends on the intensity and wavelength of light, but the exact way these properties influence the number is quite surprising if you're thinking of light as a wave.

If electrons are to be knocked out by absorbing energy from a light wave, you'd probably expect to knock out an *occasional* electron, regardless of the light wavelength, even for low intensities. You'd probably also expect higher-energy electrons for higher-intensity light.

But what is observed in experiments is somewhat different. First, if the wavelength is larger than some particular value—that

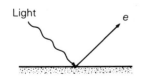

Figure 16.32 In the photoelectric effect, light incident upon a solid surface ejects electrons (e).

is, the frequency is less than some particular value—no electrons are ejected, no matter how intense the light. The particular wavelength depends on what the substance is. For copper, for example, the longest wavelength is 272 nm (frequency 1.1×10^{15} Hz). Second, if the wavelength is less than that particular amount, you immediately get *some* electrons even in extremely dim light. Furthermore the *number* of electrons increases with increasing intensity, but the *energy* of the electrons *does not*.

In 1905 Albert Einstein was able to explain these effects which seemed rather odd at that time. It was for this discovery that he won the Nobel Prize in 1921. Based on Max Planck's earlier discovery that light is *emitted* in discrete quantities, Einstein interpreted the photoelectric effect to mean that light also *propagates* as discrete quanta, which we now call *photons*, each with energy *hf*. He pointed out that *each photon acts independently,* and he applied the principle of conservation of energy to each photon as follows:

energy of photon = (energy to remove electron) +

(kinetic energy of emitted electron),

or, in symbols,

$$hf = \phi + \mathrm{KE}.$$

In other words, a given photon carries energy *hf*. When it strikes the surface, a certain amount of energy goes to remove the electron from the substance. The *minimum* energy needed to remove an electron depends on the substance, and is called the *work function* ϕ. The energy left over goes to kinetic energy KE of the electron. When the photon frequency f is so low that its energy is less than the work function, no electrons are emitted regardless of the intensity of the light. If the energy is high enough to eject electrons, increasing the intensity—increasing the number of incoming photons—gives more emitted electrons, but *not* more kinetic energy for any one electron.

There are lots of practical applications of the photoelectric effect. One widely known application is the so-called "electric eye" (Fig. 16.33). Light incident on the photosensitive surface of a *phototube*, or *photocell*, ejects electrons. As long as the light shines on the surface, electric current can flow in the circuit containing the battery, the phototube, and the electromagnetic relay. When the light stops, no current can get across the phototube and thus the electromagnet is not activated. The relay can either close or open a switch that controls some other device.

A common application is in automatic door openers for supermarkets and other stores. A light beam shines across the entrance in front of the door and hits the phototube. When a person

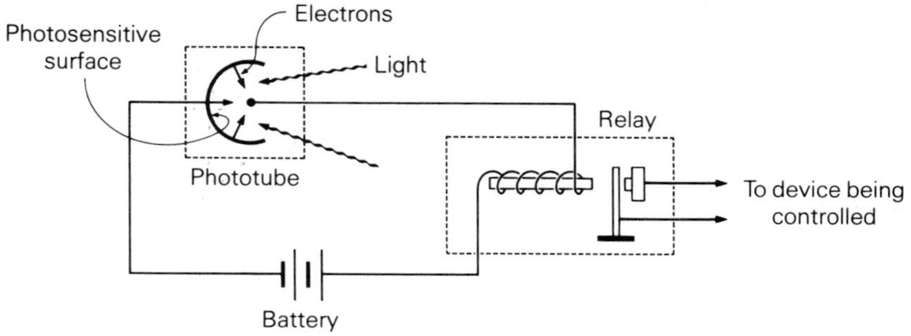

Figure 16.33 A circuit diagram showing how a phototube—an "electric eye"—can open and close a relay controlling some other device.

walks through the beam, current no longer flows in the phototube, and the relay closes the switch that opens the door. In a similar way, some burglar alarms use phototubes sensitive to nonvisible ultraviolet light. When the burglar passes through the light beam, the alarm sounds. Photocells also form the heart of photographic exposure meters, and they turn the street lights on when the sun goes down. In one type of smoke detector, photocells detect smoke particles that scatter light; and they serve as the eyes of a television camera by converting the optical image focused onto many tiny photocells into an electrical signal for transmission to your home.

16.7 The Dual Nature of Light

What goes on here? Some sections of this chapter went to great lengths to try to convince you that light is a wave. It has all the characteristics that waves have; it does all the things that waves do. Yet in other places we talked about light being emitted from sources in discrete quanta, and propagating as individual photons. In the last section, we even found that it sometimes interacts with matter as if it were made up of particles. The question arises: Just what is light? Is it a wave, or is it made up of particles? Surely it has to be one *or* the other. The properties of waves are not compatible with those of particles.

The debate among physicists over the true nature of light raged from Isaac Newton's time until the beginning of the twentieth century. Newton believed light to be made up of particles. But in the nineteenth century people accepted the wave nature because light beams were found to exhibit constructive and destructive interference, and only waves do that. In the early twentieth century we found light to be doing things—the photoelectric effect for example—that waves cannot do but particles can. The nature of light seemed to depend on the type of experiment being done—a rather disconcerting situation.

Physicists finally had to accept the idea that light has a **dual** nature: **both wave and particle.** In those situations where light

483

interacts with individual atoms, it acts as a *particle;* when light interacts with macroscopic quantities of matter, it acts as a *wave.* No experiment will show both the wave and particle nature at the same time. These particle and wave ideas are merely *models* that represent certain aspects of light but do not alone tell the whole story of its nature.

Key Concepts

Light is a **transverse electromagnetic wave** with wavelengths in air between 400 and 700 nm. Its propagation speed in air is almost the same as in vacuum: 3.00×10^8 m/s. The color of light depends on its frequency, and is the brain's interpretation of the eye's response to a particular frequency.

Light is emitted in discrete wavelengths from gas atoms as they give up energy. One **photon** is emitted as one atom de-excites. The *energy* of the photon is Planck's constant h multiplied by its frequency f, and exactly equals the energy lost by the atom. A photon is sometimes called a **quantum** of light; the light energy is **quantized** in units of hf. An atom cannot emit light if it is in its **ground state**—the lowest energy state. Atoms can be raised to **excited** states by collisions with electrons or other atoms and by absorption of photons of energy exactly equal to the change in the atom's energy.

Fluorescence is the process in which an atom absorbs radiation of one wavelength and emits radiation of longer wavelength (lower energy photons).

In solids and liquids, electrons can exist in a *continuous range* of energies. **Incandescence** is the emission of light from such a substance. Since an electron in a solid or liquid can lose any fraction of its energy, the light emitted has a continuous range of wavelengths.

The **luminous efficiency** of a light source is the ratio of light output rate in *lumens* to power input in watts.

A **laser** is a source of *single-wavelength, coherent* light produced by *stimulated emission.*

The **color** of a light source is determined by the wavelengths emitted. The color of illuminated objects is determined by the wavelengths of the light reflected, scattered, or transmitted, but not absorbed.

Light is **polarized** if the electric field oscillations are all in the same direction. Light can be polarized by absorption, as in Polaroid, and by reflection and scattering.

Light waves can **interfere** *constructively and destructively,* depending on whether the waves are in phase or out of phase, respectively.

Light has a *particle nature,* in that its energy is carried in photons of energy *hf.* This nature is demonstrated by the *photoelectric effect.*

Even though light has a **dual** nature, both wave and particle, in any one event it can act *either* as a wave *or* a particle, but not both.

Questions

Light Waves

1. The first known attempt to measure the speed of light was made by Galileo. He and a partner located themselves on hilltops about a mile apart. Each had a lantern. One would uncover his lantern. When the second person saw the light he would uncover *his* lantern. The idea was that the time between the first person uncovering his lantern and seeing the light from the other lantern would be twice the time for the light to travel from one person to the other. Why do you think this attempt failed?

2. Rank the following forms of electromagnetic radiation in order of increasing energy of its photons: red light, x rays, FM radio, AM radio, ultraviolet radiation, infrared radiation, green light. How would the order change if you ranked them in order of increasing wavelength?

Light Sources

3. One of the end-of-chapter references in this chapter says that "all light is the radiation of surplus energy by some substance." Do you agree with this statement?

4. How does the energy needed to excite an atom from state E_1 to state E_2 compare with the energy of a photon given off when the atom de-excites from E_2 to E_1?

5. Why do the heating elements of an electric range get red when they get hot? Does the light contain only one wavelength?

6. Suppose your electrical power company imposes a "brown-out"—a reduction in operation voltage to reduce power use during peak-load periods. What would that do to the color of the light from your incandescent bulbs?

7. When you strike certain kinds of stones together you get sparks. The sparks are particles of stone knocked off that are heated by the collision. Do you expect the light from the sparks to have a continuous range of wavelengths, or to consist of discrete wavelengths?

8. Is the light from the flame of a gas range due mostly to atomic and molecular de-excitation or to incandescence, as in a candle flame? (Hint: What colors dominate most incandescent objects?)

9. As advertisements, businesses sometimes give away little yellow strips (containing their ad) that you can tape to your telephone. The strip glows in the dark to show you where the phone is after the lights are off. How does it work? (Some clock dials glow in the dark for the same reason.)

10. A laser is a device for "light amplification." It runs on electrical energy. Does this mean you get more light energy out than electrical energy put in?

Seeing and Color

11. We can speak of *single-wavelength* red, green, or blue light, for example. Can we legitimately speak of *single-wavelength* white light? Why?

12. If a particular type of photographic film has a light sensitivity that makes it give correct colors when the pictures are taken in daylight, why are the colors in the picture not correct if taken under ordinary incandescent or fluorescent lights?

13. Suppose that when you look at a red traffic light and then a green traffic light, the same amount of light *energy* entered your eye per second from each. Did your eye receive more photons per second from one light than the other? If so, which?

14. If sunlight is white, why don't you see *white* when you look through a region of sunshine? What does it mean to "see" something?

15. What color is a red shirt illuminated in red light? in blue light?

16. Do you see a black cat because the light coming from it is black? What does it mean to say that a shirt is white?

17. Why do some of your pants appear to have a strange color under city street lights?

18. The technical crew for a stage performance can change the color of the performers' clothes in the middle of a performance. How do they do this? What is the physics of this change?

19. Does a red rose petal absorb long-wavelength or short-wavelength visible light more effectively?

20. Considering that we see objects that aren't light sources only by reflected light, explain why during the day it's hard to see the inside of a room through a window from the outside, even if the window is open. Remember, light can only get to the room through the window, and is likely to have many reflections inside the room.

21. Explain why a black horse gets hotter in sunlight than a white one. What is the relative hotness of a brown horse?

22. Think about the cause and effect stated in Question 21. Should we really say that an object is dark-colored because it gets hotter, rather than gets hotter because it's a dark color?

23. Deep clear ocean water gets its color mostly from incident light scattered by water molecules. As with scattering by air, short-wavelength light scatters more effectively. What color do you expect (and observe) clear ocean water to be?

24. At some beaches, clear water appears very green, rather than what you might predict in response to Question 23. This coloring is thought to be due to fluorescence in certain organic substances concentrated in the water. Explain how this might occur in the same way that the coating on a fluorescent bulb gives off light.

25. My wife, a member of a relatively rare species called *native* Floridian, likes to impress me when we go to the beach by estimating the depth of the water and the location of sandbars by the color of the water. At least three factors are different where the water is shallower: the bottom is nearer the water surface and hence likely to scatter more incoming light; more solid particles are stirred up in the water in these regions; and the nature of the waves and ripples on the surface varies from that in the deeper regions. Explain how all three of these can influence the water color.

26. If the earth had no atmosphere, what color would the sky be? Does the moon have a blue sky?

Polarization

27. If you have two pairs of sunglasses, how can you tell if they're both polarized? (Assume they are not marked as such.)

28. When you wear polarized sunglasses, why can you often see the following more clearly:

  a. A person underwater when you look from above?

  b. Cars or obstacles on a highway when the sun is up but somewhat in front of you?

  c. The true colors of water at the beach?

  d. The scene through the back window of a car you're following?

29. If you have papers on the dashboard of your car, their reflection in the windshield sometimes makes it hard to see what's ahead of the car. Why does wearing polarized sunglasses reduce this interference? Why is the dashboard of most cars a dark color?

30. Viking sailors of the ninth century navigated their ships during the day by the position of the sun. By using "magical sun stones," they reportedly could locate the sun in any weather (see *Time,* July 14, 1967, p. 58). A natural material found in Scandinavia called cordierite has polarizing properties very similar to those of Polaroid. From the fact that sunlight scattered from the atmosphere is polarized, can you formulate a theory of what these "sun stones" were and how they were used?

Interference

31. Some people mount their color slides between two pieces of glass for protection of the slides. When these are projected on a screen, why do you sometimes see curvy lines with all the colors of the rainbow superimposed in places on the picture?

32. Oily spots on a wet road often show all the colors of the visible spectrum. Why?

Photoelectric Effect

33. Many cameras have built-in exposure meters that set the diaphragm opening to let the proper amount of light reach the film. Figure 16.34 is a diagram that shows how a photocell based on the photoelectric effect can be used to indicate the intensity of incoming light. Explain why there is a current through the meter that is proportional to the light intensity.

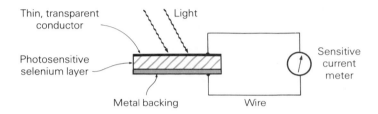

Figure 16.34   Diagram of a light exposure meter.

34. Draw a schematic wiring diagram for a circuit that could use a photocell (such as shown in Fig. 16.33) to turn on an outdoor light after sunset.

35. The sound track of a movie film is a band on the side of the film that varies in darkness in proportion to the pressure variations of the sound wave it represents. When a bright lamp shines through the track, the transmitted light varies in intensity in the same proportion. What phenomenon that we've studied in this chapter could be used to monitor these intensity variations and convert them to electric current variations, which can be amplified and used to operate a speaker system?

36. Draw a schematic diagram for the steps described in Question 35.

37. Sunburn occurs when photons of radiation cause chemical changes that produce cell damage in the skin. Which *one* of the three causes sunburn: infrared, visible, or ultraviolet radiation? Justify your choice.

38. If an electric eye is used to open a supermarket door when a person walks through a beam of white light, could you make the door open faster if you used a beam of blue light?

Problems

1. The *Apollo* astronauts placed a special reflector on the moon. A laser beam was aimed at this reflector and reflected back to earth. The time between the sending and return of a light pulse gave an accurate measure of the distance to the moon. If, at that instant, the reflector on the moon were 380 000 km away, what was the time interval between sending and receiving the light pulse? (See Question 1.)   **Ans:** 2.5 s

2. What is the frequency of the 640-nm wavelength component of the red light of a neon sign?   **Ans:** 4.69×10^{14} Hz

3. From the energies of the states in Fig. 16.6, verify the wavelengths shown for light emitted in E_3-to-E_2 and E_5-to-E_2 transitions.

4. How many photons per second are sent out by an FM radio station transmitting 50 000 W at a frequency of 100 MHz?

   Ans: 7.5×10^{29}

5. How much kinetic energy must be given up by an electron to excite a hydrogen atom in the ground state so that it can emit 656-nm light?   **Ans:** 12.09 eV

6. If you burn table salt, the sodium will give off yellow light of wavelength 589 nm. What is the energy difference between the energy levels in sodium that give rise to this light?

   Ans: 2.11 eV

7. From Fig. 6.12, estimate the energy of the photons most often emitted from an incandescent light bulb whose filament operates at 3000°C. What is the energy of the most abundant photons from a "red hot" fireplace poker at 700°C?

   Ans: 1.4 eV; 0.4 eV

8. The *absolute* efficiency of a 100-W incandescent bulb is 2.5%. How much light energy is produced in 8 h of operation of such a bulb?   **Ans:** 20 Wh

9. Based on the absolute efficiency given in Problem 8, how many visible photons are emitted per second from a 100-W bulb? Take the average wavelength of the visible radiation to be 560 nm.   **Ans:** 7.0×10^{18}

10. If you convert your 100-W kitchen incandescent light to a 40-W fluorescent, how much more light will you get? How

much electrical energy will you save per month if you use the light for an average of 6 h per day?   **Ans:** 46%; 11 kWh

11. A *helium-neon* laser emits light of wavelength 633 nm. What is the energy of the photons emitted from such a laser?

    Ans: 1.96 eV

12. About 4.8 eV of energy are needed to carry out a photosynthesis reaction in a plant. If eight 700-nm photons are needed to supply the energy for the reaction, what is the efficiency of light use in this process?   **Ans:** 34%

13. If the speed of light in water is ¾ that in air, and that in window glass is ⅔ that in air, what are the wavelengths of light of frequency 5.0×10^{14} Hz in air, water, and glass?

    Ans: 600 nm; 450 nm; 400 nm

14. Suppose a sheet of glass has a coating on the front in which light has a wavelength of 0.70 of that in air. How thick should the coating be to minimize the reflection of 550-nm light—that to which the eye is most sensitive?   **Ans:** 190 nm

15. If you were designing a light exposure meter to be sensitive to 550-nm light, what is the largest work function the photosensitive material in the photocell could have?   **Ans:** 2.2 eV

Home Experiments

1. View the spectrum of a white light source with a water prism. Put a dressing table or shaving mirror at an angle underwater in your wash basin as shown in Fig. 16.35. Back away from the mirror and hold a flashlight near your head so that you can see its reflection transmitted through the water as shown. The

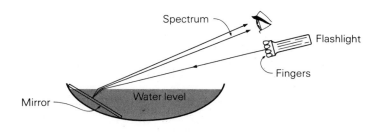

Figure 16.35 A water prism.

situation is like half of the prism of Fig. 16.4, with a mirror in the middle so that light is transmitted back through the front as it would be through the second half of the prism. One side of the reflected flashlight beam will appear bluish, the other reddish. Put your hand over the flashlight with a small *horizontal* crack between your fingers. You'll then see the entire spectrum of the white light that emerges through the crack. Why does the crack need to be horizontal? The farther back you are, the more the spectrum spreads out. You might see a spectrum from other light sources or reflectors that are parallel to the water surface: for example, the reflection of the ceiling light from a shower-curtain rod.

2. If you have a strong magnifying glass, look closely at the screen of an operating color TV. You can see that the screen is covered with small dots that are red, blue, and green. Any color is formed by different intensities of these dots. If you can get a stationary picture, such as a test pattern, you should be able to predict which dots are most intense for particular colors, and then test your prediction.

3. Look through polarized sunglasses at the glare from various horizontal smooth surfaces. Notice the difference without the sunglasses, and when they are rotated through 90° from the normal orientation. Compare the effect with sunglasses that are merely darkened glass.

4. Add a couple of drops of milk to a clear glass of water. Shine a flashlight beam through the glass. Looking from above at 90° to the beam, you see the beam by the light scattered from the milk particles. Look at the scattered light through polarized sunglasses. Then rotate the glasses 90° and notice the change in the amount of light you see. Put the sunglasses between the flashlight and milky water, rotate through 90°, and observe the effect on the amount of light scattered. Reconcile what you observe with the polarization expected from light scattering. Notice the bluish tint of the scattered light. Relate your results to the color of the sky.

5. Mix up a very soapy solution with dishwashing detergent, so that bubbles form on top. Look at the bubbles under white light. They first appear clear, then with all the colors of the visible spectrum, then clear again within a fraction of a minute. These effects happen because the film over the bubbles starts relatively thick, and then gets thinner with time. About how thick is it when you see the rainbow colors?

References for Further Reading

Goswami, A., *The Concepts of Physics* (D.C. Heath & Co., Lexington, Mass., 1979). Chapters 19 and 20 are an interesting, readable account of Einstein's theory of relativity.

Hafele, J. C., and R. E. Keating, "Around-the-World Atomic Clocks: Predicted Relativistic Time Gains," *Science,* 14 July 1972, p. 166. "Around-the-World Atomic Clocks: Observed Relativisitic Time Gains," p. 168. These two papers discuss why flying and nonflying clocks should show different times, and the results of the experiments to test this phase of relativity theory.

Hewlett-Packard Optoelectronics Division Staff, *Optoelectronics Applications Manual* (McGraw-Hill Book Co., New York, 1977). Discusses light emitting diodes and other optoelectronic devices, starting with a readable introduction and progressing to considerable detail.

Kock, W. E., *Radar, Sonar, and Holography: An Introduction* (Academic Press, New York, 1973). Parts of the book discuss the making and properties of holograms.

Küppers, H., *Color* (Van Nostrand Reinhold Ltd., London, 1972). Describes qualitatively the various aspects of color, beautifully illustrated with color plates throughout.

Meier, G., E. Sackmann, and J. G. Grabmaier, *Applications of Liquid Crystals* (Springer-Verlag, New York, 1975). Properties and applications of liquid crystals.

Minnaert, M., *The Nature of Light and Color in the Open Air* (Dover Publications, Inc., New York, 1954). The physics of hundreds of everyday observations on light.

Walker, J., "The Physics and Chemistry Underlying the Infinite Charm of a Candle Flame." The Amateur Scientist Column, *Scientific American,* April 1978, p. 154. An interesting description of the various processes going on in a candle flame, as well as experiments to support this description.

Walker, J. "The Bright Colors in a Soap Film Are a Lesson in Wave Interference," The Amateur Scientist Column, *Scientific American,* Sept. 1978, p. 232.

Wehner, R. "Polarized-Light Navigation by Insects," *Scientific American,* July 1976, p. 106. An interesting application of the polarization of light.

17

Light Rays and Optical Devices

We have talked about the wave nature of light, and its behavior as a result of that nature. We've also talked about the particle nature of light: The fact that its energy is carried in discrete bundles called photons. There is another way of describing much of the behavior of light that doesn't involve the details of either light's wave or particle nature. This viewpoint merely treats light as *rays* that travel in straight lines in a given medium. We usually represent these rays by lines with arrows on them to show the direction of light travel. At the boundary between two media, light usually changes direction somewhat. That change in direction is described by certain rules that were determined from experiments hundreds of years ago.

The study of light from the viewpoint of rays, we call geometric optics. In this chapter, we will be using that viewpoint. We will be concerned mainly with two general effects: the *reflection* of light by smooth surfaces such as mirrors; and the *refraction*, or bending, of light rays as they go from one medium to another. We'll see how lenses, that depend on refraction effects, focus light in optical instruments.

17.1   Reflection

The reflections of light that you see probably range from those created for you by nature (Fig. 17.1) to the delightful sight you see

Figure 17.1 One of nature's own mirrors.

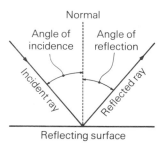

Figure 17.2 When light reflects from a surface, the angle of reflection equals the angle of incidence. The *normal* is perpendicular to the surface.

in your bathroom mirror each day as you prepare to face the world. Light reflections follow one basic **law of reflection** illustrated in Fig. 17.2: *The angle of reflection is equal to the angle of incidence.* By tradition, we usually measure angles relative to the **normal**—a line perpendicular—to the surface rather than to the surface itself.

Most surfaces are rough, as far as light reflection is concerned. That is, the surface irregularities are large compared with the wavelength of light. The law of reflection is obeyed at any one spot, but a parallel beam of light is reflected in all directions from such a surface, as Fig. 17.3(a) shows. That's the way you *see* the object—by *diffuse* reflection.

On the other hand, if the surface is smooth enough that surface irregularities are small compared with the wavelength of light, *regular* reflection occurs, as illustrated in Fig. 17.3(b). There, the entire beam is reflected so that the angle of reflection equals the angle of incidence. Many surfaces—for example, a shiny red apple—give both diffuse and regular reflection. Our remaining

Figure 17.3 (a) The scattering from most surfaces is *diffuse*. (b) If the surface is smooth, the scattering is *regular*.

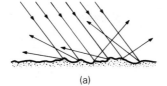

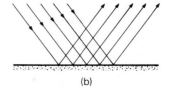

(a)   (b)

discussion will be concerned with regular reflection, the kind you get from a mirror.

Reflection from Plane Mirrors

When you look at yourself in a mirror, the face you see is an *image* that appears exactly the same distance behind the mirror as your face—the *object*—is in front of the mirror. To see that this statement is generally true for a **plane** (flat) **mirror,** imagine that you're looking at the reflection of a flower, as in Fig. 17.4. Concentrate for now on the light you see coming *only* from the center. The light that gets to either eye is reflected so that the angle of reflection equals the angle of incidence. Drawing the rays so that this situation is true, we see that the rays to each eye are at slightly different angles. As the dashed lines show, the rays appear to be coming from a point behind the mirror a distance equal to the distance from the mirror to the center of the flower. We call this point the **image.** If you draw rays similarly for other points on the flower, you can see that the image of the entire flower is as shown. The point from which the rays originate *before* hitting the mirror, we call the **object.**

Figure 17.4 The image of an object—the point from which the rays appear to come—viewed in a plane mirror is the same distance behind the mirror as the object is in front. The mirror reverses the sides: Notice the missing petal.

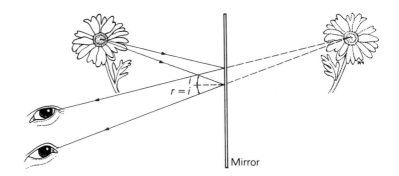

$r = i$

Mirror

Notice that the missing petal is on the *left* side of the *object* but is on the *right* side of the *image*. The "mirror image" is reversed relative to the object. If you draw ray diagrams for each side, you can see why this reversal happens. That's why you might think your photographs don't look quite like you. The left side of your face

Figure 17.5 (a) The writing on the ambulance is reversed. (b) The writing appears in the right order when looked at through a rear-view mirror. (Photographs by Loren Long.)

may be slightly different than the right, or maybe you part your hair on the side. In the mirror you see things reversed from the way others see you, and from the way you appear in a photograph.

Writing is reversed when viewed through a mirror. The writing on the front of some vehicles is backwards (Fig. 17.5-a) so that when viewed through a rear-view mirror (Fig. 17.5-b), it appears in the right order.

We can use the law of reflection to find out what size a mirror needs to be for proper viewing. Let's find the length of the shortest mirror in which Sandy Allan, the tallest living woman at 232 cm, can see herself completely. Suppose the mirror is mounted vertically on her bedroom wall in Shelbyville, Indiana, and she stands so that the tips of her toes are directly under her eyes. Assume her eyes are 12 cm below the top of her head. Fig. 17.6 shows the situation.

Light from any point that gets to her eyes travels so that the angle of reflection equals the angle of incidence. Light from her toes hits the mirror at a level halfway between the floor and her eyes. Light from the top of her head hits halfway between head level and eye level. Light from other points hits between these two extremes. The minimum length needed is then 110 cm for viewing from the eyes down, plus 6 cm for viewing from the eyes up, or a

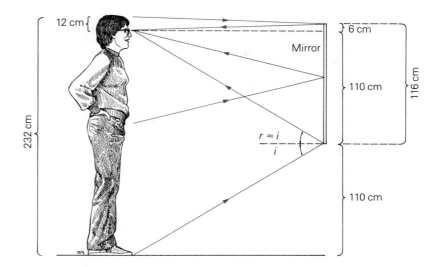

Figure 17.6   If placed vertically at the right height, a mirror needs to be only ½ the height of the person for a full view. For a given ray, the angle of reflection *r* equals the angle of incidence *i*.

total of 116 cm: ½ her height. The mirror needs mounting with the bottom 110 cm from the floor: halfway between floor and eye level. Using a "floor-length" mirror merely lets her see more of the floor and other surroundings—to give added perspective—but doesn't show any more of her body.

Question   (1) If Sandy's size 19½ EEE shoes extend forward of a point directly below her eyes, how will this change the minimum length the mirror must be to show her entire shoes?

(2) Does the rear-view mirror in Fig. 17.5(b) need to be ½ as large as the front of the ambulance for the driver to see the entire ambulance front? (Hint: The ambulance is farther from the mirror than the viewer's eyes. Draw ray diagrams to verify the answer you know from experience.)

Reflection from Curved Mirrors

Even though plane mirrors are more common, we do use curved mirrors as magnifiers (Fig. 17.7), as reflectors for various kinds of lights, as rear-view mirrors for trucks (Fig. 17.9-a), and in many other ways.

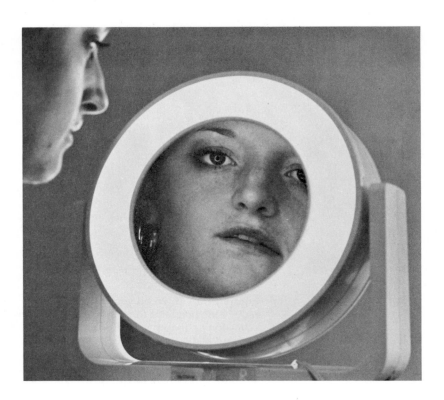

Figure 17.7 Dressing-table magnifiers are curved mirrors.

First we'll find out how a curved mirror can act as a magnifier. Suppose we put our flower near the concave side of a mirror with a slight spherical curvature. Figure 17.8 shows two rays from the center of the flower, reflected so that the angle of reflection at the point of reflection equals the angle of incidence. After reflection, these rays diverge as if they were coming from the center of the dashed image. *Any* ray from the center of the flower that makes a fairly small angle of incidence on the mirror will reflect so that it appears to come from the same point on the image. (Notice that the normal to any point on the mirror passes through the center of curvature.) Drawing this diagram as described locates the image of the center of the flower. Doing the same thing for other points on the flower, locates the entire image. We see that the curved mirror causes the image to be *magnified:* The image appears larger than the object. This is the principle of the magnifying mirror (Fig. 17.7), and it works whether you're viewing a flower or your own face.

If you draw lots of ray diagrams such as the ones in this figure, you can demonstrate that for any object that is small compared with the radius of curvature of a spherical mirror, and that is lo-

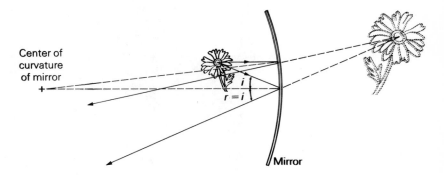

Figure 17.8 A ray from any point on the object reflects according to the law of reflection. The apparent intersection of two such reflected rays locates the image of that point. Doing the same thing for other points on the object shows that the overall image is magnified.

cated anywhere between the mirror and ½ the distance to the center of curvature, you get a magnified image.* This fact can be shown mathematically, using a little geometry and trigonometry, but we'll settle for the ray-drawing approach here.

Exercise   Draw diagrams such as Fig. 17.8 for an object near the mirror and for an object *almost* ½ the distance to the center of curvature. Which way gives more magnification? If you have a magnifying mirror, check your answer by trying it.

Sometimes it's useful to have a curved mirror that reflects from the convex side. Figure 17.9 illustrates one such application.

Figure 17.10 shows why the convex mirror gives a greatly expanded field of view. Light from an object at point O appears after reflection to be coming from point O'. That from an object at point P appears to come from point P'. Each ray obeys the law of reflection. The image is reduced in size and badly distorted, but you can tell that the object is there.

---

*To do this, you can measure with a protractor the angle between the incident ray and the normal where the ray hits the mirror. The normal is a straight line from that point on the mirror through the center of curvature. Then draw the reflected ray at the same angle on the other side of the normal. The seeming intersection of any two such rays locates the image.

Figure 17.9 A rear-view mirror used by trucks. (a) The plane mirror reflects light toward the driver's eyes from only one direction, showing the driver what is behind as seen directly beside the trailer body. The small spherical mirror at the bottom shows a greatly expanded, though distorted, view of what is nearby. (b) The actual location of the car that shows up in the spherical, but not the plane, mirror. With this mirror the driver can see cars, or rather the images of cars, that are almost beside the cab. (Photographs by Loren Long.)

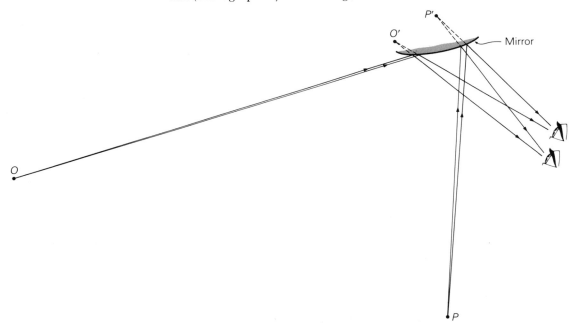

Figure 17.10 A convex spherical mirror reflects light from a broad range of angles into the direction of the viewer. The image of point O is at point O'; that of point P is at P'.

17.2   **Refraction**

When light traveling in one medium enters a new medium in which its speed differs, some rather strange things can happen. One example is shown in Fig. 17.11(a). This photograph is of *one* fish, but because of the shape of the boundaries of the different media transferring the light, it can look like *two* fish, one larger than the other. You may have noticed some other strange-looking effects: the apparent bending or separation of a straw at the surface of a transparent drink (Fig. 17.11-b); the short squatty appearance of a person standing underwater; and the shallower look of the far end of a swimming pool, no matter which end is deeper. Some of the things that happen when light enters a different medium are fortunate: for example, one of them gives you the ability to see, as we'll soon describe.

Figure 17.11   Because of the refraction of light rays, (a) the *one* fish in this photograph looks like *two;* (b) straws seem to do strange things.

To see why light should bend as it enters a medium in which its speed is different, consider an analogy. Suppose you're coasting on skates down a smooth sidewalk, and you go off the sidewalk into the grass at an angle so that one skate goes into the grass before the other. The skates meet more resistance, and therefore move slower, in the grass. Since one skate slows down before the other, you will gradually change direction until the other skate gets into the grass. You then move in a fixed *new* direction. The analogy isn't perfect, because a given skate doesn't instantaneously change speed at the grass boundary and then move at constant speed thereafter. But you can imagine what would happen if it did.

The bending of a light beam in entering a new medium is similar. Figure 17.12 shows parallel light rays incident on glass from air. The light's speed in glass is about ⅔ that in air. The lines perpendicular to the rays are *wavefronts,* representing crests in the wave. A crest of ray *A* gets to the glass before one of *B*. It slows down, causing the wavefront to bend at the boundary. Once the entire crest has reached the glass, it moves off in the new direction.

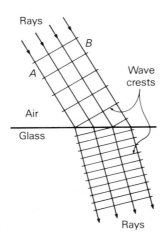

Figure 17.12   Light rays slow down as they enter the glass, so that the part that enters first is retarded relative to the other side of the beam. This retardation causes the beam to bend, or *refract.*

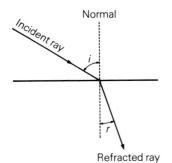

Figure 17.13   We usually measure angles relative to the normal. The angle *i* is the angle of incidence; *r* is the angle of refraction.

We call this bending due to speed difference **refraction.** It can occur in any kind of wave. In discussing refraction of sound we considered the gradual bending caused by a continuous variation in speed with position, rather than an abrupt change at a boundary.

As we did in considering reflection, we usually specify the angle a ray makes with the *normal* to the surface, as shown in Fig. 17.13. When the ray hits perpendicular to the surface, there is no bending. The greater the angle with the normal, the more the ray is refracted. For a given angle, the amount of bending increases with increase in the ratio of initial to final speeds.

We've been discussing the case where the speed in the second medium is lower than in the first. In that case the ray bends *toward* the normal: The angle of refraction is *less* than the angle of incidence. If the speed in the second medium is higher, the ray bends *away* from the normal.

As an example, we'll see how refraction causes a pool to seem shallower at the far end, no matter which end is deeper (Fig. 17.14). For light from the far end to get to the observer, it has to bend so that it appears to come from just below the water surface.

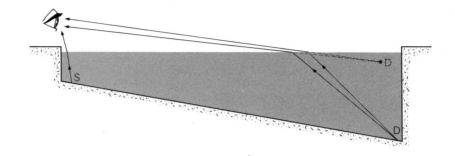

Figure 17.14 The far end of a pool appears shallower than the near end, no matter what its depth.

Exercise   By drawing a diagram like Fig. 17.14, show why a person standing underwater in a pool looks much shorter to someone on the side than he actually is. That is, show that light from his feet seems to come from a point higher than his feet.

Total Internal Reflection: Fiber Optics

Imagine you are under water with a flashlight, and you shine the beam upward. You shine it at various angles between straight-up and about 60°, as shown in Fig. 17.15. At each angle some light is reflected and some transmitted. As the incident angle of the beam

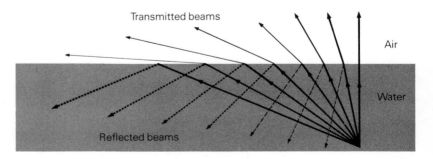

Figure 17.15   Light entering a medium with higher speed of travel bends *away* from the normal. When the incident angle between ray and normal exceeds that for which the refracted ray travels along the surface (about 48° for water), the light is totally internally reflected.

increases, the fraction of the beam that is transmitted decreases, and it bends farther from the normal as it emerges. Eventually (at about 48° in water) the emerging ray is bent right along the surface. For larger angles, the ray cannot get out and is totally internally reflected at an angle equal to the angle of incidence. We call the angle of incidence beyond which total internal reflection occurs, the **critical angle.**

Total internal reflection has many practical applications. In a prism, it can reflect light back parallel to its original direction (Fig. 17.16). This process comes in handy in instruments such as binoculars and periscopes.

Figure 17.16   Total internal reflection can be used in a prism to reverse the direction of a light beam.

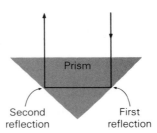

A more recent development that is probably going to have far more impact on our lives is the use of total internal reflection in a field called **fiber optics.** Suppose you have a thin glass fiber bent into a curved path as shown in Fig. 17.17. Light entering one end is totally reflected at various points inside the fiber and eventually emerges from the other end. Bundles of these fibers can be made flexible enough to curve around in fairly sharp bends.

One important use of fibers is in medicine (Fig. 17.18). Another is in communication: television, telephone, etc. It is likely that, within a few years, optical fibers will take over a large part of the job now done in communication by metal wires. Rather than sending a signal down a wire by means of an electrical current, we can send it down a fiber by means of a pulsed light beam—the message being coded in the pulses. A receiver then decodes the signal back into a picture or sound or whatever is being transmitted. Because of the extremely high frequency of light waves, the beam can be pulsed at a very high rate. Systems under test carry about 45 million pulses per second, a rate sufficient to carry 672 one-way voice signals. Therefore, a cable of 24 fibers, separately pulsed, can carry about 8000 two-way telephone conversations—vastly more than can be carried in the same sized copper wire conductors.

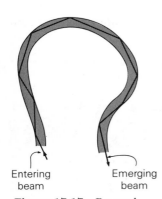

Figure 17.17   By total internal reflection in a small glass fiber, a light beam can be bent in any direction.

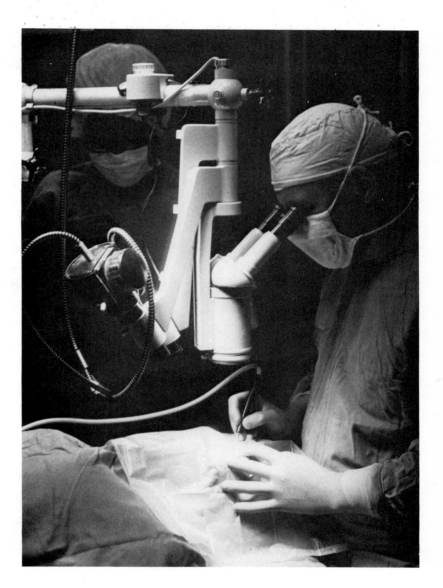

Figure 17.18   Fiber optics is being widely used in microsurgery.

There are still a few problems to be solved before fibers take over the job. One is long-distance transmission without too much loss in light intensity. Another is the problem of connectors: It's easy to plug a telephone into a jack and have it work reliably after many pluggings and unpluggings; a comparably reliable and sturdy optical connector hasn't been invented yet.

Dispersion and the Rainbow

We mentioned in Chapter 16 that, when white light passes through a prism, it *disperses* into its various colors (see Fig. 16.4). We can now see why. Figure 17.19 shows a beam of *yellow* light incident on a glass prism. The incident beam bends toward the normal in entering the glass, which means it bends downward in the figure. When the ray reaches the other side, the slope is such that, in bending away from the normal, the ray again bends downward.

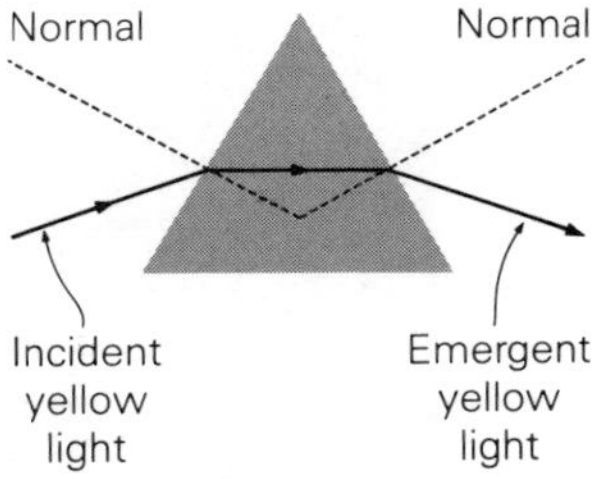

Figure 17.19
Single-wavelength light incident on a transparent prism refracts downward at both surfaces.

Question   What happens to a beam of yellow light that hits at an oblique angle on a piece of glass with *parallel* sides?

Since the speed of light depends on its wavelength, the amount of refraction depends on the wavelength. Therefore, white light breaks up into its component colors as in Fig. 16.4.

Something similar happens when you see a rainbow. A rainbow appears when the sun shines on drops of water in the atmosphere (Fig. 17.20). A ray of white light enters a water drop at the point shown, and is refracted at the surface. Each color is refracted a slightly different amount. The rays partly reflect at the back of the drop, and refract as they leave. The violet light leaves at about $40°$

Figure 17.20   A rainbow is formed by the refraction, reflection, and refraction of sunlight in many, many water drops. The various colors are dispersed in different directions because they refract different amounts.

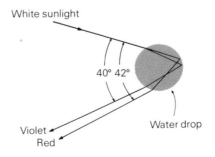

to the incoming beam, the red at about 42°. Other colors are in between. Your eyes would intercept only a small portion of one color from any given drop. By looking in a slightly different direction, you see the other colors from other drops. The drops that reflect red into your eyes lie on a circular path that gives you the correct 42° angle with the incident sunlight.

17.3   **Lenses**

Lenses abound in our lives. They form the heart of magnifying glasses, eyeglasses, cameras, projectors, microscopes, telescopes, and most important of all, our own eyes. Lenses depend on the *refraction* of light. We'll now see how they work, and how they are used in various optical devices.

The Focusing of Parallel Rays

Suppose we put two prisms base-to-base (as in Fig. 17.21), and bring in a beam of parallel light. The rays will refract at each surface, and bend toward the line formed by the bases. We'll call this line the *axis*. Any two rays coming in at the same distance on each side of the axis will cross at the same point on the axis.

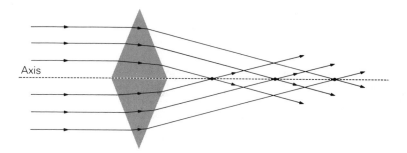

Figure 17.21 Base-to-base prisms cause parallel rays *equal distances from the axis* to cross the axis at the same place.

If we could arrange to have *all* the rays cross at the *same* point on the axis, after passing through the refracting device, we would have a **lens.** Look at the way the rays bend, and you will see that to reach this goal, our prisms need to slope *less* near the bases, *more* near the points. That way, the rays near the axis will bend less, those away from the axis more, so that all will cross at an intermediate point.

It turns out that if we make the refracting surfaces spherical in shape, and keep the rays near the axis, they will all cross at the same point (as shown in Fig. 17.22). We call this point at which the rays *focus* the **focal point.*** We call the distance from the lens center to the focal point the **focal length.**

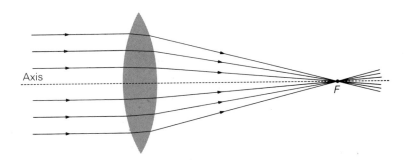

Figure 17.22 A lens with spherical surfaces brings all parallel rays near the axis to a focus at the same point F called the *focal point.*

The path of a given ray is the same, no matter which direction along the path the ray is traveling. As Fig. 17.23 shows, if we have a small light source placed at the focal point, those rays passing through the lens will move off parallel to the axis.

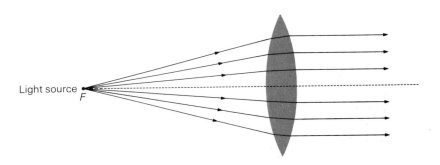

Figure 17.23 Light rays coming from the focal point travel parallel to the axis after traversing the lens.

Next, notice what happens if we bring a parallel beam of light rays onto a lens with spherical surfaces arranged so that the lens is thin in the middle and thick at the edges (Fig. 17.24). Because of the way the edges are curved, the rays bend outward and *diverge* away from the axis. They appear to be coming from a point on the incoming side of the lens, as shown by the dashed lines. The focal

---

*When the distance from the ray to the axis is large enough that the angles of bending are more than about 10°, a spherical surface no longer brings the rays to a sharp focus.

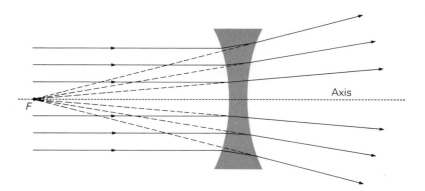

Figure 17.24 If a lens with spherical surfaces is thinner in the middle than at the edges, parallel rays will appear to diverge from the focal point on the incoming side of the lens.

point is, therefore, on the *opposite* side of this lens from that for the previous lens.

We call a lens that causes parallel rays to converge, a **converging** *lens*. If parallel rays diverge after passing through the lens, it is a **diverging** *lens*. Converging lenses are always thicker in the middle than on the edges, but diverging lenses are always thinner in the middle. Both surfaces may curve the same way (as for most eyeglasses) but, depending on which surface has the greater curvature, the lens may be diverging or converging.

Notice that if you turn one of our lenses completely around, it would have the same effect. This statement means that there is a focal point on both sides of the lens, equal distances from the center of the lens. Since *both* surfaces determine the distance to *each* focal point, both focal lengths are equal, even if the curvature of the two surfaces is different.

Focusing Nonparallel Rays

Next look at a situation such as shown in Fig. 17.23, but with the light source farther than the focal length away from the lens. In Figure. 17.25(a) a small light source S is located on the axis of a converging lens. Compare this case with that where the source is at the focal point. Because a given ray hits the lens at a smaller angle than if it had come from the focal point, it bends beyond the "parallel direction." The rays all converge to a point on the axis. In Fig. 17.25(b), the source is located off-axis. The rays then come to focus at a point off-axis. Notice that, as you should expect from our discussion so far, the particular ray that comes in parallel to the axis, goes off through the focal point. Also, the one that comes in through the other focal point, goes off parallel to the axis. We'll make special use of these two particular rays.

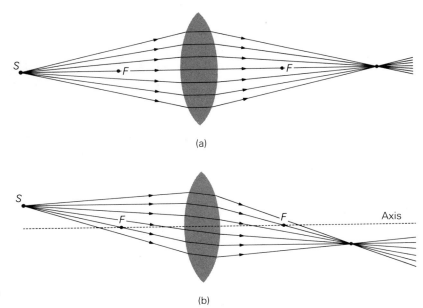

(a)

(b)

Figure 17.25 (a) Rays from a small light source on the axis in front of the focal point converge to a focus beyond the other focal point. (b) Rays from an off-axis source come to focus off-axis.

Image Formation by Lenses

In talking about mirrors, we described the *image* formed by the reflected light (the image being the point from which the light appears to come after reflection). A similar process happens for lenses. Light from an **object** forms an **image** upon passing through a lens. The transmitted light *appears* to come from this image.

To simplify the drawing, we'll use an upright tower-shaped form to represent the object, as we've done in Fig. 17.26. This tower symbolically represents anything that sends light through the lens, which is what we mean here by the term "object."

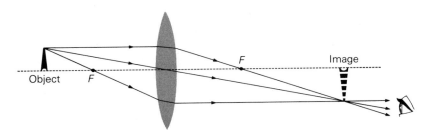

Figure 17.26 Any two of the three principal rays shown will locate the image (dashed) of the object (solid).

In this figure, the object is shown *in front of* the focal point of a converging lens. We have located the image by drawing three rays, all of which can be drawn easily without measuring any angles. Here's why.

First, notice that the rays drawn locate the image of only the tip of the tower. Once it's located, the rest can be drawn in. But you should locate a few other points on the image the first time around, just to convince yourself.

Second, let's see what is special about these three rays. *One travels parallel to the axis, so that it then passes through the focal point. Another passes first through a focal point, so it then must move off parallel to the axis. The third passes through the center of the lens and is almost undeviated,* since the lens is thin and the sides are nearly parallel there. Since *all* rays (within the small-angle limit) must pass through the same point, we don't need to draw any more. Actually, the crossing point of *any two* of these rays is enough to locate the image. We call the three rays we've described **principal rays.**

If you look from any position to the right of the image at the light from the object, the object will *appear* to be at the image position. It will appear upside down, and the size you see will be that of the image, not that of the object.

Exercise   Draw two diagrams such as Fig. 17.26, one with the object near but in front of the focal point, one with it far from the lens. Demonstrate that the nearer the object is to the focal point, the farther away is the image.

By comparing the triangle formed by the object, the axis, and the straight-through ray with the triangle formed by the image, the axis, and the straight-through ray, you can see that the ratio of *image size* to *object size* is the same as the ratio of *image distance* to *object distance:*

$$\frac{\text{image size}}{\text{object size}} = \frac{\text{distance from lens to image}}{\text{distance from lens to object}}.$$

For example, if the image is three times as far from the lens as the object, it will be three times as big as the object.

In drawing ray diagrams, you can usually ignore the thickness of the lens and draw the bending as if it all occurred at the center. We will continue to draw the bending at each *surface,* since that's where it actually occurs. If you know the focal length of the lens and how far the object is from the lens, you can find out exactly where the image falls by drawing the ray diagram to scale. We'll do that in the next section when we discuss specific instruments.

The Magnifying Glass   Look at the case where an object is put *between* a converging lens and its focal point. Figure 17.27 shows the image location. The rays never cross after passing through the lens, but appear to cross at a point in front of the lens, on the left side of the focal point. Since the image is farther from the lens than the object, it appears magnified.

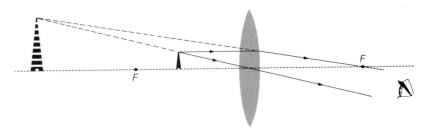

Figure 17.27   A simple magnifying glass is a converging lens with the object placed *inside* the focal length.

This is the principle of a simple magnifying glass, which is merely a converging lens placed so that the object you're magnifying is *inside* the focal length.

Diverging Lenses   We find the image of a diverging lens the same way, but it's a little trickier (see Fig. 17.28). A ray that comes in parallel diverges so that it appears to come from the focal point on the left. A ray headed for the focal point on the *right* is refracted by the lens so that it moves off parallel to the axis. The ray through the center is undeflected as before. In other words, the role of the two focal points is reversed from that for the converging lens.

Again, any two of the three rays locate the image. No matter where you place the object to the left of the diverging lens, the rays will *always* diverge and appear to come from an image on the left side, reduced in size compared with the object.

Exercise   Draw ray diagrams for diverging lenses with several different object locations. Show that you always get the image on the same side as the object, upright, and reduced in size.

Notice that for the magnifying glass (converging lens with object within focal length) and for all diverging lenses, the focused

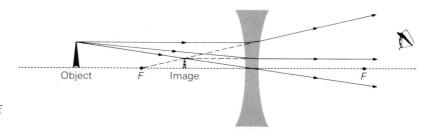

Figure 17.28 Rays diverge from a diverging lens so that the image is located on the same side of the lens as the object.

rays do not actually pass *through* the image. They only *appear* to come from the image. We call such an image a **virtual** image, as opposed to a **real** image, through which the focused rays actually pass. If you have a real image, you can put a screen there and see the image formed on the screen. You cannot do this for a virtual image, since the rays do not actually pass through the point of focus.

The Lens Equation

It's sometimes convenient to have a mathematical relationship between the image distance, object distance, and focal length. This relationship, called the *lens equation,* is

$$\frac{1}{f} = \frac{1}{o} + \frac{1}{i},$$

where o and i are the object and image distances, respectively, and f is the focal length.

In using the lens equation, remember that the focal length of a diverging lens is negative. If you find that an image distance comes out to be negative, this means that the image is on the same side of the lens as the object. For a single lens, you can always take the object distance to be positive.

**17.4   Optical Instruments

In discussing optical devices that use lenses, we'll start with the camera. The camera directly uses the image-formation ideas we've been discussing, and offers a good background for discussing the human eye.

The Camera

A camera operates by bringing the rays from an object to focus—that is, by forming an image of the object—on the film (Fig. 17.29). Suppose you're photographing a tree. When you open the shutter, you want the light from *one spot* on a leaf on the tree to hit only *one spot* on the *film*. No other light should hit that spot. In other words, the image of the tree should fall exactly on the film. Light in the image causes chemical changes in the film that, after processing, reproduce the colors in the original light.

How, then can we be sure the image is always on the film? The focal length of the lens is fixed; yet you can take pictures of the tree from various distances: The tree can be at various *object distances*. From the previous section, you know that changing the object distance for a lens changes the image distance. When you focus the camera for a particular object distance, you are changing the distance between the lens and the film, making this distance equal the image distance. Usually, the object distance is large compared with the lens focal length, making the image fall near the focal point for all object distances. As a result, little lens movement is needed in focusing.

Question   If a camera has a 10-cm focal length, how far should the film be from the lens when taking a picture of a distant mountain? Hint: Rays hitting the camera from a point on a distant object are almost parallel. For nearby objects, would the lens be closer to, or farther from, the film?

Ans: 10 cm; farther

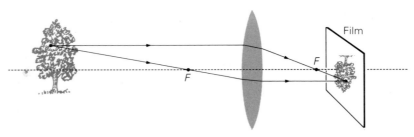

Figure 17.29   A camera uses a converging lens to bring the image to focus on the film.

Camera and projector lenses are rated not only by their focal length but by an "f-number," for example f/2. (Don't confuse this "f" with the "*f*" we used earlier for focal length.) The f-*number is the ratio of focal length to lens diameter.* If our f/2 lens has a focal length of 48 mm, the effective lens diameter is 24 mm. Reducing the diaphragm opening to let less light enter the camera gives a higher f-number setting, and vice versa.

The Eye

The eye is essentially a high-speed "instant" camera that processes the picture quickly and sends the result to the brain rather than printing it out on paper. But Polaroid and Kodak have a long way to go to catch up with the technology in the eye!

Figure 17.30 is a diagram of the human eye. Light from an object enters the eye through the pupil and is brought to a focus on the *retina.* The pupil, the opening in the iris, varies in size, depending on the brightness of the object. The retina, which plays the analogous role of the film in a camera, contains the photosensitive cells that translate the visual image into the electrical signals sent to the brain by way of the optic nerve. A clear message is sent only when the image is focused clearly on the retina.

The focusing is done not only by the lens, but by the curved surface of the cornea, the outer protective layer of the eye. With varying object distances, something has to change if the image is always to fall on the retina. The camera accomplishes this feat through movement of the lens relative to the film. But the distance from the lens to retina in the eye is fixed. The eye adjusts to varying object distances by adjusting the focal length of the lens, letting the image distance stay constant. When a normal eye looks at a distant object, the muscles controlling the lens relax, and the lens takes on

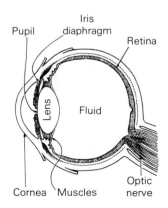

Figure 17.30 The human eye.

the shape needed to focus the image on the retina. When the eye looks at a closer object, the muscles pull the lens into a fatter shape so that the focal length is shorter and the image still falls on the retina.

The process of adjusting the focal length is called **accommodation.** Typically at about the age of 40, a person's lens starts to stiffen, and can no longer accommodate as effectively. For his eyes to focus, he then has to hold things farther away than normal. By the time a person's arm gets too short to read the newspaper, he usually gives up and visits the ophthalmologist. The accommodation ability continues to decrease with age, and usually vanishes by about age 75.

Eye Defects   A normal eye brings images to a focus on the retina (Fig. 17.31-a). But there are certain ordinary eye defects that can be corrected by eyeglasses, such as *nearsightedness* (myopia) and *farsightedness* (hyperopia). In a nearsighted person (Fig. 17.31-b), a distant object is focused in front of the retina. Glasses with diverging lenses cancel some of the converging effect of the eye and make the image fall on the retina. In a farsighted person (Fig. 17.31-c), nearby objects focus behind the retina. A converging lens corrects the problem.

Loss of accommodation may cause a form of farsightedness for close-up objects. Thus, an older person's "bifocals" may have a diverging region for seeing distant objects and a converging region for reading. You might call the use of bifocals "artificial accommodation."

Another eye defect that can be corrected with glasses is *astigmatism.* The cornea and/or lens of an astigmatic eye has more curvature, and therefore a shorter focal length, in one direction than another—for example, more curvature in the vertical plane than in the horizontal. The remedy is glasses whose lenses are more curved in the plane that the eye is least curved. Figure 17.32 shows the focusing effect of lenses that are purely converging compared with that of lenses for correction of astigmatism.

Visual Acuity   Most people know that 20/20 vision is considered good and that some other ratio of numbers is not as good. But few seem to know just what these numbers mean. This ratio, called *visual acuity*, deals with your ability to view certain test letters or other symbols. It is defined as:

$$\text{visual acuity} \quad \frac{\text{distance at which test was made}}{\text{distance for which the letter was designed}}.$$

A letter is designed (rather arbitrarily) to be read at a distance for

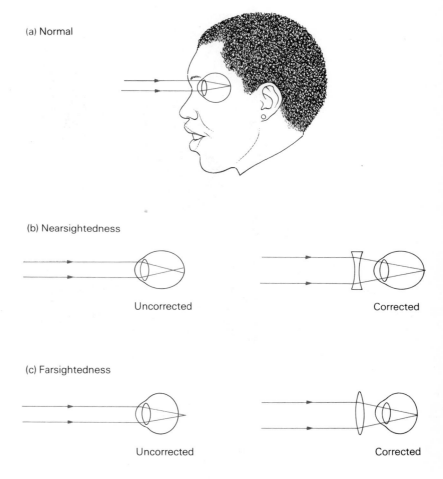

(a) Normal

(b) Nearsightedness

Uncorrected   Corrected

(c) Farsightedness

Uncorrected   Corrected

Figure 17.31   (a) Normal vision. (b) Nearsightedness is corrected by a diverging lens. (c) Farsightedness is corrected by a converging lens.

which it subtends an angle of 5 minutes. This "design" agrees with what a person with good eyesight can see. For example, if a certain letter is designed to be read from a distance of 9 m (30 ft), and Myron Myopia cannot read it farther away than 6 m (20 ft), his visual acuity is 6/9 or 20/30.

A Closer Look at Eyeglasses

We can use the ray drawing techniques we learned earlier to find the specific focal lengths needed for particular nearsightedness and farsightedness problems.

Figure 17.32   The shadow of eyeglasses in a parallel beam of light. The upper lenses, mine, are purely converging with a focal length of 1.3 m. The lower lenses, my wife's, are slightly diverging and are astigmatic in addition. As you can tell, both the axis of the astigmatism and the amount of astigmatism are different for each of my wife's eyes.

what a person with good eyesight can see. For example, if a certain letter is designed to be read from a distance of 9 m (30 ft), and Myron Myopia cannot read it farther away than 6 m (20 ft), his visual acuity is 6/9 or 20/30.

A Closer Look at Eyeglasses

We can use the ray drawing techniques we learned earlier to find the specific focal lengths needed for particular nearsightedness and farsightedness problems.

First, consider nearsightedness. A normal eye can focus on distant objects clearly. But a nearsighted person cannot see clearly beyond a certain distance. This distance is called the *far point*.

Let's find the focal length of lenses needed by a woman whose far point is 100 cm. For her, light from a distant object needs to *appear* to come from 100 cm away. Remember that rays intercepted by the eye from a point on a distant object are almost parallel.

Figure 17.33 shows how these rays can appear to come from 100 cm away. Parallel rays appear to diverge from the focal point. Thus, a diverging lens of focal length 100 cm is what is needed.

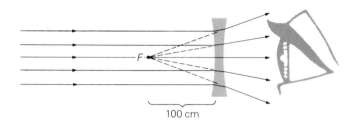

Figure 17.33 For a person with a far point of 100 cm, a diverging lens of focal length 100 cm makes a distant object have an image 100 cm away.

Now, consider farsightedness. A normal eye can focus on objects *up to* a distance of about 25 cm. Thus, 25 cm is the normal *near point*. A farsighted person has a near point longer than this.

What lens is needed to correct the farsightedness of a man with a 60-cm near point? This man needs to fool his eyes into thinking an object at 25 cm is at 60 cm. In other words, an object 25 cm away needs to have an image 60 cm away. We get the object and image located appropriately (as in Fig. 17.34) only if the lens is converging with a focal length of 43 cm. (The fact that one focal point of the lens is behind the observer's eye is not important.)

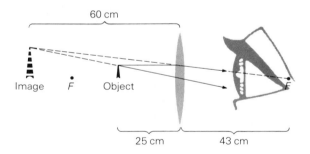

Figure 17.34 For a person with a near point of 60 cm, a converging lens with a focal length of 43 cm makes an object 25 cm away have an image at 60 cm.

Diopter Units

When you get a prescription for glasses, the magnifying power for each lens is specified by the opthalmologist in units of **diopters.** The *power* in *diopters* is the *reciprocal* of the *focal length in meters.* Thus, if your prescription calls for a "+0.75" lens, that means a 0.75-diopter lens: one with a focal length of 1/0.75 or 1.33 m. The *positive* "+" sign means a *converging* lens. A *diverging* lens is specified as having a *negative focal length* and therefore a *negative power.*

Complex Instruments

Optical instruments such as microscopes, telescopes, and binoculars use two or more lenses to do their magnifying jobs. The light transmitted through one lens forms an image, and this image serves as the object for the lens that follows. In many instruments, "the lens" is not a *single* lens but several lenses grouped together as a unit with an appropriate overall focal length.

Key Concepts

Geometric optics treats light as rays that travel in straight lines in a given medium. In general, the rays change direction at the boundary between two media.

The **law of reflection:** The angle of reflection is equal to the angle of incidence.

The **image** formed by a *plane* mirror is as far behind the mirror as the **object** is in front of the mirror. A **concave** spherical mirror acts as a *magnifier* if the object is within a distance of ½ the radius of curvature of the mirror. A **convex** spherical mirror increases the field of view of the observer.

When light travels from one medium into one in which its speed is different, it **refracts.** The angle the rays make with the normal is lower in the medium in which the speed is lower.

Light incident upon a medium in which its speed increases undergoes **total internal reflection** if the angle of incidence exceeds the **critical angle:** the angle at which the refracted ray would travel parallel to the boundary.

Dispersion is caused by the fact that light of different wavelengths travels at different speeds in a material medium.

Light rays parallel to the axis of a lens refract so that they pass through the **focal point** of a **converging** lens; they appear to have come from the focal point of a **diverging** lens. The **image** of an **object** that sends light through a lens can be located by the point of intersection of any two of the three **principal rays.**

A **magnifying glass** is a *converging lens* located so that the object is *within* the focal length. A camera is focused by changing the image distance, and the human eye is focused by changing the lens' focal length. This process in the eye is called **accommodation.**

Nearsightedness is corrected by a diverging lens, *farsightedness* by a converging lens, and *astigmatism* by a lens that focuses more strongly in one plane than in another. Lens prescriptions are written in units of **diopters,** the inverse of the focal length expressed in meters.

Telescopes, binoculars, and microscopes use two or more lenses (or groups of lenses that act as a unit). The image of the first lens serves as the object for the second.

Questions

Reflection

1. Show, by drawing diagrams that obey the law of reflection, that the distance you stand from a *vertical* mirror has no influence on how much of yourself you can see in the mirror? Why is this not true if the mirror is over a dresser, and the dresser top shows in the mirror? Does the answer to either of these questions change if the mirror is not vertical?
2. I once rented a "U-Haul" trailer that had painted across the *front* of the trailer: HꟼM Ƨ4 ᗡƎƎꟼƧ MUMIXAM. Was that because the sign painter had something stronger than milk for breakfast the day he painted the letters?
3. Figure 17.35 shows a top view of a clothing-store mirror used for seeing yourself on all sides. Trace the approximate path of a ray from your back to your eyes. Since you stand sideways as shown to see your back, does the center section of the mirror contribute anything to that view?

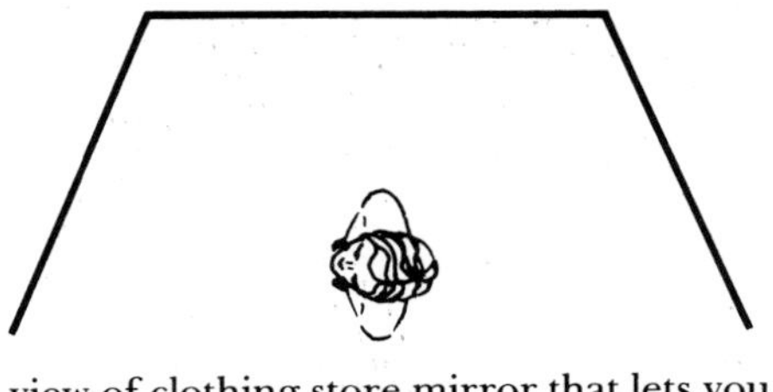

Figure 17.35 Top view of clothing store mirror that lets you see all sides of yourself.

4. Bike riders sometimes use small rear-view mirrors attached to the frames of their eyeglasses. If the mirror is 2 cm in diameter, does the rider see only a 2-cm-diameter circle of the car following behind? Why more?
5. A short person driving a car moves the seat forward to better reach the steering wheel and pedals, while a tall person moves the seat farther back. Which person will see a wider field of view in a fixed rear-view mirror?

6. After taking a night physics class at Flat Mountain Community College, the manager of the local clothing store installed a convex spherical mirror in the corner of his store so he could keep an eye out for shoplifters (Fig. 17.36). Why did he use that type of mirror rather than a plane one?

Figure 17.36 Convex spherical mirrors are often used in retail stores for monitoring.

Refraction

7. As you and your rich aunt converse beside her pool (Fig. 17.37), you spot her lost diamond ring on the pool bottom. Why would you not have been able to see it in that spot from where you're sitting if there had been no water in the pool?

8. The tube of liquid in a liquid-in-glass thermometer looks *much* bigger than it really is. Why? (In a clinical thermometer, the mercury column looks almost as large as the glass tube itself.)

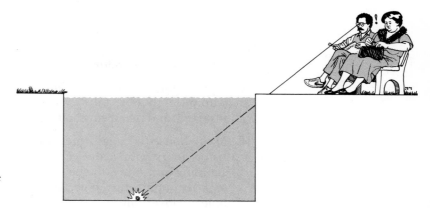

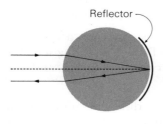

Figure 17.37 A straight line from the diamond to your eyes goes through the concrete pool wall.

Figure 17.38 Reflecting tape and road signs have tiny spheres embedded in a reflector. The reflected light is directed back parallel to the incoming beam.

9. A hot highway far ahead looks "glassy" or wet. You're really seeing light from the sky refracted (bent) into your eyes because the air near the road is hotter than that a little higher, giving light a slightly higher speed near the surface. Explain.

10. Highly reflecting road signs and bumper stickers consist of many tiny glass spheres embedded on a reflecting surface. Figure 17.38 shows one such sphere. The refractive properties of the glass bend a given ray of light so that it hits the back surface on a line through the center of the sphere and parallel to the incoming ray. Explain how such materials reflect any incident light back parallel to its incoming direction. A similar approach is used for the retroreflectors of some liquid crystal displays we discussed in Sec. 16.2.

11. Cats' eyes shine brightly at night, but only if you beam a light at them, and then most strongly right back along the path of the incoming light. Read the background part of Question 10 and then formulate a theory of shining cat eyes.

12. Optical cables for communication are made up of many tiny glass or plastic fibers in a bundle, rather than one larger fiber. This construction makes them more flexible. Why does this also make them work better optically? (Hint: Draw what happens to a ray in a sharply bent small fiber and in a large fiber bent to the same radius.)

13. By drawing a side view of the fishbowl of Fig. 17.11-a, explain why you see two fish when there is only one. (Hint: Draw two rays from the fish to your eye—one through the top surface of the water and one through the curved side of the water and bowl.) By drawing a top view, explain why the "bottom fish" looks larger than the fish really is.

Lenses

14. If the sun is shining, how can you easily measure the focal length of a converging lens?

15. By drawing ray diagrams for converging and diverging lenses with different object distances, show that a real image is always inverted, and that a virtual image is always upright.

16. Many houses and apartments have little "peep-holes" in their front doors with a device inside that lets you see someone standing *anywhere near* the front of the door, even though you look through a tiny hole. What kind of lens is in this peep-hole—converging or diverging? If you have one, check your answer by shining a flashlight through it from the inside after dark, and watching what happens to the beam outside.

17. Draw a *converging* lens for which one surface has a radius of curvature of 20 cm, and one 40 cm. Draw a *diverging* lens for which one surface has a radius of curvature of 20 cm, and one 40 cm.

18. Since different colors of light refract different amounts, all colors do not have the same focal point in a lens. Explain why this effect, called *chromatic aberration,* can give you rainbow-colored regions around white spots on a slide projected onto a screen. (That's one reason "the lens" of a device is often two or more lenses of different materials that compensate for this aberration.)

Optical Instruments

19. Why does a photograph come out blurred if the camera is not properly focused when the picture is taken?

20. Why do you have to put slides upside down in a slide projector to have the image upright on the screen?

21. If your optometrist prescribes glasses with −0.50-diopter magnifying power, what will be the focal length of the glasses? Would that indicate you are nearsighted or farsighted?

22. During an eye examination, an ophthalmologist usually adds drops that dilate your eyes. In addition to opening your pupils so that she can see inside your eye, the fluid temporarily paralyzes the muscles that control accommodation. Why would she want to eliminate your accommodation ability in order to accurately evaluate your eyesight? Would you expect this

paralyzation to be more important for older or for younger people?

23. Light passing near the edges of a lens crosses the axis slightly closer to the lens than light passing near the center. The blurring effect that results is called *spherical aberration.* Does this explain why you might get clearer pictures with an inexpensive camera if you use it in bright light so that the diaphragm opening can be set smaller, rather than in dim light that needs a large diaphragm opening?

24. Would you expect spherical aberration, defined in Question 23, to have anything to do with why you can read more clearly in bright light than in dim? (Remember what happens to the pupil when the light level changes.)

25. When people first get new glasses, why do they sometimes step too soon or too late when they get to a staircase? (Hint: The eye sees the image formed by the glasses, not the object.)

Problems

1. For a mirror mounted vertically on a wall, what must be its minimum height for you to see yourself completely? How high must the bottom be from the floor in order to get by with this minimum height? Would the same mirror at the same place work for a shorter person? Why not?

Ans: ½ your height; ½ distance from floor to eyes.

2. If, for bike riding, you clip a 2-cm-diameter flat mirror onto your sunglasses so that it is 4 cm from your eye, what's the tallest object you can completely see 10 m behind you?

Ans: 5 m

3. Suppose you look at a dime under a magnifying glass of focal length 5 cm. If the dime is 3 cm from the magnifier, how far away will it appear to be? How big will it appear?

Ans: 7.5 cm; 2.5 times its actual size

4. A personalized slide projector uses a screen 60 cm from the projection lens. If the lens focal length is 10 cm, how far should the slide be from the lens? **Ans:** 12 cm

5. If you're 10 m from Queen Elizabeth and your camera has a focal length of 50 mm, how far should your film be from the center of the lens for proper focus? (Because of the big difference between object distance and focal length, the ray-drawing method won't be very accurate. You'll need to use the lens equation).   **Ans:** 50.2 mm

6. If John Travolta discovers at age 43 that he can no longer read his lines, and his ophthalmologist prescribes lenses of +1.0 diopter, what will be his near point, assuming he is fitted correctly?   **Ans:** 33 cm

7. A nearsighted person cannot clearly see anything farther than 3 m away. What magnifying power should this person's glasses have to correct the problem?   **Ans:** −0.33 diopter

8. A 35-mm color slide has film dimensions of 24 mm by 36 mm. You want to show the picture on a 1.0-m by 1.5-m screen. If the slide is 120 mm from the proejction lens when the image is focused, how far away should the screen be?   **Ans:** 5.0 m

Home Experiments

1. Verify that you need a mirror only ½ your height to see your entire body, if there are no obstructions such as a dresser top. Show that if the mirror is vertical, your distance from it makes no difference in how much you see.

2. Get in a swimming pool or pond with no one else in it to ripple the water. Wearing a skindiver's viewing mask, swim to the bottom and look up. Why do things look as they do?

3. If you have a converging lens available (for example, the projection lens of many slide projectors can be taken out along with the tube that contains it), hold it near a nonfrosted light bulb such as a high-intensity reading lamp. It should be slightly farther away than the focal length of the lens. Hold a paper out from the lens until the image of the filament comes to focus on the paper. You can easily see the filament coils. Vary the distance to the lens (the object distance) and see what happens to the image distance. If you know the lens focal length, measure the distances involved, and draw ray diagrams to check the correctness of the ideas in the chapter. If you don't know the focal length, you can measure it this way.

4. If you wear eyeglasses, find out from your optometrist or optician what your prescription is. See if you can roughly verify this prescription by the way your glasses refract the *parallel* rays of the sun.

References for Further Reading

Boyle, W. S., "Light-Wave Communications," *Scientific American,* Aug. 1977, p. 40. Discusses the techniques in the first test of telephone communication via light waves.

Gregory, R. L., "Visual Illusions," *Scientific American,* Nov. 1968, p. 66. Discusses many of the interesting optical illusions that we see.

Kaufman, L., and I. Rock, "The Moon Illusion," *Scientific American,* July 1962, p. 120. An explanation of why the moon looks larger when it first "rises."

Lynch, D. K., "Atmospheric Halos," *Scientific American,* April 1978, p. 144. Describes how the misty ring we sometimes see around the sun and moon is caused by reflection and refraction of light by ice crystals in the air.

The Way Things Work, an Illustrated Encyclopedia of Technology (Simon and Schuster, New York, 1967). An equivalent publication: *How Things Work, The Universal Encyclopedia of Machines* (George Allen and Unwin Ltd., London, 1967). Details of many optical instruments including cameras, telescopes, microscopes, etc.

Part V

The Atom

Humanity certainly needs practical men, who
get the most out of their work, and, without
forgetting the general good, safeguard their
own interests. But humanity also needs
dreamers, for whom the disinterested
development of an enterprise is so
captivating that it becomes impossible for
them to devote their care to their own
material profit.

Marie Curie

Marja Sklodowska, the youngest child of a
physics teacher in Warsaw, was born in 1867.
After her marriage to Pierre Curie in 1895, the
two shared not only a happy marriage but an
extremely tedious and intensive search that
culminated in the discovery of the elements
radium and polonium. Though saddened by
her husband's death in 1906, she continued not
only her detailed studies of radioactivity, but
pioneered a mobile x-ray unit and founded a
radiological school for nurses. She rejected the
money, comfort, and many advantages offered
by her great fame, crowned by Nobel Prizes in
both physics and chemistry. She died of
leukemia in 1934, apparently from prolonged
exposure to radiation during her scientific
investigations. (She is shown here with her
daughters Eve and Irene.)

18

The Atom and Its Nucleus

Our modern idea that all matter consists of tiny particles called *atoms* is actually not so modern. The Greek philosophers Leucippus and Democritus introduced the first atomic theory in the fifth century B.C. After considerable debate among the Greek philosophers, atomism became associated with atheism and fell into disrepute from Aristotle's time until reintroduced by the English chemist John Dalton in the early nineteenth century.

The word **atom** meant "uncuttable" or "indivisible" in Greek. As we study the properties of the atom and its nucleus in this chapter, we'll see that the word's origin doesn't quite fit the way the atom behaves.

We'll first look at some models of the atom, starting with early models and working up to a description of our present understanding of its structure. We'll then consider the nucleus of the atom: its structure, its radioactivity, the reactions it participates in, and the ways we are able to use its energy. As we do, we'll point out many of the practical applications of nuclear physics, and its effects on our lives. As background for this chapter, you need to have read the material on atomic structure and electric charge in Section 10.2 of Chapter 10, and that on light emission from atoms and the dual nature of light in Sections 16.2 and 16.7 of Chapter 16.

18.1   The Rutherford-Bohr Model of the Atom

In earlier chapters when we needed to talk about the atoms that make up all matter, we used a "planetary" model for the atom. In this model, the atom is pictured as consisting of electrons orbiting the nucleus much as planets orbit the sun. Each time we did this, we tried to point out that this crude picture is useful for discussing *some* atomic behavior, but it has some serious drawbacks. We'll now look at the origin of this model, its drawbacks, and what it takes to give a better description of the atom.

Rutherford and the Nuclear Atom

How would you try to learn about something so small that it takes about 10^{24} of them to make up this book—something about 10^{-10} m (0.1 nm) across? You certainly couldn't look at it with light, no matter *what* the magnification of your microscope. Something only 1/10 000 the wavelength of light is not about to scatter any light into your microscope. The atom is so small that, in a sense, you can't get "close" enough to see inside it.

Well, what do you do when you're standing beside a pond in a mountain stream, but you can't get close enough, or in it, to find out how deep it is or what the bottom is like? You throw rocks at the pond and watch what happens to them. If you can see what happens to the rocks, you can tell something about the pond. That's exactly how we can study the atom: We throw "rocks" at it and watch what happens to them. These "rocks" have to be a special kind—particles of atomic or smaller sizes—and we have to "watch" with equipment that can tell what these particles do.

Around 1910 the prevailing view was that the atom had a sort of "plum pudding" arrangement, with electrons spread out like plums in a sphere of positively charged "pudding." Between 1909 and 1911 the English physicist Ernest Rutherford and his associates found out differently. The "rocks" they used were *alpha* (α) particles—clumps of two neutrons and two protons—emitted from certain *radioactive* substances. (More about radioactivity later.) They aimed these α particles at a gold foil in an arrangement such as that shown in Fig. 18.1. If the foil were made of "plum pudding" type atoms, these high-speed particles would move through almost undeflected from a straight line. Much to their surprise, many of the α particles were scattered at large angles, some even straight backwards. In Rutherford's words, "It was almost as incredible as if you fired a 15-inch shell at a piece of tissue paper and it came back and hit you."

Rutherford correctly interpreted the results to mean that *most* of the atom's mass, and *all* of its *positive* charge, are concentrated within an extremely tiny region—the **nucleus.** You can visualize the reason for the difference in scattering pattern by the following

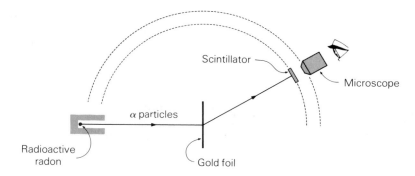

Figure 18.1   The apparatus that Rutherford and his associates used to study the scattering of alpha particles by atoms of a gold foil. An α particle striking the scintillator caused it to give off a small amount of light that could be seen in the microscope. The scintillator and microscope could be moved through 180° to find the number scattered at each angle.

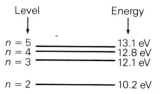

Figure 18.2   An energy level diagram for the hydrogen atom. Taking the *ground state* (level 1) to have *zero* energy, the energies of the lowest four *excited* states are given in electron volts (1 eV = 1.60 × 10⁻¹⁹ J).

thought experiment. Imagine many, many billiard balls to be thrown in the general vicinity of a bowling ball hanging from a rope. Most would miss the bowling ball, but some would be scattered at large angles. Now imagine the billiard balls to be thrown at a *cloud* having the same total mass as the bowling ball. Most would hit the cloud, but you'd be very surprised to see any billiard balls bouncing backward. That was Rutherford's line of reasoning, but he also was able to calculate mathematically the correct number of α particles scattered at each angle. His insights brought him knighthood and a Nobel Prize.

Nuclear dimensions are 10^{-15} to 10^{-14} m, 1/100 000 to 1/10 000 the size of the atom as a whole. Let's look at the atom on a much larger scale: If a typical atom were expanded, with everything keeping the same proportions, until the nucleus became the size of our sun, the outer electrons would be 500 to 1000 times as far away as the earth is from the sun.

The Bohr Model

Remember that, in our discussion of light emission from gas atoms, we stated that an atom can exist only in certain discrete energy levels. For the hydrogen atom we found the lowest few energy levels to be those shown in Fig. 18.2. These levels correspond to certain states of the electron in the atom. Each level is identified by a quantum number (n) starting with $n = 1$ for the ground state.

Let's assume a planetary model, with the electron held in orbit by the attractive electrical force between it and the nucleus. If we were to calculate the total energy—the sum of the kinetic and potential energies—of the electron in the orbit, we would find that the only orbits that would give the correct energy levels for the hydrogen atom are the ones whose orbital radii are shown in Fig. 18.3. Notice that you can get the radius of any orbit by multiplying the smallest (*ground state*) radius by n^2.

We have two conditions here: (1) a planetary model with electrons in orbit around a massive nucleus; and (2) discrete energy levels with their accompanying requirement of only certain orbits being allowed. Neither of these requires the other to exist, although we know from the light emission spectra (Chapter 16) that the second condition is a *fact of life*.

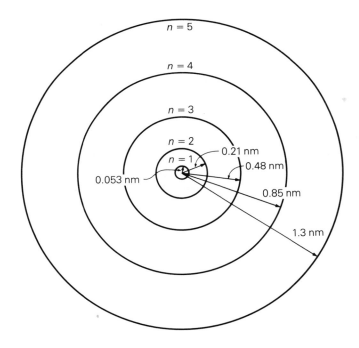

Figure 18.3   Electron orbits in the Bohr model for the lowest five states of the hydrogen atom. The radii of the orbits are given in nanometers (1 nm $= 10^{-9}$ m).

In 1913 Niels Bohr tied these two ideas together by *postulating* an additional condition on the electron orbits. He postulated that the angular momentum of an electron in orbit—its linear momentum multiplied by the orbit radius—could not take on any arbitrary value. Rather, the angular momentum could have values of *only* some **integer** (1, 2, 3, 4, . . .) multiplied by Planck's constant h divided by 2π. That is,

$$\text{angular momentum} = n \times \frac{h}{2\pi},$$

where n is an integer. Remember from Chapter 16 that Planck's constant is a universal constant with the value of 6.62×10^{-34} joule-second. The particular orbits that satisfy this postulate give the exact energy levels for the hydrogen atom—those of Fig. 18.2. We call this planetary atomic model, with quantized angular momentum values, the **Bohr model** of the atom.

The Bohr model quite accurately predicts the energy levels of the hydrogen atom: The simplest atom we have. But for more complicated atoms this model doesn't quite give the right answers. We'll see in the next section what is needed to do that. The model is useful, however, because it gives us an intuitive picture of the atom, and helps us to visualize many atomic properties.

From *our* historical perspective, it might seem that it should not have taken much insight to come up with the ideas that Rutherford and Bohr gave us. They seem perfectly obvious. But, for the state of atomic knowledge in 1910, their contributions were immense.

18.2 The Quantum Theory of the Atom

The mid-1920s brought the introduction of the **quantum theory** that we now use to describe the atom and its behavior. As background for discussing this theory, we'll summarize the status in about 1920 of the understanding of the atom and its parts.

1. From the fact that light is emitted in *discrete wavelengths* from individual atoms, the atom must have *discrete energy levels.* The energy of the photon emitted equals the *change* in energy of the atom emitting it.

2. The Bohr planetary model with quantized orbits could explain the energy levels of hydrogen. But it did this by a somewhat arbitrary assumption about angular momentum, and it didn't work for more complicated atoms. Physicists, including Bohr, knew that something new and different was needed.

3. Light was known to have a *dual* nature: Sometimes it acted as a wave; sometimes it acted as a particle.

In 1923 a young French graduate student named Louis de Broglie, in searching for a Ph.D. thesis topic, began thinking about the dual nature of light and about the symmetry that usually exists in nature. De Broglie's line of reasoning was something like this. Light, which is usually thought of as a wave, also has a particle nature. Since things in nature are normally not lopsided, it is reasonable that electrons, which we usually think of as particles, would also have a wave nature. In his thesis, de Broglie proposed just that. Furthermore, he was able to predict that the wavelength associated with the electron was given by Planck's constant divided by the electron momentum.

If we combine the de Broglie wavelength hypothesis with the Bohr model, a surprising coincidence appears. You find that for those orbits allowed by the Bohr model, the orbit circumference is exactly a whole number of electron wavelengths. (Fig. 18.4-a). This implies a sort of standing wave similar to those of a plucked guitar string. For other orbits, the wave does not close on itself (as Fig. 18.4-b illustrates).

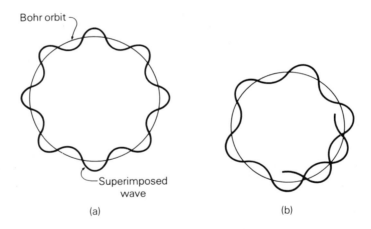

Figure 18.4   (a) For the orbits allowed by the Bohr model, the orbit circumference is a whole number of electron wavelengths. (b) For other orbits, the waves overlap randomly and cancel each other's effect.

What this standing wave effect says to us is that there must be *something to* the idea that the electron has a wave nature. It still doesn't give us the picture of the atom that we need, but it points us in the right direction.

Three years *after* de Broglie made his hypothesis, Clinton Davisson and Lester Germer at Bell Telephone Laboratories showed that electrons undergo *diffraction*—something that only waves can do. Within a few years other experimenters showed that the other entities we usually consider to be particles also sometimes behave as waves.

Quantum Mechanics à la Schrödinger

In 1926 the Austrian physicist Erwin Schrödinger invented a model that serves as the basis of our present understanding of the atom. Using de Broglie's idea of the wave nature of electrons, he developed a **wave equation** that is the starting point in deriving a model for the atom. This model is a *mathematical* one and, unfortunately, is not something that can be visualized easily. We will describe in qualitative terms some of the main features of the picture this equation gives of the atom and of atomic particles in general.

In case you'd like to see what the Schrödinger wave equation looks like, I'll write it down:

$$-\frac{\hbar^2}{2m}\nabla^2\Psi + V\Psi = i\hbar\frac{\partial\Psi}{\partial t};$$

$$\underbrace{\text{kinetic energy}} + \underbrace{\text{potential energy}} = \underbrace{\text{total energy.}}$$

I won't bother you with the mathematical details of what those strange symbols mean. But, as implied by the line below the equation, it is in principle just an expression of the conservation of energy. In a mechanical system without friction, the total energy stays constant, and is the sum of the kinetic and potential energies. Each term of the Schrödinger equation, as shown, represents one of these forms of energy for atomic particles.

When you apply the wave equation to the electrons of a particular kind of atom, what you get out is something called a **wave function,** that we usually represent by the Greek letter psi (Ψ). This wave function is a mathematical expression that has in it all the information that it is possible to know about the electrons.

The wave function cannot tell you precisely *where* an electron *is* in the atom. It can only tell you the *probability,* or relative chance, of finding the electron in a certain region of the atom. Instead of having a precise electron orbit, we have a *probability distribution* for the electron that tells the probability of finding it at various points in space. The closest we can come to a visual picture of this is to draw a cloudlike picture in which the density of the cloud is proportional to the probability of finding the electron there. Figure 18.5 shows this "probability cloud" for the ground state and one of the higher energy states of the hydrogen atom.

Figure 18.5 The "cloud" density is proportional to the probability of finding the electron at that location. These "probability clouds" are for the (a) ground state and (b) one of the excited states of the hydrogen atom.

(a)   (b)

This probability distribution business is not nearly as satisfying as being able to say that the electron is at "such-and-such" a place in "thus-and-so" orbit. However, the creator of the atom has not allowed us the satisfaction of being able to have such precise information. So far as physicists can tell at this time, it is not *in principle* possible to precisely determine the status of any given electron, but only to know the probability of finding it at a certain place with a certain velocity. But, within that limitation, we can calculate the structure of all the atoms.

Our knowledge of atomic electrons is something like the knowledge statisticians have of how many people will be killed on the highways over the Fourth of July weekend. They can tell you in advance with considerable accuracy how many people will be killed. But they can predict *nothing* about what will happen to Mary Elizabeth Johnson-Whippendorfer, or you, or any other individual. The *quantum mechanics* of the Schrödinger equation can tell you *how many* electrons are *where* in a given group of atoms, but cannot tell you what a particular electron is doing.

A quantum mechanical calculation using the Schrödinger wave equation predicts the *highest probability* for the electron of the hydrogen ground state to be at the position of the first Bohr orbit. But the electron can also be, with high probability, at other locations. And using quantum mechanics, we can calculate the positions of the energy levels of *any* atom. In the ground state of a many-electron atom, electrons occupy the lowest possible energy states.

The Heisenberg Uncertainty Principle

If you're like many people, you probably visualize the electron as some tiny round speck. But as we've seen, the electron on an atomic scale acts more like a wave than a particle. When you *really really* get down where the electron "is," how do you locate it?

Early in the days of quantum mechanics, Werner Heisenberg showed that you cannot simultaneously know an electron's position and momentum with infinite precision. If the momentum mv is *uncertain* by an amount $\Delta(mv)$, and the position x is *uncertain* by Δx, then $\Delta(mv)$ multiplied by Δx must be equal to or greater than Planck's constant divided by 2π. In symbols,

$$\Delta(mv) \cdot \Delta x \geq \frac{h}{2\pi}.$$

The Δ here means uncertainty, and $\geq$ means greater than or equal to. We call this statement the **Heisenberg uncertainty principle.** It says that we can measure either position *or* momentum precisely, but not both at the same time. Or, for example, suppose you know the electron is within an interval of 0.1 nm along some line. (That's about the diameter of a hydrogen atom.) The closest you can then determine its momentum is

$$\Delta(mv) \geq \frac{h}{2\pi \cdot \Delta x} = \frac{6.62 \times 10^{-34} \, \text{J} \cdot \text{s}}{2\pi(0.1 \times 10^{-9} \, \text{m})} = 1 \times 10^{-24} \, \text{kg} \cdot \text{m/s}.$$

Because the electron mass is so small, 9.1×10^{-31} kg, the speed is uncertain by about 1000 km/s.

Question   Look at the size of the numbers involved, and explain why the Heisenberg uncertainty principle does not limit Pete Rose's ability to locate and hit a baseball pitched by Catfish Hunter.

Quantum Mechanics and You

You may be wondering what all this quantum mechanics has to do with anything relevant to you. The truth is that a person going about his daily routine does not *directly* see or experience anything quantum mechanical. That's one reason quantum mechanical ideas are hard to grasp. They are not part of our ordinary experiences that give intuition about things. But *indirectly* the quantum mechanical behavior of atoms influences *everything* we do or think. Everything is composed of atoms, and these atoms interact by the rules of quantum mechanics.

Quantum mechanics does more than just satisfy our intellectual curiosity about the nature of the atom. For example, there is a rather large area of *applied physics* called **chemistry.** Quantum mechanics lets the chemist in on the laws of how atoms interact with each other, and how to combine them to form substances that are very useful to you and me.

The Quantized Nature of Things

Even though most physical properties seem to be able to exist in any arbitrary amount, quantization into discrete small bundles is an inherent feature of nature. Let's review the things we've talked about in this book that are *quantized*, that come in discrete amounts.

1. All matter is quantized into atoms.

2. The energy levels of atoms are quantized.

3. The charge on the atom's constituents is quantized into units of electron charge.

4. Light is quantized into photons.

Quantum mechanics treats this quantization in a natural way, without artificial assumptions as in the Bohr model. Some physicists have even theorized that *time* is quantized—that we can choose time intervals only in whole number multiples of some very small time interval. No one has yet given any experimental verification of this hypothesis.

18.3 X Rays

You've probably had some part of your body—your teeth, your chest, your broken foot—x-rayed at some time in the past. The radiologist beams these highly penetrating rays at the part to be examined—let's say your foot. The film beyond the foot is sensitive to the radiation coming through. A crack or other abnormality absorbs a different amount of radiation and thus shows up on the film (Fig. 18.6). Various industries use x rays in the same way, to detect cracks that cannot be seen visually in pipes or structural materials.

Aside from medical and industrial uses, x rays are a very useful tool for understanding the structure and energy levels of atoms. Remember from Chapter 16 that atoms give off photons of electromagnetic radiation when they de-excite from a higher to a lower energy level (Fig. 18.7). If the energy change, and thus the photon energy, is 2 or 3 eV, the photon has a wavelength in the visible region. At 100 eV, the photon is in the ultraviolet region. If the energy change exceeds a few hundred electron volts, the photon is an x ray.

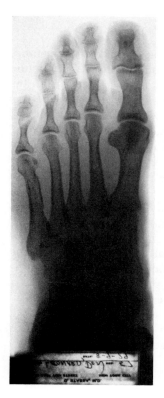

Figure 18.6 An X-ray photograph of a broken foot.

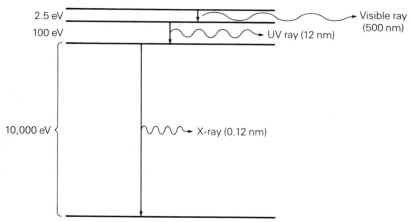

Figure 18.7 An atomic energy level diagram showing energy changes in the visible, ultraviolet, and x-ray regions. Wavelengths in nanometers for each transition are in parentheses. Energy changes are shown in electron volts. The diagram is not to scale; if it were, you couldn't see the separation between the three highest levels.

Figure 18.8 shows how an x-ray tube produces x rays. Electrons emitted from a heated filament are accelerated by a voltage of typically 40 to 140 kV. This high-energy electron beam hitting a heavy-metal target excites some of the atoms, causing them to emit x rays. The x-ray beam emerges through a narrow opening—a collimator—that confines the beams to hit only the part of the body that needs x-raying.

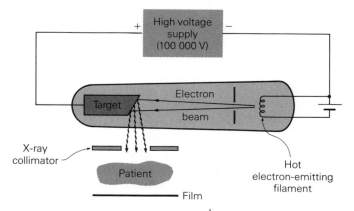

Figure 18.8 The basic principle of an x-ray tube.

18.4   The Nucleus of the Atom

The **nucleus** of the atom—that tiny speck at its center—is far more influential than its small size might lead you to believe. The potential for release of vast amounts of energy from atomic nuclei became evident to the whole word when "the bomb" was dropped on Hiroshima and Nagasaki. The social controversy that centered around "the bomb" has, in recent years, extended to the use of nuclear energy for generation of electricity. Even though nuclear power plants have the spotlight, there are many noncontroversial nuclear processes that are extremely useful and important to our way of life.

Nuclear applications include medical diagnosis and treatment for cancer and various other diseases, establishing the date at which particular ancient cultures existed, detection of smoke by household smoke alarms, identification of small quantities of pollutants in our environment, and lighting of some LCD watches for nighttime viewing. Industries apply nuclear processes in hundreds of useful ways. We'll employ some of these applications to illustrate the physics of the nucleus.

In order to understand basic nuclear processes and their uses, we first need to understand nuclear structure.

The Structure of the Nucleus

The nucleus consists of two types of particles: **protons** and **neutrons.** We often use the word **nucleon** to mean *either* a neutron *or* a proton—one of the nuclear particles. Protons have a positive electrical charge equal to the negative charge of the electron, 1.60×10^{-19} C. In a neutral atom, there are as many electrons surrounding the nucleus as there are protons inside, making the *net* charge of the overall atom *zero*. The mass of the proton is 1.67×10^{-27} kg, about 1850 times that of the electron.

The other nuclear constituent, the neutron, has no electrical charge, and its mass is just slightly greater than that of the proton. Most stable nuclei have at least as many neutrons as protons.

We now see why most of the mass of the atom is in the nucleus. Nucleons have a mass of about 1850 times that of the electron, and there are usually at least twice as many nucleons as electrons. Result: About 99.97% of the mass of the atom is in the nucleus.

We describe particular nuclei by symbols such as $^{60}_{27}$Co. The Co is the *chemical symbol* for the element cobalt, which means the nuclei belong to cobalt atoms. The 27 is the **atomic number:** the number of protons in the nucleus. This number gives redundant information and is often left off, since cobalt is *the* element with 27 protons in its nuclei, and 27 electrons in its neutral atoms. In general, the atomic number identifies the element. Known nuclei range in their number of protons from hydrogen with only one proton up to element 107, which hasn't been named yet.

The 60 in $^{60}_{27}$Co is the **mass number.** This gives the number of protons *plus* neutrons in the nucleus. It's called "mass" number because it tells you the approximate value of the mass in a particular unit, the atomic mass unit (abbreviated amu). The amu, which is equivalent to 1.66×10^{-27} kg, is defined so that the mass of a nucleon is *approximately* 1 amu, and that of any nucleus is *approximately* equal in value to its mass number.* Actually, the neutron mass is 1.008665 amu, and that of ^{60}Co is 59.933813 amu. We get the number of neutrons by subtracting the atomic number from the mass number. Thus, there are $60 - 27$, or 33 neutrons in ^{60}Co.

A particular nuclear species has the name **nuclide.** Thus, $^{235}_{92}$U is a nuclide with 92 protons and 143 neutrons (U = uranium). Nuclides of the same element can have different numbers of neutrons. For example, $^{88}_{38}$Sr has 50 neutrons, $^{90}_{38}$Sr has 52 neutrons (Sr = strontium). Nuclides of the same element having different numbers of neutrons are called **isotopes.** Hydrogen, the simplest element, has three isotopes: $^{1}_{1}$H; $^{2}_{1}$H; and $^{3}_{1}$H. The ^{1}H nucleus is just a single proton. The ^{2}H and ^{3}H nuclei have the name deuteron and triton, and bulk quantities of these materials are called deuterium and tritium, respectively. Isotopes of other elements do not have special names.

Question   How many protons and how many neutrons are in each of the nuclides $^{2}_{1}$H, $^{3}_{1}$H, $^{4}_{2}$He, $^{35}_{17}$Cl, $^{37}_{17}$Cl, $^{137}_{55}$Cs, $^{238}_{92}$U, and $^{239}_{94}$Pu? (H = hydrogen, He = helium, Cl = chlorine, Cs = cesium, U = uranium, Pu = plutonium.)

The Nuclear Force

The nucleons in a nucleus form a cluster somewhat like eggs in a basket (Fig. 18.9). That is, they don't overlap but each nucleon takes up roughly the same amount of space. This means that a ^{234}Th nucleus occupies about 78 times the volume of a ^{3}He nucleus.

---

*The amu is defined as 1/12 the mass of the carbon-12 atom. Thus, ^{12}C is the only atom with mass in amu exactly equal to its mass number.

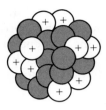

Figure 18.9 In an approximation that is about as valid as the Bohr model of the atom, you can think of the nucleus as a cluster of protons (labeled "+") and neutrons.

Nucleons are typically about 1 to 2 fm apart in a nucleus. You can interpret fm to mean either femtometer (10^{-15} m) or *fermi*, the name often used for this unit in honor of Enrico Fermi, a renowned nuclear physicist about whom we'll have more to say later. Because it's about the right size, a fermi is a handy unit for talking about nuclear dimensions.

What do you suppose holds this cluster together? Remember that the electrons are held in the *atom* by the electrical force between the negative charge of the electron and the positive charge of the nucleus. But the *nuclear* particles are either uncharged or positively charged. The electrical force is either zero (between neutron and neutron or between neutron and proton) or repulsive (between proton and proton). As a matter of fact, if you use Coulomb's law (Chapter 10) and calculate the *electrical* force between two protons 1.5 fm apart, you get 100 N (22 lb). That doesn't seem like much force, until you think about the mass of the particles on which the force is acting. The following exercise may put this force in perspective. You can see that there is a tremendous electrical repulsion in a heavy nucleus like uranium, with 92 protons.

Exercise   Use Newton's second law of motion ($F = ma$) to calculate the acceleration of a proton free to move, and acted on by a net force of 100 N.

Ans: 7.0×10^{28} m/s² (7 000 000 000 000 000 000 000 000 000 times your acceleration if you fall off your rocker).

What about the other ordinary force—the gravitational force? Could it hold the nucleus together? It would act to attract the nucleons to each other. But if you calculate its strength between our protons 1.5 fm apart, you get a force of only 8×10^{-35} N. That would do about as much good in overcoming the electrical repulsion between protons as your trying to hold back a tidal wave with a surfboard.

Obviously, there must be *some* very strong force that acts between nucleons, attracting them to each other enough to overcome the electrical force. This force, called the **nuclear force** or sometimes just the **strong force,** exists only between nuclear particles. Experiments show that this force is the same between any two kinds of nucleons, and that it doesn't act if the nucleons are separated by more than a few fermis. For example, if a proton, or a group of nucleons including at least one proton, gets separated from the nucleus by a small amount, it will be repelled by the electrical force and move off at an extremely high speed.

This nuclear force is one of nature's fundamental kinds of force, and is distinct in nature from the other two we've discussed in some detail: the electromagnetic and the gravitational forces.

One of the functions of the neutrons is to act as a sort of nuclear "glue." Since the nuclear force doesn't care what kind of nucleon it acts on, the neutrons can provide some added nuclear force to overcome the electrical repulsion. Figure 18.10 is a plot of the number of neutrons versus number of protons for the nuclei that exist stably in nature. As the graph shows, for light nuclei up to about the calcium region (atomic number 20), the number of neutrons and protons is nearly the same. For heavier nuclei, the percentage of neutrons steadily increases to compensate for the higher electrical repulsion.

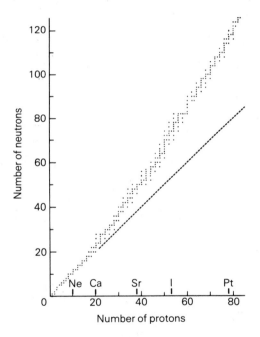

Figure 18.10   Number of neutrons versus number of protons in stable nuclei. The approximate location of some of the elements is shown (in order, neon, calcium, strontium, iodine, platinum). The dashed line shows where the number of neutrons equals the number of protons.

Since nuclear forces are at least strong enough to overcome the electrical repulsion between protons, we can get a *general feel* for the minimum strength of the nuclear force. From our earlier calculation, it must be at least about 100 N between nucleons. Let's compare this with the force holding an electron in orbit around the nucleus. A Coulomb's-law calculation for an electron in the first Bohr orbit of hydrogen gives the force to be about 10^{-7} N, one-billionth that for the protons in the nucleus. That gives us a hint at why nuclear energies are so much larger than chemical energies. The burning of coal—a chemical reaction—involves the rearrangement of orbital electrons as the atoms form different combinations. The nuclear reactions that occur in the reactor of a nuclear power plant involve the rearrangement of nuclear particles. Considering the vastly stronger forces holding the nuclear particles in place, you might expect vastly more energy to be involved in their rearrangement. In fact, single nuclear reactions typically involve about a million times more energy than single chemical reactions.

18.5   Radioactivity

It's possible *in principle* to combine *any* number of neutrons with *any* number of protons. However, that does *not* mean that *any* combination of nucleons will form a stable, or near-stable, nucleus.

Figure 18.10 shows the combinations that we know to be stable. If by some means, you managed to create a nucleus with relative neutron and proton numbers far from this "stability line," it would simply fly apart. But what if we have a nucleus with only slightly too many or too few neutrons for its number of protons: a nucleus that is *almost* stable? Then, that nucleus merely *decays* to a more stable arrangement by the emission of some particle. We call this emission of particles **radioactivity.**

You'll notice that uranium, and the other heavy elements with atomic number higher than 83, are not on the graph of Fig. 18.10. That's because there are *no* known *stable* nuclides with atomic number higher than 83. All heavier elements are radioactive.

The term "radioactivity" may seem bothersome because its use to describe particle emission doesn't seem to fit with the word "radio." The dictionary, though, defines the prefix "radio-" to indicate the emission of radiation, and defines "radiation" as the emission and propagation of waves *or* particles. Radiation, then, can include both the radio waves broadcast from your favorite FM station *and* the particles emitted from uranium nuclei. (You already know that electromagnetic waves sometimes act as particles, and particles sometimes act as waves.) Radioactivity *is* just what its name implies.

Types of Radioactivity

Radioactivity was first discovered in uranium by Henri Becquerel in 1896. Within a couple of years, Marie and Pierre Curie discovered the radioactive elements thorium, polonium, and radium.

By making the radiation pass through a magnetic field, early investigators showed the radiation to be of three types. They labeled them α (*alpha*), β (*beta*), and γ (*gamma*) rays. The direction of the bending of the paths of α and β rays as they passed through a magnetic field indicated that they are charged positively and negatively, respectively. Since the γ rays pass through undeflected, they must be uncharged. More detailed studies showed that the α rays are actually $^4_2$He (helium) nuclei: clusters of two neutrons and two protons very tightly bound together. The β rays are *electrons,* and γ rays are *photons* of electromagnetic radiation. We'll look at each type decay in more detail.

α **Decay**   If α decay involves the emission of two neutrons and two protons, the nucleus doing the emitting is obviously *transmuted* into a nucleus of a different nuclide. Consider the α decay of radium-226, symbolized in Fig. 18.11. A ^{226}Ra nucleus spontaneously breaks up into an α particle and a nucleus with two less each of

Figure 18.11   The α decay of a ^{226}Ra nucleus. The arrows indicate the relative velocities after decay, that of the α being higher in the ratio of masses: 222/4.

Before decay $\qquad\qquad\qquad\qquad\qquad\qquad$ After decay

$^{226}_{88}$Ra $\qquad\qquad\qquad\qquad\qquad\qquad$ $^{222}_{86}$Rn $\qquad$ $^4_2\alpha$

neutrons and protons: radon-222. We write this in a reaction equation, somewhat as chemists write chemical reactions:

$$^{226}_{88}\text{Ra} \longrightarrow {}^{222}_{86}\text{Rn} + {}^4_2\alpha.$$

Sometimes people write the $^4_2\alpha$ as $^4_2$He; the two symbols mean the same thing. After the decay, the α moves off at high speed, being repelled by the electrical force between it and the nucleus. But the principle of momentum conservation requires the ^{222}Rn to recoil in the opposite direction just as a rifle recoils—kicks—when you fire it. The relative speeds are inversely proportional to the particle masses.

Radioactive decay reactions have to pay their own way; you cannot get something for nothing *there* any more than anywhere else. That is, the principles of conservation of energy and momentum apply. In addition, there are two more conservation laws that apply in any nuclear interaction:

1. *The total number of nucleons remains constant.*

2. *The total charge remains constant.*

The reaction equation we wrote includes these laws. The first law makes the total number of nucleons on the right, $222 + 4$, equal the total on the left, 226. The conservation of charge is taken care of in the protons: the total number on the right, $86 + 2$, equals the total number on the left, 88. From these two laws, you can find out the third member for any reaction in which you know two of the nuclides involved. For example, if you know that some nuclide α-decays to $^{209}_{82}\text{Pb}$ (lead), the initial mass number has to be $209 + 4$, and the initial proton number (atomic number) has to be $82 + 2$. The element with atomic number 84 is polonium. Thus, the reaction is

$$^{213}_{84}\text{Po} \longrightarrow {}^{209}_{82}\text{Pb} + {}^{4}_{2}\text{He}.$$

Question Uranium-234, $^{234}_{92}\text{U}$, α-decays to Th, which α-decays to Ra, which α-decays to Rn, which α-decays to Po, which α-decays to $^{214}_{82}\text{Pb}$. Write the reaction equation for each reaction in this decay "chain."

We often call the initial nuclide before decay the **parent,** the one left after decay the **daughter.** This terminology applies to all types of radioactivity, not just α decay.

β **Decay** Did you notice anything strange about the definition of β rays—*electrons* emitted from the nucleus? There are *no* electrons in the nucleus! That sounds like picking potatoes from an orange tree.

What effectively happens is that a neutron in the nucleus converts to a proton, kicking out an electron in the process. In reaction terms,

$$^{1}_{0}\text{n} \longrightarrow {}^{1}_{1}\text{p} + {}^{0}_{-1}\text{e}.$$

Notice that the number of nucleons is the same—1—on both sides of the equation, and the total charge is 0 on both sides. Since the electron has a *negative* charge, we write its "atomic" number, or *charge number,* as -1. Since it is not a nucleon, its mass number is 0.

One naturally occurring β emitter is lead-212, $^{212}_{82}\text{Pb}$. Its reaction equation is

$$^{212}_{82}\text{Pb} \longrightarrow {}^{212}_{83}\text{Bi} + {}^{0}_{-1}\text{e}.$$

Notice that, since a *negative* charge is given off, the charge of the nucleus *increases* by 1 unit: the nucleus becomes a bismuth-212 nucleus with the same *total* number of nucleons. But the daughter has one less neutron and one more proton than the parent.

You might wonder what could cause a neutron to convert to a proton with the accompanying *creation* of an electron. To account for this conversion, there is a fourth *general* type of force that is distinct from the other three we've discussed: the gravitational, the electromagnetic, and the strong (nuclear) interactions. This force, that goes by the name **weak interaction,** is responsible for the β-decay process. So far as we know, these four general types of forces account for every specific force that can exist in the universe.

γ **Decay** We said that γ rays are photons of electromagnetic radiation. They have no electric charge, and contain no nucleons. A typical reaction is

$$^{209}_{22}\text{Pb*} \longrightarrow \, ^{209}_{82}\text{Pb} + \, ^{0}_{0}\gamma.$$

The nuclide does not change in γ decay. Rather, this is the way a nucleus in an *excited* state (indicated by the symbol *) can get rid of some excess energy. The process is identical in principle to the way the orbital electrons of an atom go to a lower energy state by the emission of light or x-ray photons. The wavelength regions of γ rays and x rays overlap, meaning that you can't tell a γ ray from an x ray. Their only difference is in where they come from: an x ray from the outer atom, a γ ray from the nucleus.

When a nucleus decays by α or β emission, this process is almost always followed by γ decay because the final nucleus in α or β decay is usually left in an energy state higher than its ground state. It gets rid of this excess energy by γ decay.

Decay Rate and Half-Life

Imagine that your doctor suspects that you have thyroid cancer. The function of your thyroid gland is to remove iodine from your blood and use it in the manufacture of thyroid hormones. The food you eat each day probably contains an average of about 150 μg of iodine. Your thyroid doesn't care which isotopes of iodine it gets; it just takes whatever iodine comes down the pipe. However, cancerous regions of the thyroid usually have a lower-than-normal uptake of iodine.

Among the other tests your doctor might perform to check out his suspicion, there is a fair chance he will feed you something containing a small amount of an isotope of iodine that is radioactive—maybe ^{131}I. After waiting a specific period for your thyroid to perform its function, he will put you in front of a scan-

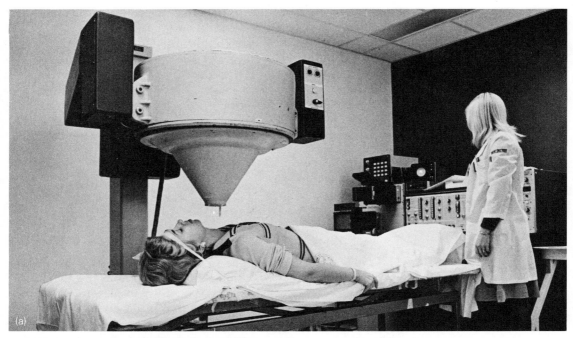

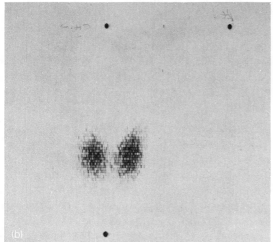
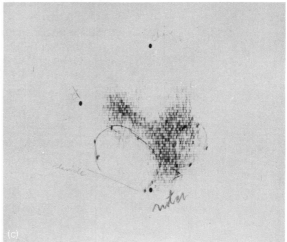

Figure 18.12   (a) A thyroid scanner in operation. (b) A normal thyroid scan. The darkness is proportional to the activity from that spot. (c) An abnormal thyroid scan.

ner (Fig. 18.12-a) that can detect the *number and direction* of γ rays from the ^{131}I. If your thyroid is normal, the map of the γ activity—the *scan*—will look something like Fig. 18.12(b). The darkness of each spot is proportional to the amount of γ radiation coming from that direction. The relatively even distribution in (b) indicates no regions of excessively high or low iodine concentration, such as those in Fig. 18.12(c).

Now look at the nuclear physics of the procedure we've just described. First of all, even though ^{131}I is radioactive, not all the nuclei decay immediately. Instead, they decay at some given rate; that is, some fraction of the ^{131}I nuclei decay each second. *The total rate of decay is proportional to the number of radioactive nuclei present.* The scanning technique depends heavily on this fact.

The number of nuclei of a substance decaying per unit of time is called its **activity.** The customary unit used to measure activity is the *curie* (Ci), defined as 3.7×10^{10} decays per s.* The SI unit of activity, the becquerel (Bq), defined as the number of decays per second, hasn't yet gained widespread use. The usual dose of ^{131}I used for a thyroid scan is from 50 to 100 μCi: 2 to 4 MBq ($\mu = 10^{-6}$, M $= 10^{+6}$).

Exercise   How many γ rays are emitted from a 3.7-MBq (100-μCi) ^{131}I sample during the 20 min of a typical thyroid scan? Note: The scanner detects only a small fraction of this number because not all the ^{131}I goes to the thyroid, and the scanner itself intercepts only a tiny portion of the γs emitted equally in *all* directions.   **Ans:** 4.4 billion

---

It's interesting that, even though radioactive ^{131}I is used in the scanning technique, the γ rays do not actually come from I nuclei. The I nuclei β-decay through the reaction

$$^{131}_{53}\text{I} \longrightarrow \,^{131}_{54}\text{Xe} + \,^{0}_{-1}\text{e},$$

leaving the xenon nuclei in an excited state. These ^{131}Xe nuclei give up their excess energy by emitting the observed γ rays.

Half-Life   As ^{131}I converts to ^{131}Xe, the supply of ^{131}I is depleted. Therefore, the activity must decrease proportionately. Suppose the technician in charge of radioisotope storage in Dr. Cureall's clinic measures the activity of his bottle of ^{131}I and finds he has 20 MBq (5.4 mCi). He comes back 8 days later, and finds he has only 10 MBq (2.7 mCi). Then 8 days later 5 MBq, and in another 8 days 2.5 MBq. A graph of this activity (Fig. 18.13) shows that the decay rate after *any* 8-day period is ½ what it was at the beginning of that period.

---

*The curie, named after the discoverers of radioactivity in radium, was originally defined as the activity of 1 g of ^{226}Ra.

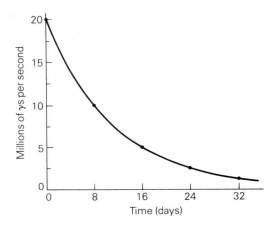

Figure 18.13   The activity of an ^{131}I sample that initially has a decay rate of 20 MBq.

Suppose Dr. Cureall's technician does the same with his supply of ^{123}I, also sometimes used for thyroid studies. He would get an identical looking graph, except that the activity would cut in half every 13 hours.

This change in decay rate is typical of *all* radioactive nuclides. There is a time period for each nuclide, called its **half-life,** during which time its decay rate decreases by 50%. Remember that the decay rate is proportional to the amount of the nuclide present. Therefore, *the half-life is the time needed for half the nuclei in a given sample of a radioactive nuclide to decay.* The half-life varies from nuclide to nuclide.

Exercise   If Dr. Cureall gives Mrs. Neversick a 3.7-MBq (100-μCi) dose of ^{131}I, how long will it take for the activity in her body to decrease to 0.037 MBq (1 μCi)? How long if she gets the same dose of ^{123}I? Can you think of an advantage of using ^{123}I rather than ^{131}I? Unfortunately, it's not as readily available.

Ans: 56 days; 3.8 days (to nearest half-life)

The measured half-lives of radioactive nuclides vary from extremely small fractions of a second to billions of years: 4.5 billion for ^{238}U and 14 billion for ^{232}Th. Table 18.1 gives the half-lives of a few selected nuclides, most of which have some practical application.

When radioactive nuclides, often called **radioisotopes,** are used in following the path of some element such as iodine in the body, they are called **tracers.** The name comes from the fact that you can *trace* their paths.

Table 18.1 Half-lives of Selected Nuclides

Nuclide	Half-life	Nuclide	Half-life
neutron	11 min	strontium-90	28.9 yr
tritium (^{3}H)	12.3 yr	iodine-131	8.1 day
beryllium-8	10^{-16} s	radium-226	1620 yr
carbon-14	5730 yr	uranium-235	7.1×10^6 yr
sodium-24	15.0 h	plutonium-239	24 400 yr
cobalt-60	5.3 yr	americium-241	433 yr

CAUTION

RADIOACTIVE MATERIALS

Figure 18.14 The symbol used universally to identify the presence of a radioactive source or a general radiation area.

An interesting industrial use of tracers is by soap manufacturers. They like to know how effective their soap is in getting dirt out. One way they measure this effectiveness is by adding a radioactive tracer to the dirt, and then checking the activity of the dirty water.

Do you, or your parents, have a smoke detector (alarm) in your house? Take the cover off and look inside. If it's the "ionization" type, the most popular, you'll see the "radiation" symbol inside (Fig. 18.14). This symbol is used to identify either the presence of a radioactive source or an area in which you may be exposed to radiation. The smoke detector contains a weak radioactive source, usually Americium-241, located as shown schematically in Fig. 18.15. The α particles from the source ionize the air; that is, they knock out electrons leaving positive ions and free electrons. The battery then causes a weak electric current through the air. When smoke enters the measurement chamber, the ions attach themselves to the smoke particles. Their very slow movement reduces the current, causing the alarm to sound. Some brands advertize *dual* ionization chambers. The second "reference" chamber lets no smoke in, but compensates for changes in temperature, air pressure, and humidity, thus reducing false alarms.

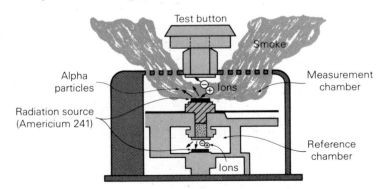

Figure 18.15 An ionization type smoke detector uses a radioactive source.

18.6   Biological Effects of Radiation

When radiation passes through matter, it knocks out electrons, thereby forming *electron-ion pairs*. This **ionization** process allows the detection of radiation in instruments such as geiger counters. But radiation also ionizes the atoms of biological cells that it traverses. This ionization can displace or chemically change the molecules of a cell, thereby doing damage to it. Radiation effects may harm the person exposed or, if the damage is to genetic cells, the offspring of the exposed person may suffer the consequences.

We are using the word "radiation" here to mean *ionizing* radiation such as x rays, γ rays, electrons, and nuclear particles. Heavy exposure to other types of radiation—infrared and ultraviolet, for example—may do surface damage, but these forms of radiation do not penetrate the body and produce ionization deep inside.

Radiation Exposure   The unit usually used for radiation **dose** is the rem, which takes into account both the energy deposited per kilogram of exposed matter and the differing effects of different types of radiation. Ordinary doses are often given in mrem (0.001 rem).

Most of the radiation exposure the average person gets is from natural unavoidable sources. These include cosmic radiation from outer space and naturally radioactive nuclides in the earth. These quantities vary drastically from place to place. For example, people in higher altitudes or who fly in airplanes get more cosmic rays; people in brick houses get more exposure than those in wooden houses because of the natural radioactivity in the sand of the bricks. Table 18.2 shows how you can compute your own radiation dose. The column on the right tells the percentage of the *average* annual dose in the United States attributable to each of the sources. If you receive medical x rays, your exposure might be twice this average annual dose of 148 mrem.

Notice that 1 mrem/yr is equivalent to moving to an elevation 100 ft higher; increasing your diet by 4%; or taking a 4- to 5-day vacation in the Sierra Nevada. The people living within a 10-mile radius of the Three Mile Island nuclear power plant near Harrisburg, Pa., received an average dose of about 10 mrem from the highly publicized accident there in 1979. Comparing that 10 mrem with the figures in Table 18.2 will help to put the accident's consequences in perspective.

Benefits of Radiation Exposure   There are numerous benefits to radiation "damage." For example, certain foods are exposed to

Table 18.2 Compute Your Own Radiation Dose.[a]

Fill in the blanks with the numbers appropriate to your situation.

	Common Source of Radiation	Your Annual Inventory (mrem)	Percent[b]
WHERE YOU LIVE	Location: Cosmic radiation at sea level Elevation: Add 1 for every 100 feet of elevation Typical elevations (feet): Pittsburgh 1200; Minneapolis 815; Atlanta 1050; Las Vegas 2000; Denver 5280; St. Louis 455; Salt Lake City 4400; Dallas 435; Bangor 20; Spokane 1890; Chicago 595. (Coastal cities are assumed to be zero, or sea level.)	44 ——	29.7
	House construction (based on ¾ of time indoors) Brick 45 Stone 50 Wood 35 Concrete 45	——	27.0
	Ground (U.S. average based on ¼ of the time outdoors)	15	10.1
WHAT YOU EAT, DRINK, AND BREATHE	Water, Food, Air, U.S. average	25	16.9
	Weapons test fallout	4	2.7
HOW YOU LIVE	X-ray diagnosis: Chest x ray —— × 9 Gastrointestinal tract x ray —— × 210	——	13.5
	Jet airplane travel: Number of 6000-mile flights —— × 4.	——	0.1
	Television viewing: Number of hours per day —— × 0.15.	——	0.1
HOW CLOSE YOU LIVE TO A NUCLEAR PLANT	At site boundary: Annual average number of hours per day —— × 0.2 One mile away: Annual average number of hours per day —— × 0.02 Five miles away: Annual average number of hours per day —— × 0.002 Over 5 miles away: None	—— —— ——	Less Than 0.1%
	Total mrem	——	100

[a] From *Nuclear Power and the Environment—Questions and Answers* (American Nuclear Society, La Grange Park, Ill., 1976).

[b] Percentage of the average annual U.S. dose of 148 mrem.

high doses of radiation to kill microorganisms and extend shelf life. Some medical supplies are sterilized by exposure to radiation.

You no doubt know of people given radiation treatment for cancer. The cancerous tissue is exposed to heavy doses of high-energy radiation in order to destroy the cancerous cells. In some cases this radiation is administered from "within," and in other cases from outside the body.

We discussed earlier the use of ^{131}I in *detecting* thyroid disorders. In some cases of thyroid cancer, the thyroid still has the ability to concentrate iodine. The doctor might then administer large doses—up to 7 GBq (200 mCi)—of radioactive iodine. The thyroid will then concentrate the iodine there and hopefully destroy the cancerous tissue by radiation exposure.

The more common radiation treatment uses a source outside the body, often referred to as "cobalt" treatment because most hospitals use the γ rays from ^{60}Co for this purpose. The source is kept in a lead box with a small opening so that the γ rays emerge in a narrow beam. The radiologist aims the beam at the tumor. Unfortunately the γ rays damage healthy cells as well as diseased ones. For this reason the beams are often brought in from different directions so that they intersect at the position of the tumor (Fig. 18.16). This procedure maximizes the energy deposited there and minimizes that delivered to surrounding healthy cells.

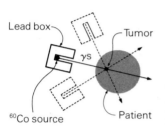

Figure 18.16   In cancer therapy, the γ-ray beam is directed from different angles to concentrate the radiation at the point of the tumor.

18.7   Nuclear Reactions

Let's get a little more "mileage" from our ^{131}I examples discussed earlier. Where do you suppose the ^{131}I comes from? We didn't mention that it does not exist naturally. Evidently, there is some way of artificially converting one element to another. You may remember that converting one element to another was the goal of the ancient alchemists; they wanted to convert base metals to gold. We haven't quite perfected that art yet, but we can transmute a fraction of the nuclei in a sample to a nuclide with a nearby mass and atomic number.

Causing Nuclear Reactions

We get this change of nuclide by bombarding appropriate nuclei with nuclear particles. For example, we can produce ^{131}I by bombarding the naturally available nuclide tellurium-130 with protons.

Before reaction   After reaction γ

^1_1p $^{130}_{52}\text{Te}$ $^{131}_{53}\text{I}$

Figure 18.17 The nuclear reaction that transmutes a ^{130}Te nucleus to an ^{131}I nucleus.

Figure 18.17 pictures what happens to one Te nucleus. The **reaction** *equation* is

$$^{130}_{52}\text{Te} + {}^1_1\text{H} \longrightarrow {}^{131}_{53}\text{I} + {}^0_0\gamma.$$

The resulting ^{131}I nuclei are left in an excited state after the proton *capture,* and quickly de-excite by γ emission, as the equation shows.

Reaction equations are similar to those for radioactive decay. The difference is the presence of the bombarding particle—in this case the proton, which is sometimes written ^1_1p. Notice that, as for radioactive decay, both charge and mass numbers are *conserved:* Their sums on both sides of the equation are equal.

We can also get ^{131}I in a two-step process by bombarding ^{130}Te with neutrons:

$$^{130}_{52}\text{Te} + {}^1_0\text{n} \longrightarrow {}^{131}_{52}\text{Te} + {}^0_0\gamma,$$

$$^{131}_{52}\text{Te} \longrightarrow {}^{131}_{53}\text{I} + {}^{\ 0}_{-1}\text{e}.$$

The ^{131}Te β-decays to ^{131}I with a half-life of 25 min. After several hours, most of the tellurium has decayed, leaving the wanted iodine isotope.

Aside from the use of nuclear reactions as a way of producing radioisotopes for everyday use, such reactions provide nuclear physicists their main tool for learning about the nucleus. We said in describing Rutherford's experiments that showed the existence of the nucleus, that the nucleus is so small that we can't get "close enough" to observe it. So we throw "rocks" at it—nuclear projectiles—and use radiation detectors to find out what comes out. We can infer what some of the nuclear features are by what comes out, and by the direction and kinetic energy of the outcoming particle relative to what went in.

There's another analogy we can use to explain how we study the nucleus by nuclear reactions. If you want to see the skunk stealing your cookies from the picnic table outside your camping tent, you shine a flashlight on him. The light scattered from the skunk into your eyes lets you identify the thief. The "light" you shine on a *nucleus* to "see" it is a beam of protons, deuterons, α particles, or some other particles. You see the effect of the scattering by radiation detectors rather than your eyes. You could not use visible light because the wavelength is much longer (100 million times longer) than the object you're trying to see. High-speed nuclear particles have a wavelength (remember de Broglie) about the right size to scatter from nuclei. Researchers use machines called *particle accelerators* to produce beams of high-speed charged particles. For bombardment by neutrons, they immerse the sample in the high neutron population of a nuclear *reactor* (Section 18.8).

**Carbon-14 Dating

Have you ever wondered how we can know the age of a 5000-year-old Egyptian mummy? One method of dating that can tell its age within 100 years makes use of both nuclear reactions and radioactive decay.

The story starts in the upper atmosphere where neutrons from cosmic-ray interactions produce carbon-14 in the reaction

$$^{14}_{7}N + ^{1}_{0}n \longrightarrow ^{14}_{6}C + ^{1}_{1}p$$

(N = nitrogen). This radioactive ^{14}C, with a half-life of 5700 years, combines with oxygen to form carbon dioxide. When the carbon dioxide migrates to earth, it constitutes about 1 out of every 10^{12} carbon dioxide molecules taken in by plants. These plants then get eaten by people and animals.

When the plant, animal, or person dies, it quits taking in ^{14}C, and the relative percentage of its carbon that is ^{14}C decreases by radioactive decay. By measuring the activity and thus the fraction of ^{14}C in the remains, we can measure the time since the organism's death. The method is not very accurate for ages beyond several 5700-yr half-lives of ^{14}C. But it works well for things that are from a thousand or so (the Dead Sea scrolls, for example) to 15 or 20 thousand years old.

**Nuclear Fingerprints

It's now suspected that Napoleon was slowly poisoned to death with arsenic, rather than dying of cancer, as some of us learned in History 1111. Some English scientists not long ago found unusual amounts of arsenic in a sample of Napoleon's hair. (For some reason arsenic taken into the body concentrates in the hair.) They irradiated the sample in a reactor, and found γ rays characteristic of ^{76}As, which you get by irradiating natural arsenic, ^{75}As, with neutrons.

Neutron activation analysis, the name for this technique, is a sensitive way of finding extremely small amounts of almost any substance in a sample. Each different nuclide, after neutron irradiation, emits γ rays with energies and half-lives characteristic of that nuclide only. The procedure does not destroy the sample; the small fraction of nuclei made radioactive merely decay back to stable nuclides.

This method has been used in thousands of broad-ranging applications—matching small bits of blood found on a murder suspect with that of the victim; identifying small flakes of paint in hit-and-run automobile accidents; identifying elements in samples of moon rocks; looking for lead concentrations in the hair from deer that live near highways; protecting against counterfeiting of products by deliberately including traces of known elements that can be easily activated for identification—to name just a few.

Energy from Nuclear Reactions?

The world seems to have divided into "pro-nukes" and "anti-nukes": those favoring and those opposing the use of nuclear energy. Let's not "choose up sides" on that issue here, but instead let's look at the conditions under which we can get energy from a nuclear reaction. That will hopefully put us in a more informed position for choosing sides.

I hope that, by now, you have come to believe in the principle of conservation of energy: that you can't create or destroy energy. If energy comes from a nuclear reaction, it is not created, but must result from the loss of *some* form of energy by the nuclei involved.

We'll examine a particular reaction that can liberate energy,

$$\mathrm{^2_1H} + \mathrm{^2_1H} \longrightarrow \mathrm{^3_2He} + \mathrm{^1_0n},$$

and see both where this energy comes from and in what form it shows up. This reaction, which incidentally might supply part of the energy to light your house and cook your food in twenty or thirty years, combines two deuterons to form a neutron and a helium-3 nucleus.

The masses of the ^{2}H, ^{3}He, and n are 2.0141 amu, 3.0160 amu, and 1.0087 amu, respectively. As illustrated *symbolically* in Fig. 18.18, this means that the total mass *before* the reaction exceeds that *after* the reaction by 0.0035 amu. So you're not impressed by a trivially small difference in the masses of things that have an unbelievably small mass to start with!

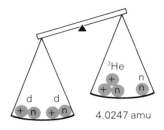

4.0247 amu

4.0282 amu

Figure 18.18 The mass of two deuterons is larger than that of a neutron plus a ^{3}He. The diagram is purely symbolic: You can't actually *weigh* nuclei with a balance. You find their mass indirectly by methods such as measuring the curvature of their paths in a known magnetic field.

In Einstein's theory of relativity, he showed that mass can be converted to energy, and vice versa. The relationship between mass and energy is given by the popularized equation

$$E = mc^2,$$

where m is the mass in question and c is the speed of light, 3×10^8 m/s. On this basis, if you could convert the ½ kg of food you probably had for dinner yesterday entirely to energy, you'd get 4.5×10^{16} J, or about 12 billion kWh. This quantity is about 1/6 of the total energy used per day by the *entire* United States for *all* purposes! This example makes it clear that small mass changes can provide significant energy changes.

Now go back to our nuclear reaction. The energy equivalent of our 0.0035 amu mass difference is 5.2×10^{-13} J, or 3.2 million electron volts (MeV). We can get this by converting the 0.0035 amu to kilograms, multiplying by c^2 to get energy in J, and converting this to MeV. If we lump all these conversion factors into one number, we find the *energy equivalent of 1 amu is 931 MeV*. In other words, if we have a mass in amu, we can get its energy equivalent in MeV by multiplying by 931:

$$E = 0.0035 \text{ amu} \times 931 \text{ MeV/amu}$$

$$= 3.2 \text{ MeV}.$$

This result tells us that 3.2 MeV of energy is released in each

$$^2_1\text{H} + {}^2_1\text{H} \longrightarrow {}^3_2\text{He} + {}^1_0\text{n}$$

reaction. *This energy shows up as kinetic energy of the reaction products.* In other words, the total *kinetic energy* of the ^{3}He and n exceeds that of the two deuterons by 3.2 MeV. This is an *extremely* small amount of energy by everyday standards: it would take 4.4×10^{18} such reactions to give enough energy to boil away a liter of water. But as we'll see, there are ways of getting many times that number of reactions.

If you start with a heavy nucleus in the uranium region and split it—*fission* it—into two parts of roughly equal size, the resulting *fission products* usually have somewhat lower mass than the original nucleus. On the other hand, if you take very light nuclei—^{2}H, ^{3}H, etc.—and *fuse* them, they form nuclei of roughly twice their mass. The overall mass is significantly decreased.

Since either of these types of reactions, **fission** of heavy nuclei and **fusion** of light nuclei, result in a lower total mass, energy is released as they occur.

18.8   **Nuclear Energy from Fission and Fusion**

Even though all types of nuclear reactions are useful for learning about the nucleus and its behavior, only two types, fission and fusion, seem useful for large-scale release of energy. We'll consider the two approaches separately.

Fission and Fission Reactors

Heavy nuclei may once in a while fission on their own, but to get any significant fraction of the nuclei of a sample to fission, we need to add energy by hitting the nucleus with some kind of particle, for example a neutron.

If a neutron hits a ^{235}U nucleus, it gives energy to the nucleus that can cause it to vibrate and oscillate much as you might expect a water drop to do if you hit it with a smaller droplet. As Fig. 18.19 depicts, these oscillations often cause the nucleus to take an elongated "barbell" shape, and then split into two parts. Usually a few extra neutrons splatter out in the process. The particular **fission fragments** and the number of neutrons emitted vary from fission to fission. One such reaction is

$$\,^{1}_{0}\text{n} \; + \; \,^{235}_{92}\text{U} \; \longrightarrow \; \,^{236}_{92}\text{U} \; \longrightarrow \; \,^{94}_{38}\text{Sr} \; + \; \,^{139}_{54}\text{Xe} \; + \; 3\,^{1}_{0}\text{n}.$$

In the intermediate stage between neutron absorption and fission, the nucleus is ^{236}U. Even though the fission fragments sometimes have nearly equal mass, usually they are somewhat unequal: typically in the 100 and 140 mass-number region.

The mass change during fission of ^{235}U has an average energy equivalence of about 200 MeV. Each fission, then, releases an average of about 200 MeV of kinetic energy to the fission fragments and neutrons. To get enough energy from fission for practical use, we need billions of such reactions.

Fission fragments

Figure 18.19   When a ^{235}U nucleus captures a neutron, the energy it gets causes it to oscillate and sometimes fission into two parts with a few neutrons left over.

This goal can be reached by using the neutrons released in each fission to interact with other heavy nuclei to cause other fissions: a **chain reaction.** Figure 18.20 symbolizes this effect. We show two neutrons from each fission going on to cause other fissions. If this situation were left uncontrolled, it would build up *very rapidly* with a very quick release of energy. We would have a bomb! This effect was just the one used in the "atomic" bombs dropped on Hiroshima and Nagasaki. As long as the average number of fission-released neutrons that go on to produce other fissions is larger than 1 per fission, the reaction rate will increase indefinitely.

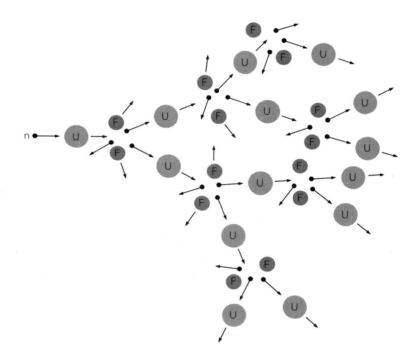

Figure 18.20 A fission chain reaction. An incoming neutron causes the fission of a uranium nucleus, which releases three neutrons. In each fission two of the neutrons (black dots) released go on to cause other fissions.

To use fission-released energy for everyday purposes, we need to control the fission chain reaction and force it to proceed at a manageable rate. The device that does this is a **nuclear reactor.** The reactor provides the ability to keep the reaction rate at a fixed level: on the average *one* neutron from each fission going on to cause another fission. The other neutrons either escape or are absorbed without producing a fission.

Figure 18.21 is a schematic diagram of the main components of a nuclear reactor. Neutrons released in the fission of uranium or

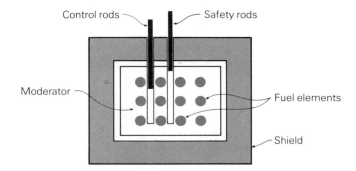

Figure 18.21 Schematic diagram showing the main components of a nuclear reactor.

plutonium in the fuel elements are slowed down by a *moderator* (usually water or graphite), so that they have a higher probability of interaction with another fuel nucleus.

The control rods are made of a material such as cadmium that easily absorbs neutrons, without fissioning. The reactor operator controls the reaction rate by inserting or withdrawing the control rods to slow down or speed up the chain reaction. The safety rods are good neutron absorbers that can be inserted quickly to shut down—**scram**—the reactor in an emergency.

When a reactor is operated at a fixed rate, with exactly one neutron per fission causing another fission, we say it is **critical.** When more than one neutron per fission causes another fission, the chain reaction is building up and we say the reactor is *supercritical.* Below the critical rate, we say it's *subcritical.*

As a source of useful energy, a reactor is just a nuclear furnace. The kinetic energy of the fission fragments and neutrons heats the materials in the reactor core. A liquid *coolant* pumped through the core absorbs and extracts the heat in the same way water does in a coal-fired boiler. In most present reactors, this coolant is water, which also serves as the moderator. The schematic diagram of a nuclear power plant would be similar to Fig. 13.1. There's one additional step: The coolant water from the reactor usually goes to a steam generator, where it gives up heat to produce the steam.

Football and the "Atomic" Age

Even though football brings out more emotion on many campuses than any other activity, its most ardent enthusiasts (excluding the coaches and players) think of it as a diversion rather than the stuff that shapes their lives. But the events that took place in 1942 under the football stadium at the University of Chicago have changed the

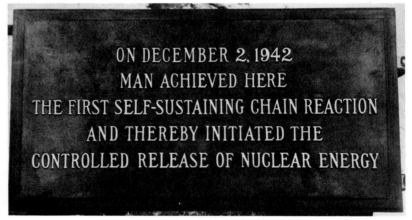

ON DECEMBER 2, 1942
MAN ACHIEVED HERE
THE FIRST SELF-SUSTAINING CHAIN REACTION
AND THEREBY INITIATED THE
CONTROLLED RELEASE OF NUCLEAR ENERGY

Figure 18.22   Enrico Fermi and his co-workers produced the first self-sustaining chain reaction in a reactor under the stands of Stagg Field at the University of Chicago.

course of world affairs. It was in that unsuspected location, beneath the stands of Stagg Field (Fig. 18.22), that Enrico Fermi and his co-workers secretly produced the world's first self-sustaining chain reaction. They accomplished this feat in a reactor containing uranium fuel embedded in graphite moderator.

If you ever have a chance to hear a description of that fascinating story by one of the members of the team involved, don't miss it! The story has been recorded by many people, including Fermi's wife Laura.*

---

* Fermi, Laura, *Atoms in the Family: My Life with Enrico Fermi* (University of Chicago Press, Chicago, 1954).

Nuclear Energy: Benefits and Risks

The hot debate and raging emotions that sometimes arise obscure the fact that nuclear energy is really a mixed bag. There are *obvious* advantages, and there are *obvious* problems associated with its use. We'll briefly describe the physical bases for both. First the advantages.

The most obvious advantage of nuclear energy lies in the use of nuclear fuels rather than fossil fuels. The energy available will be many times the world's total supply from fossil fuels, and its use will not deplete the finite supply of fossil fuels.

Studies show the cost of nuclear power to be appreciably less than from fossil fuels, especially as the technology is improved and fossil fuels become more scarce. The "burning" of nuclear fuels does not release the vast amount of smoke and fumes into the air that fossil fuel plants release. And the devastation of land per year for uranium mining is only a small fraction of that stripped in coal mining.

Now, for the problems. A question that probably occurs to everyone is: Can a runaway reactor become an atomic bomb? We can answer that question with an emphatic **no.** Reactor fuel contains only 3 to 5% fissionable nuclides such as ^{235}U. This percentage is much lower than that needed for a bomb.

However, in the event of an accident resulting in loss of coolant, the reactor core could become hot enough to melt everything inside. Reactors are always built with several redundant independent safety systems to shut down—"scram"—the reactor and prevent such a *meltdown*. Nevertheless, every reactor core is surrounded by two containment vessels to contain the radioactive hodgepodge that would result in the unlikely event of a meltdown. There is, of course, the small possibility that the vessels will rupture and release radioactive material into the environment.

Pollution from nuclear power plants is of two forms: radiation and thermal. Small amounts of radioactive gases, mostly krypton-85, are released from some reactors. Even though the total radiation produced is small compared with natural sources, any increase in radioactivity in the environment is undesirable. Thermal pollution comes from the heat deposited in the atmosphere or nearby body of water in condensing the steam from the turbines. For a nuclear plant, this thermal pollution is about 1.5 times that for a fossil fuel plant.

The problem of radioactive waste disposal is a nagging one whose final solution has not yet been found. Let's see why we have this problem.

Remember that the more massive the nucleus, the higher the ratio of neutrons to protons. Therefore, when a heavy nucleus fissions, the fission fragments have too many neutrons to be stable. They get rid of these excess neutrons by β-decaying until they reach stability. For example, the ^{139}Xe in the fission reaction we wrote earlier β-decays to lanthanum-139 through the reactions

$$^{139}_{54}\text{Xe} \xrightarrow[40\text{ s}]{\beta} {}^{139}_{55}\text{Cs} \xrightarrow[9.3\text{ min}]{\beta} {}^{139}_{5}\text{Ba} \xrightarrow[83\text{ min}]{\beta} {}^{139}_{57}\text{La}.$$

In this more compact notation, the half-life for each decay is shown below the arrow. Notice in these reactions something that is typically, but not always, true: The closer the nucleus is to stability, the longer its half-life.

Most of the fission fragments and their decay products have fairly short half-lives, and decay to stability within days. A few hang around for quite a long time. A couple of the more notorious ones are strontium-90 and cesium-137, which have half-lives of 29 and 30 years, respectively. Some fragments have half-lives of thousands of years, but their relative abundance is small.

After spent fuel elements are reprocessed, and all useful materials such as unused fuels are removed, there remains some radioactive garbage that will be dangerously radioactive for *hundreds* of years. This material has to be maintained and monitored to be sure it doesn't contaminate the environment. Various agencies, both private and governmental, are studying alternatives for long-term handling of this material. It's now being kept in large heavily shielded underground tanks.

Finally, there's the *plutonium* problem. Because the uranium in most reactors is at least 95% nonfissionable ^{238}U, a significant amount of ^{239}Pu is produced by neutron absorption in the uranium. This plutonium can be recovered by chemical separation when the fuel is reprocessed, and later used for nuclear fuel.

Plutonium is a concern because it is highly toxic. It's chemically poisonous, but its main toxicity is associated with the radiation: Long before you get enough to hurt you chemically, you've been "gotten" by the radiation. The most likely source of exposure in working with plutonium is from breathing plutonium-contaminated air, some of which attaches itself to lung tissue where it may later cause lung cancer. The plutonium that gets into the blood usually deposits in the skeleton where it can produce bone disease, including cancer.

For these reasons, workers need to be extremely careful in handling plutonium. Experts consider the maximum acceptable *lifetime* intake to be about 0.7 μg of plutonium.

The Big Question The decision of whether or not our country should commit itself to additional large-scale use of nuclear energy from fission depends upon two questions:

1. Is the potential benefit of nuclear power worth the risk involved in using it?

2. Are the alternatives any less risky?

Taking risks is part of living. Most of us regularly ride in cars, participate in athletic events, work in factories, and commit ourselves to be involved with other people, all of which require some measure of risk. We do these things because we feel the potential benefit is worth the risk.

Several of the end-of-chapter references address the problem of the risks of using nuclear energy. Some compare probability of injury from nuclear power plants with the probability of injury in other risks that we regularly accept. Other references compare the risk of nuclear versus other energy sources. I urge you to read these references.

Fusion and Fusion Reactors

We pointed out in Section 18.7 that when light nuclei are fused to form heavier ones, the final mass is usually less than the initial. Some reactions that meet this condition are:

$$^2_1H + {}^2_1H \longrightarrow {}^3_2He + {}^1_0n \ (3.2 \text{ MeV}),$$

$$^2_1H + {}^3_2He \longrightarrow {}^4_2He + {}^1_1H \ (18.3 \text{ MeV}).$$

The energy released per reaction by loss of mass is listed in parentheses.

These can go only if the initial particles come together with enough kinetic energy to overcome their mutual electrical repulsion. The only known way to do this on a large scale is to raise the gas to an intensely high temperature, at least 40 million °C. At these temperatures, thermal energy provides the needed kinetic energy, and collisions knock off all electrons from the atoms. We call this completely ionized gas a *plasma*. This high-temperature plasma presents a "slight" problem: What kind of pot do you put it in? Any material medium would immediately vaporize as soon as the plasma touched it.

Reactions in which thermal energy provides the needed kinetic energy for the nuclear reactions to occur are called **thermonuclear**

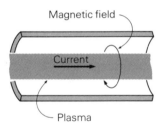

Figure 18.23 The "pinch effect" used in confinement of a plasma.

reactions. Efforts to produce a controlled thermonuclear reaction that gives off more energy than it takes to get it going center on three main objectives: (1) achieving high enough plasma temperatures; (2) compressing and confining the plasma long enough for a net energy output; and (3) recovering the energy in a useful form.

There are two general approaches being used to solve these problems: **magnetic confinement** of the plasma, and **laser bombardment** of the fuel pellets. In magnetic confinement, the plasma carries a large electric current and is confined by magnetic fields away from any material objects. Figure 18.23 shows *one* way of achieving this confinement. The current in the plasma produces a magnetic field around it as shown by the curved arrow. The charged plasma particles moving through this field experience a force toward the center. The plasma thus gets "pinched" into a small region, which both confines it and heats it by compression.

In the *laser fusion* scheme, pellets about 1 mm in diameter of frozen deuterium-tritium are bombarded with high-intensity laser beams from many directions simultaneously. The intense pulse of light energy heats the pellet to thermonuclear temperatures before it has time to expand and disperse. The approach is sometimes called **inertial confinement** because the inertia of the pellet particles themselves confines them there long enough to be heated. The idea is that pellets are continually dropped and hit with laser pulses, thereby releasing energy in pulses.

A **fusion reactor** for producing useful energy from fusion would need some means of getting the energy out. A likely approach is to surround the reactor core with a blanket of liquid coolant. Neutrons released in the fusion process would be slowed in the coolant, giving up their kinetic energy to heat the coolant. The coolant could be circulated to a heat exchanger to produce steam and turn a turbine to generate electricity in the usual way.

As of this time (1980) neither magnetic confinement nor laser fusion have succeeded in getting more energy out than is put into the machines. Nevertheless, steady progress is being made in both areas. Fusion researchers have a fairly detailed timetable, according to which larger and larger machines are to be built and operated, each one being based on the knowledge gained from its predecessors. They *expect* to have working fusion reactors by the early twenty-first century.

One big plus for using fusion for energy is that its fuel supply—deuterium—is virtually limitless. About one in every 6000 hydrogen atoms in sea water is deuterium. That doesn't sound like much, but the potential fusion energy in each liter of water is about 100 times the combustion energy in a liter of gasoline. A second plus: Fusion reactions won't leave the long half-life radioactive fission fragments left by fission reactors.

Some drawbacks of fusion include: (1) The extreme neutron density in the reactor will irradiate all components of the reactor structure, making them radioactive and probably weakening them considerably. This problem is an engineering one that presumably can be solved. (2) Large supplies of tritium will be generated in a fusion reactor. Tritium is the radioactive isotope of hydrogen of mass number 3, with a half-life of 12 years. Tritium behaves chemically like ordinary hydrogen, and can form water and other compounds easily taken into the human body. It will be absolutely necessary to design fusion reactors to minimize the leakage of tritium. (3) Feasibility of fusion reactors has not yet been proven, even though researchers in the field are quite optimistic about success.

Key Concepts

All matter consists of **atoms**—about 0.1 nm across—in which the entire positive charge and most of the mass is concentrated in a **nucleus** of about 1/100 000 to 1/10 000 this size. The **Bohr model** of the atom—a planetary model with **quantized** orbits—correctly describes the energy levels of the hydrogen atom, but not of more complicated atoms. **Quantum mechanics** uses the *wave nature of the electron* to determine a wave function that gives all the information possible about the atom. This information includes the *probability distribution* for the electron. X rays are photons of electromagnetic radiation given off when atoms de-excite with energy changes of a few hundred electron volts or higher.

The atomic *nucleus* consists of **neutrons** and **protons** held together by the **strong,** or **nuclear,** force. For light stable nuclei, the number of neutrons is approximately equal to the number of protons; the ratio of neutrons to protons increases with total number of **nucleons.** Unstable nuclei decay to more stable forms by **radioactivity:** the emission of particles from the nucleus. In naturally occurring radioactive nuclides, these particles are α (^{4}He nuclei), β (electrons), and γ (photons) rays. In radioactive decay, *total charge and total number of nucleons remain constant.* The **half-life** is the time for half of the nuclei in a radioactive sample to decay.

An important application of radioactivity is in the use of *tracers* to follow the path of a particular element.

When radiation passes through matter, it *ionizes* the atoms. This ionization allows the detection of radiation, and it can destroy living cells. The largest radiation exposure for most people is from natural sources.

Nuclear reactions release energy if the total mass after the reaction is less than that before. Both **fusion** of light nuclei and **fission** of heavy nuclei result in lower mass and thus a release of energy.

A *nuclear reactor* using *fission* depends on a chain reaction from the neutrons released in the fission. *Fusion reactors* will, if expectations of researchers are fulfilled, use thermonuclear reactions for the release of energy.

Questions

The Atom and Its Radiation

1. What is the relation between the colors you see at night from a lighted Coca Cola sign and the energy levels of the atoms of the gases in the lights?

2. When you're playing pool, does the Heisenberg uncertainty principle influence your ability to get the eight ball into the side pocket? Why so or why not?

3. Baggage at airports is inspected by x-ray beams that give an exposure of no more than 1 mrem per examination. It takes about 5 mrem to noticeably fog your photographic film. What does this say about the number of times you should let a given film go through airport inspection?

4. In xeromammography, a process developed by Xerox Corporation, the film used in breast x-ray examinations is replaced by a sheet of selenium backed by photographic film. X rays hitting the selenium screen cause it to fluoresce, giving off visible light that exposes the film. With this method, about 1/20 the x-ray dose gives the same film exposure as when it is exposed directly by x rays. Because of the two-step process, do you expect some loss in sharpness of the image?

Nuclei and Radioactivity

5. Nuclear physicists use the symbol N for the number of neutrons and Z for the number of protons in a nucleus. There are 160 stable nuclides in which both N and Z are even numbers,

53 with N odd and Z even, 49 with N even and Z odd, and only 4 with both N and Z odd. From these data, what would you suggest about the strength of the nuclear force when nucleons can pair off two-by-two of *like* particles?

6. If you were to pull a nucleus apart into its individual nucleons, work would be needed to pull each nucleon out. Anytime you do work, you produce some form of energy; in this case, it goes to mass. Would you expect the mass of all the separated nucleons to be more *or* less than the mass of the nucleus? This difference in mass energy is called the *binding energy* of the nucleus.

7. X rays and γ rays are the same stuff; they just come from different places. Explain how a ^{60}Co γ-ray source could be used to "x-ray" inaccessible equipment—such as the pipes of a boiler—to check for cracks.

8. Complete the following radioactive decay equations (use an X for any unknown chemical symbols):

   (a) $^{235}_{92}U \longrightarrow {}^{231}_{90}Th +$

   (b) $\longrightarrow {}^{237}_{93}Np + {}^{4}_{2}He$

   (c) $^{60}_{27}Co \longrightarrow \quad + {}^{0}_{-1}e$

   (d) $^{60}_{28}Ni \longrightarrow \quad + {}^{0}_{0}\gamma$

   (e) $^{14}_{6}C \longrightarrow {}^{14}_{7}N +$

Applications of Radioactivity

9. Is there any difference in what ^{131}I does for a doctor in searching for thyroid cancer and what a cowbell does for a farmer?

10. Emogine Embolism complains to her doctor that her left foot is cold and numb. The doctor suspects that a blockage in an artery is impeding circulation. To pinpoint it, he injects into Emogine's blood a small amount of salt solution (sodium-chloride) in which a minute fraction of the sodium is radioactive ^{24}Na. What would be his next step in the search?

11. A fractured or diseased area of a bone takes in calcium more rapidly than does normal bone. Strontium behaves chemically almost exactly the same as calcium. How could physicians use radioactive ^{85}Sr (half-life 65 days) or ^{87}Sr (half-life 2.8 h) to locate bone lesions before they are large enough to show up on x-ray pictures? What is wrong with using ^{90}Sr (half-life 29 yr)?

12. When a new batch of oil is started down the Alaskan pipeline right behind an old one, how can radioactive tracers be used to let people at the other end know when the new batch gets there?

13. Some brands of LCD digital watches have a tritium lighting system that continually gives off enough green light that you can read the display at night. A small amount of tritium (^{3}H that β-decays with a half-life of 12 yr) is tightly sealed in a glass capsule coated on the inside with a phospher that glows when hit by electrons. What form of energy produces this light? If you bought such a watch today, how would the light output 12 years from now compare with that now?

14. A *nuclear battery* could be constructed as in Fig. 18.24. Electrons coming off the β source on the inner cylinder travel out and hit the outer cylinder, charging it negatively. A voltage of 10 000 to 100 000 V can develop and can produce a few microamperes of current in an external circuit. If strontium-90 is used as a β source, how would you expect the output of this battery to change with time?

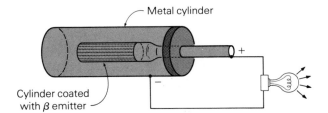

Figure 18.24   A nuclear battery.

15. When a γ-ray beam passes through a layer of matter, the fraction of the γ rays absorbed by the matter increases with the thickness of the layer. Draw a diagram showing a scheme that uses a γ-ray source and a radiation detector to continuously monitor the thickness of plastic sheet coming off a production line.

Biological Effects of Radiation

16. The radiation dose to the exposed area during a chest x ray should not exceed about 30 mrem. Since this is considerably less radiation than you get from natural sources every year, is there any reason for not routinely having a chest x ray each year?

17. Certain foods are pasteurized by a radiation dose of 200 000 to 500 000 rem and sterilized by a dose of 2 to 5 million rem. Since radiation is known to cause genetic damage, is there any way you could genetically damage your offspring by eating irradiated food?

Nuclear Reactions and Applications

18. Complete the following nuclear reaction equations, using an X for any chemical symbols you don't know.

    (a) $^{59}_{27}Co + ^{1}_{0}n \longrightarrow \quad + ^{0}_{0}\gamma$

    (used in producing a nuclide for therapy)

    (b) $^{6}_{3}Li + ^{1}_{1}H \longrightarrow ^{3}_{2}He +$

    (a potential fusion reaction for fusion reactors)

    (c) $^{235}_{92}U + ^{1}_{0}n \longrightarrow ^{236}_{92}U \longrightarrow \quad + \quad + 2^{1}_{0}n$

    (a reaction in a fission reactor)

    (d) $^{7}_{3}Li + ^{1}_{1}H \longrightarrow ^{4}_{2}He +$

    (the first reaction ever produced using a particle accelerator)

19. Tracer methods are very useful in wear studies. For example, gasoline engine manufacturers like to know how fast their piston rings wear. They can use radioactive piston rings and then, after a certain period of running the engine, look for radioactivity in the lubricating oil. How can they make the piston rings radioactive?

20. McGarrett of Hawaii Five-0 has uncovered an illegal gunpowder manufacturing gang. He wants to know if *their* gunpowder was involved in the recent murder of one of his officers. Knowing the exact composition of their powder, he wipes the murder weapon with a cloth, and sends the cloth to the Honolulu Nuclear Research Reactor for neutron activation analysis. Do you think he will get his answer?

21. Natural sodium is all ^{23}Na. How could you use a nuclear reactor to produce the ^{24}Na used by Emogine's doctor in locating her arterial blockage (Question 10)?

22. An oil slick is found floating in the Gulf of Mexico. Investigators are not sure where it came from. How can they use neutron activation analysis to find out whose oil it is?

23. An alleged Rembrandt painting is suspected of being a twentieth-century forgery. How could neutron activation analysis be used for evidence one way or the other?

Nuclear Energy; Reactors

24. If you calculate the energy equivalent of the mass difference before and after fission of ^{235}U, you get about 170 MeV for a typical fission. (See, for example, Problem 7.) The *average total* energy released per fission is about 200 MeV. The rest is released gradually after fission. Where does it come from? (Hint: Remember what happens to the fission fragments.)

25. When a reactor is critical, is it necessary that exactly one neutron released in *every* fission cause another fission?

26. A loss of coolant in a reactor can cause a reactor core *meltdown*. Even if the fission chain reaction stops immediately, there is still enough heat being produced to cause a meltdown. Why? (Hint: What happens to the fission fragments? That's why reactors have an emergency cooling system to inject coolant if the normal coolant system ruptures.)

27. Here is a potentially realistic chain of events: Lightning knocks out a power line from a nuclear power plant; the line no longer carries current; the generator no longer produces current and therefore uses little steam; the steam generator doesn't have to produce steam and therefore doesn't cool the reactor coolant; the reactor coolant doesn't cool the reactor. Explain this sequence from a *conservation of energy* viewpoint. In designing a nuclear power plant, nuclear engineers have to provide a system for preventing meltdown from such an event.

28. We sometimes use the word "hot" to describe the fact that a substance is radioactive. Radioactive wastes from a nuclear reactor also stay very hot *thermally.* Why?

29. Discuss the feasibility of providing each home in the country with a well-encased spent reactor fuel element to put in the middle of the family room in winter and provide heat like a pot-bellied stove. (See Question 28.)

Problems

1. A clinical technician measures the acitivity of a ^{24}Na sample to be 200 MBq. What is the activity of the sample a week later? (The half-life of ^{24}Na is 15 h.)   **Ans:** 100 kBq

2. Suppose you take a sample of ^{130}Te (Te = tellurium) to your friendly neighborhood reactor operator and convince her to make you some $^{131}_{53}$I. What she will make is $^{131}_{52}$Te, which has a half-life of 25 min. How long will it take for 99% of the ^{131}Te to decay to ^{131}I?   **Ans:** about 2 hr, 55 min

3. A hardworking physics graduate student, while skin-diving at Wakula Springs, Florida, discovers an old mastodon bone. If he takes it back to his campus and finds that the ratio of ^{14}C to stable carbon is ¼ what it is in the backbone of his goldfish that died that morning, how long ago did the mastodon drink from Wakula Springs?   **Ans:** 11 500 years

4. In the radioactive α decay

    $$^{226}_{88}\text{Ra} \longrightarrow \, ^{222}_{86}\text{Rn} + \, ^{4}_{2}\text{He},$$

    what is the total kinetic energy of the Rn and α particle after decay? Since the α is much lighter, it carries away most of the kinetic energy. (The masses are 226.02536 amu, 222.01753 amu, and 4.00260 amu, respectively.)   **Ans:** 4.87 MeV

5. Find the energy released in the fusion reaction

    $$^{2}_{1}\text{H} + \, ^{3}_{1}\text{H} \longrightarrow \, ^{4}_{2}\text{He} + \, ^{1}_{0}\text{n}.$$

    (The masses in order of appearance are 2.0141 amu, 3.0160 amu, 4.0026 amu, and 1.0087 amu.)   **Ans:** 17.5 MeV

6. In order to make the first-observed nuclear reaction

    $$^{14}_{7}\text{N} + \, ^{4}_{2}\text{He} \longrightarrow \, ^{17}_{8}\text{O} + \, ^{1}_{1}\text{H}$$

    go, how much mass energy must be made up by the kinetic energy of the incoming α particle? (The masses in order of appearance are 14.00307 amu, 4.00260 amu, 16.99913 amu, and 1.00782 amu.)   **Ans:** 1.19 MeV

7. The following fission reaction releases 5 neutrons:

    $$^{1}_{0}\text{n} + \, ^{235}_{92}\text{U} \longrightarrow \, ^{136}_{53}\text{I} + \, ^{95}_{39}\text{Y} + 5(^{1}_{0}\text{n}).$$

    If the kinetic energy of the incoming neutron is negligible, what is the total kinetic energy of the fission fragments and neutrons immediately after fission? (The masses in order of appearance are 1.00866 amu, 235.04392 amu, 135.91474 amu, and 94.91254 amu.)   **Ans:** 169 MeV

References for Further Reading

Bushong, S. C., "Radiation Exposure in our Daily Lives," *The Physics Teacher,* March 1977, p. 135. Discusses the sources of radiation exposure to United States citizens.

Cohen, B. L., "Health Risks of Nuclear Power," *The Physics Teacher,* Nov. 1978, p. 526. An authoritative summary of the various risks of large-scale nuclear power generation.

Gough, W. C., and B. J. Eastlund, "The Prospects of Fusion Power," *Scientific American,* Feb. 1971, p. 50. The experimental programs and techniques for thermonuclear fusion reactors. Concentrates on magnetic confinement techniques.

Marion, J. B., *Energy in Perspective* (Academic Press, New York, 1974). Chapters 5 and 6 deal specifically with nuclear power and radiation.

Maynard, C. D., *Clinical Nuclear Medicine* (Lea and Febiger, Philadelphia, 1969). Describes clinical uses of nuclear physics.

McBride, J. P., R. E. Moore, J. P. Witherspoon, and R. E. Blanco, "Radiological Impact of Airborne Effluents of Coal and Nuclear Plants," *Science,* Dec. 1978, p. 1045. Shows that the radiation dose from radioactive nuclides in the emissions from coal-fired power plants may be greater than those from nuclear plants.

Nuclear Power and the Environment—Questions and Answers (American Nuclear Society, La Grange Park, Ill., 1976). Answers many common questions relative to nuclear power and radiation.

Schwitters, R. F., "Fundamental Particles with Charm," *Scientific American,* Oct. 1977, p. 56. Discusses the search for particles thought to be the building blocks for ordinary particles such as neutrons and protons.

Seaborg, G. T., and J. L. Bloom, "Fast Breeder Reactors," *Scientific American,* Nov. 1970, p. 13. Describes the principles of breeder reactors, which produce more nuclear fuel than they use.

Stickley, C. M., "Laser Fusion," *Physics Today,* May 1978, p. 50. The principles of the "laser fusion" approach to thermonuclear reactions.

U.S. Nuclear Regulatory Commission, *Reactor Safety Study: An Assessment of Accident Risks in U.S. Commercial Nuclear Power Plants—Executive Summary*, WASH-1400 (NUREG 75/014), (National Technical Information Services, Springfield, Va., 1975). This report summarizes the results of a panel of experts, headed by Professor Norman Rasmussen of M.I.T., enlisted in the mid-1970s by the Nuclear Regulatory Commission to study the risks involved in potential nuclear power plant accidents. The report includes quantitative numerical comparisons between the risks of nuclear power and other everyday risks—including automobile travel, falls, air travel, and lightning.

Appendix I

Handling Numbers

If we follow certain general guidelines when we use numbers in calculations, the calculations will be easier and the answers will have more meaning. We'll discuss guidelines in two areas: properly rounding off numbers, and using powers-of-ten notation.

A. **Rounding Off Numbers**

In the numerical calculations we do in this book, we round off almost all answers to two "significant digits." (Zeros that only locate the decimal point—for example those in 0.0013—are not significant digits.) To illustrate why this rounding off is appropriate, consider our fourth-grader in Section 1.7 of Chapter 1, who wanted to know how far she rode in 2 hours traveling at 80 kilometers/hour (km/h).

Suppose she had measured the time with an accurate digital stopwatch that showed she had traveled for 7200.15 seconds, but had measured the speed by watching the speedometer. When she converted seconds to hours and figured out the distance on her calculator, she got 160.00333 kilometers. However, her speed measurement was not very accurate. It might, for example, have been anywhere between 78 and 82 km/h. That would mean the distance was somewhere between 156.00325 and 164.00342 km. In other words, there is no real significance to those last six digits in the distance. The best we can do is say the distance is 160 km.

This example illustrates two points. First, the number of significant digits tells you roughly how accurate the number is: The six digits in the time measurement tell you that you know the time to one or two hundredths of a second; the two digits in the speed measurement (assuming the zero here is significant) tell you that you know the speed to one or two km/h.

Second, this example shows that *when you multiply two numbers together, there is no significance to leaving more digits in the answer than you have in the multiplier with the fewest number of significant digits.* Since our examples usually have the input information given to two digits, it's wrong (or at least misleading) not to round off answers obtained by multiplication or division to two digits. For intermediate answers that are used in subsequent calculations, we sometimes carry along another digit to avoid ambiguities later.

B. Powers-of-Ten Notation

The sizes of physical quantities that are important in our lives vary over a *tremendous* range. For example, your height is about 300 000 000 000 000 times the diameter of the nucleus of one of the carbon atoms in your left little toe. Rather than carry along a bunch of zeros in large numbers such as this one, or small numbers such as the nuclear diameter in meters, we usually use *powers-of-ten* notation, in which the above number would be 3×10^{14}. We won't go into the mathematical meaning of the notation, but will merely summarize how to use it.

If the *power,* or *exponent,* of ten is a positive number, you get the usual notation by moving the decimal point to the right the number of places given by the exponent. For example,

$$3.2 \times 10^8 = 320\ 000\ 000,$$

$$94 \times 10^2 = 9400,$$

$$6.8 \times 10^0 = 6.8.$$

If the power of ten is negative, you move the decimal point to the left that many places. For example,

$$3.2 \times 10^{-8} = 0.000\ 000\ 032,$$

$$84 \times 10^{-3} = 0.084,$$

$$2.8 \times 10^{-1} = 0.28.$$

Appendix II

Units and
Conversion Factors

A. Metric Units

Metric units have a big advantage over the traditional "British" units because, in the metric system, there is only one basic unit for each quantity being measured. The basic metric units for length, mass, and time are the meter (m), gram (g), and second (s), respectively. We specify larger or smaller units of each quantity by attaching a prefix to the basic unit that merely moves the decimal point a prescribed number of places. These prefixes and their abbreviations are given in Table A.1.

Table A.1 Metric Prefixes

Prefix	Abbre- viation	Multiplier	Prefix	Abbre- viation	Multiplier
exa	E	10^{18}	deci	d	10^{-1}
peta	P	10^{15}	centi	c	10^{-2}
tera	T	10^{12}	milli	m	10^{-3}
giga	G	10^{9}	micro	μ	10^{-6}
mega	M	10^{6}	nano	n	10^{-9}
kilo	k	10^{3}	pico	p	10^{-12}
hecto	h	10^{2}	femto	f	10^{-15}
deca	da	10	atto	a	10^{-18}

We use these prefixes to express a quantity in units of the appropriate size. Some examples: The radius of a carbon nucleus is 2.7 fm (2.7×10^{-15} m); the average distance of the earth from the sun is 149.6 Gm (1.496×10^{11} m); the mass of a typical flea is 450 μg (4.5×10^{-5} g); and it takes about 12 ms (0.012 s) for a fastball to pass over home plate.

B. SI Units*

When doing a calculation, each quantity in the calculation usually must be expressed in the same system of units. The subgroup of metric units used internationally for scientific purposes is the SI system. Table B.1 summarizes the SI units used in this book.

Table B.1   The International System of Units

Unit	Abbreviation	Quantity Measured
meter	m	length
kilogram	kg	mass
second	s	time
newton	N	force
joule	J	work, energy
watt	W	power
kelvin	K	temperature
coulomb	C	electric charge
ampere	A	electric current
volt	V	electric potential difference
ohm	Ω	electric resistance
tesla	T	magnetic field
hertz	Hz	frequency
lumen	lm	light output
becquerel	Bq	radioactivity

C. Conversion Factors

You can convert non-SI metric units to SI units merely by moving the decimal point according to the appropriate prefix. Notice that the SI unit of mass—the kilogram—is not a basic metric unit, but is 1000 of the base units, the gram. In converting other metric units of mass to SI units, you need to allow for this factor of 1000.

---

*The International System of Units.

Table C.1 gives conversion factors from other often-used units to SI units.

Table C.1   Conversion Factors between Units

time:	1 second (s)	= 1/60 minute (min)
		= 1/3600 hour (h)
length:	1 meter (m)	= 39.37 inches (in)
		= 3.281 feet (ft)
	1 inch (in)	= 2.54 cm
	1 kilometer (km)	= 0.621 mile (mi)
	1 mile (mi)	= 1609 meters (m)
		= 1.609 kilometer (km)
volume:	1 cubic meter (m^3)	= 1000 liters (l)
	1 liter	= 0.001 m^3
		= 1.057 quart
mass:	1 atomic mass unit (amu)	= 1.66×10^{-27} kilogram (kg)
	1 kilogram (kg)	*weighs* 2.20 pounds (lb)
force:	1 newton (N)	= 0.2248 pound (lb)
		= 1×10^5 dyne (dyn)
	1 pound (lb)	= 4.448 newton (N)
pressure:	1 N/m^2	= 1.45×10^{-4} lb/in^2
	1 lb/in^2	= 6895 N/m^2
	1 atmosphere (atm)	= 101.3 kN/m^2
		= 76.0 cm of mercury
energy:	1 joule (J)	= 0.239 calorie (cal)
		= 2.39×10^{-4} kcal
		= 2.778×10^{-7} kilowatt-hour (kWh)
		= 6.24×10^{18} electron volts (eV)
		= 9.48×10^{-4} British thermal unit (Btu)
		= 0.737 foot-pound (ft·lb)
	1 kilocalorie (kcal)	= 4186 joules (J)
		= 1.16 watt-hour (Wh)
	1 electron volt (eV)	= 1.60×10^{-19} joule (J)

D.   Juggling Units

When we do a calculation to convert from one unit to another, or to work out a problem, the units themselves can give us a strong hint about whether or not we have done the right thing. We can handle units as if they were algebraic symbols. If we come out with the right units after this manipulation, we can be confident we've gone about the calculation correctly. Some examples will illustrate this point.

Example A2.1   Convert 65 in to centimeters. Noting that there are 2.54 cm/in, we get

$$65 \text{ in} = 65 \text{ in} \times 2.54 \text{ cm/in}$$

$$= 165 \text{ in} \cdot \text{cm/in}.$$

Notice that just as $x \cdot y/x = (x/x) \cdot y = 1 \cdot y = y$, then in $\cdot$ cm/in $=$ cm. Thus,

$$65 \text{ in} = 165 \text{ cm}.$$

Since we wanted the distance in cm, we know we have done the conversion correctly. If we mistakenly divide by the conversion factor instead of multiplying, we get units of in/(cm/in) = in²/cm. Since this is not the unit we wanted, we know we made a mistake.

Example A2.2   Convert a speed of 55 mi/h to meters/second.

Since 1 mi = 1609 m and 1 h = 3600 s,

$$55 \text{ mi/h} = \frac{55 \text{ mi/h} \times 1609 \text{ m/mi}}{3600 \text{ s/h}}$$

$$= 24 \frac{\text{mi/h} \times \text{m/mi}}{\text{s/h}} = 24 \text{ m/s}.$$

Notice that the miles and hours cancel, leaving meters/second.

Exercise   Show that if you either divide by the distance conversion factor or multiply by the time conversion factor in Example A2.2, you will not get the right units, meters/second.

You can use the same technique to find out if you're solving a problem correctly. We'll illustrate.

Example A2.3   How long does it take you to drive 320 km at a speed of 80 km/h? From your experience you know that

$$\text{time} = \frac{\text{distance}}{\text{speed}}$$

$$= \frac{320 \text{ km}}{80 \text{ km/h}} = 4.0 \text{ km/(km/h)}$$

$$= 4.0 \text{ (km/km)h} = 4.0 \text{ h}.$$

If we had mistakenly divided speed by distance, the resulting units—1/hour rather than hour—would have told us we did not solve the problem correctly.

The examples we have worked illustrate how to manipulate units algebraically in any problem, no matter how simple or how complicated.

Index